Athena Vakali and Lakhmi C. Jain (Eds.)

New Directions in Web Data Management 1

Studies in Computational Intelligence, Volume 331

Further volumes of this series can be found on our homepage: springer.com

Vol. 310. Dipti Srinivasan and Lakhmi C. Jain (Eds.)
Innovations in Multi-Agent Systems and Applications – 1, 2010
ISBN 978-3-642-14434-9

Vol. 311. Juan D. Velásquez and Lakhmi C. Jain (Eds.)
Advanced Techniques in Web Intelligence, 2010
ISBN 978-3-642-14460-8

Vol. 312. Patricia Melin, Janusz Kacprzyk, and Witold Pedrycz (Eds.)
Soft Computing for Recognition based on Biometrics, 2010
ISBN 978-3-642-15110-1

Vol. 313. Imre J. Rudas, János Fodor, and Janusz Kacprzyk (Eds.)
Computational Intelligence in Engineering, 2010
ISBN 978-3-642-15219-1

Vol. 314. Lorenzo Magnani, Walter Carnielli, and Claudio Pizzi (Eds.)
Model-Based Reasoning in Science and Technology, 2010
ISBN 978-3-642-15222-1

Vol. 315. Mohammad Essaaidi, Michele Malgeri, and Costin Badica (Eds.)
Intelligent Distributed Computing IV, 2010
ISBN 978-3-642-15210-8

Vol. 316. Philipp Wolfrum
Information Routing, Correspondence Finding, and Object Recognition in the Brain, 2010
ISBN 978-3-642-15253-5

Vol. 317. Roger Lee (Ed.)
Computer and Information Science 2010
ISBN 978-3-642-15404-1

Vol. 318. Oscar Castillo, Janusz Kacprzyk, and Witold Pedrycz (Eds.)
Soft Computing for Intelligent Control and Mobile Robotics, 2010
ISBN 978-3-642-15533-8

Vol. 319. Takayuki Ito, Minjie Zhang, Valentin Robu, Shaheen Fatima, Tokuro Matsuo, and Hirofumi Yamaki (Eds.)
Innovations in Agent-Based Complex Automated Negotiations, 2010
ISBN 978-3-642-15611-3

Vol. 320. xxx

Vol. 321. Dimitri Plemenos and Georgios Miaoulis (Eds.)
Intelligent Computer Graphics 2010
ISBN 978-3-642-15689-2

Vol. 322. Bruno Baruque and Emilio Corchado (Eds.)
Fusion Methods for Unsupervised Learning Ensembles, 2010
ISBN 978-3-642-16204-6

Vol. 323. Yingxu Wang, Du Zhang, and Witold Kinsner (Eds.)
Advances in Cognitive Informatics, 2010
ISBN 978-3-642-16082-0

Vol. 324. Alessandro Soro, Vargiu Eloisa, Giuliano Armano, and Gavino Paddeu (Eds.)
Information Retrieval and Mining in Distributed Environments, 2010
ISBN 978-3-642-16088-2

Vol. 325. Quan Bai and Naoki Fukuta (Eds.)
Advances in Practical Multi-Agent Systems, 2010
ISBN 978-3-642-16097-4

Vol. 326. Sheryl Brahnam and Lakhmi C. Jain (Eds.)
Advanced Computational Intelligence Paradigms in Healthcare 5, 2010
ISBN 978-3-642-16094-3

Vol. 327. Slawomir Wiak and Ewa Napieralska-Juszczak (Eds.)
Computational Methods for the Innovative Design of Electrical Devices, 2010
ISBN 978-3-642-16224-4

Vol. 328. Raoul Huys and Viktor K. Jirsa (Eds.)
Nonlinear Dynamics in Human Behavior, 2010
ISBN 978-3-642-16261-9

Vol. 329. Santi Caballé, Fatos Xhafa, and Ajith Abraham (Eds.)
Intelligent Networking, Collaborative Systems and Applications, 2010
ISBN 978-3-642-16792-8

Vol. 330. Steffen Rendle
Context-Aware Ranking with Factorization Models, 2010
ISBN 978-3-642-16897-0

Vol. 331. Athena Vakali and Lakhmi C. Jain (Eds.)
New Directions in Web Data Management 1, 2011
ISBN 978-3-642-17550-3

Athena Vakali and Lakhmi C. Jain (Eds.)

New Directions in Web Data Management 1

Springer

Prof. Athena Vakali
Department of Informatics
Aristotle University
54124 Thessaloniki
Greece
E-mail: avakali@csd.auth.gr

Prof. Lakhmi C. Jain
School of Electrical and Information Engineering
University of South Australia
Adelaide
Mawson Lakes Campus
South Australia SA 5095
Australia
E-mail: Lakhmi.jain@unisa.edu.au

ISBN 978-3-642-17550-3 e-ISBN 978-3-642-17551-0

DOI 10.1007/978-3-642-17551-0

Studies in Computational Intelligence ISSN 1860-949X

Typeset & Cover Design: Scientific Publishing Services Pvt. Ltd., Chennai, India.

Printed on acid-free paper

9 8 7 6 5 4 3 2 1

springer.com

Preface

In the past few years we have all witnessed tremendous changes and innovations in the way we deal with data on the Web. The major shift from navigating to content regulating has posed new challenges in a way data is circulated, disseminated and publicized. New technologies have emerged under the Web 2.0 umbrella and we're already progressing on a Web 3.0 reality. It certainly is a rather active and emerging period for Web data management and it is important to understand and clarify current practices and methodologies towards moving to a more effective and productive Web future.

This book addresses the major issues in the Web data management in order to highlight issues dealing with technologies and infrastructures, methodologies and techniques as well as applications and implementations. Emphasis is placed on Web engineering and technologies, on Web graph managing, on searching and querying. The importance of social Web is also acknowledged with inclusion of both social and semantic aspects.

The editors express their gratitude to all the authors of this book for their insights and excellent contributions to the book. Moreover, the editors acknowledge the help of all involved in the collation and review process, without their support this project could not have been satisfactorily completed. Certainly special thanks are due to the Springer-Verlag for their assistance during the preparation of the manuscript. Finally, we thank our families for their love and support throughout this project.

We hope that the readers will this research book informative and enlightening. Comments from readers will be greatly appreciated. Please contact us at Athena Vakali <avakali@csd.auth.gr> and Lakhmi Jain <Lakhmi.Jain@unisa.edu.au>.

Athena Vakali
Greece

Lakhmi C. Jain
Australia

Editors

Athena Vakali is an associate professor at the Department of Informatics, Aristotle University of Thessaloniki, Greece. She is the head of the Operating Systems Web/INternet Data Storage and management research group and her research activities focus on topics of Web information systems such as Web data management (clustering techniques), content delivery on the Web, Web data clustering, Web caching, text mining and multimedia data management. Her publication record is now at more than 100 research publications which have appeared in several journals, book chapters and in scientific conferences and she is also co-editor of the book "Web Data Management Practices: Emerging Techniques and Technologies" published by Idea Group Publishing. She is a member of the editorial board of the Computers and Electrical Engineering Journal (Elsevier) and since March 2007, she is the coordinator of the IEEE TCSC technical area of Content Management and Delivery Networks and she has scientifically leaded more than 15 European and national research projects.

Professor Lakhmi C. Jain is a Director/Founder of the Knowledge-Based Intelligent Engineering Systems (KES) Centre, located in the University of South Australia. He is a fellow of the Institution of Engineers Australia.

His interests focus on the artificial intelligence paradigms and their applications in complex systems, art-science fusion, e-education, e-healthcare, unmanned air vehicles and intelligent agents.

Contents

Chapter 1
Innovations and Trends in Web Data Management

Athena Vakali

Department of Informatics
Aristotle University
54124 Thessaloniki, Greece
avakali@csd.auth.gr

The growing influence and resulting importance of the Web 2.0 applications has changed the daily practices in the areas of research, education, finance, entertainment and an even wider range of applications in work and personal life. Such a development in the roles of users such as navigators, content creators and regulators has had a major impact. This impacts on the amount and type of data and the sources that are now circulated and disseminated over the Web. It has posed new and interesting research questions and problems in Web data management.

1 Communities and Open Problems in the Web 2.0 Environment

Considerable interest has developed both in the analysis and description of Web user activities. Such applications, allow users to create their own content and to form communities having these interests. Graph structures which can use these techniques have been widely used to model users problems and their relationships. The amount of data generated by today's Web systems produces very large and complex graphs that are difficult both to analyze and to interpret. Consequently, several methods for community detection and graph compression have been developed. These can identify groups of users which are based on the topology and the properties of graphs that contain them as nodes.

In Web usage the term community is used to describe a group of pages that can depict common interests and are able to share many hyperlinks (Kumar, Raghavan, Rajagopalan, and Tomkins, 1999). The community has been defined as a set of Web objects both users and pages similar with their own logical and semantic structures. This is done in order to facilitate Information Retrieval and Data Management (Zhang, Xu Yu, and Hou, 2006). In the context of Web 2.0, communities and often a community are defined with reference to certain sets of users and resources. These may involve factors such as articles, images, videos and tags. Combinations (Cattuto and others, 2008) would also be included.

Considering the structure of the graphs, a community has been defined as a set of nodes having common properties and playing similar roles within the graph. For example, groups of websites having similar topics such as Fortunato and

A. Vakali & L.C. Jain (Eds.): New Directions in Web Data Management 1, SCI 331, pp. 1–18.
springerlink.com

Castellano, 2008. In order to model the concept of communities and to identify communities using algorithmic methods, it is necessary to utilize concepts of graph theory. These include areas such as clique, cycle, k-clique (Palla, Derényi, Farkas, & Vicsek, 2005), N-clique (Alba, 1973), k-core (Batagelj & Zaveršnik, 2002) and k-plex (Seidman & Foster, 1978). These correspond to coherent sub-graph structures and have also been used. In the development of techniques to model the concept of a community based on these structures, and concepts which include (i) *density*, which indicates the coherence of the subgraph, (ii) *Degree Centrality*, which refers to the number of edges connecting a given node to other nodes, (iii) *Betweenness Centrality*, which has high values for nodes that are included in many the determination of the shortest paths between nodes, and (iv) the *edge "betweenness"*, which refers to the number of shortest paths between the different pairs of nodes containing an edge, are all of particular significance. These concepts are used to describe the significance of a node or an edge in a graph. They have been used in various techniques used to detect communities in networks.

Web 2.0 applications and social networks allow users to create virtual personalities that are represented by the users' profiles, and to create social connections. The user profiles include information about: Their characteristics and interests (e.g. *Facebook*[1]), Sites that interest them (e.g. *Delicious*[2]), or about any other type of information that is described in a given Web 2.0 application and reflects their preferences. The social connections can be represented either by explicit friendship links or they can be inferred through the analysis of the users' activity. This may be comments in a blog service or references to a research repository. The analysis of the above types of links contributes to the identification of communities in a network. This is important for analyzing the features of individual communities, and for understanding the structure and properties of the overall network (Milo & Itzkovitz, 2002) (Vázquez, Dobrin , Sergi, Eckmann, Oltvai, and Barabási, 2004).

The problem of community identification, despite having been investigated for many years, requires finding effective methods for graph partitioning which can be applied to complex and large-scale networks. Some open questions still requiring further investigation are the following:

- **Data Collection problems which can occur and are related to the common reference of data:** Users can interact with existing Web 2.0 applications such as Wikis, Blogs, Social Tagging Systems, and Social Networks. This can provide information about participants interests, dependant on the particular activity or the data in question. A problem which is as yet unsolved is the collection and integration of data from different data sources. This is necessary for the uniform reporting and analysis of data. The solution of this particular problem can lead to radical conclusions concerning the interests of users as well as to provide a more complete image of their data profile. It is possible to identify any multiple relationships that may exist between two or more users. These may be friends or associates.

[1] `http://www.facebook.com/`
[2] `http://delicious.com/`

- **Size and heterogeneity of the data:** Different types of data collected from existing Web 2.0 applications and their huge volume require the definition of new representations and indexing structures that will assemble as much information as possible about the data described and contribute to the development of efficient and scalable methods of analysis. Due to the huge volume of data, new data representation approaches are required. These should contemplate the use of an external memory (Chiang, Goodrich, Grove, Tamassia, Vengroff, and Vitter, 1988). In addition, distributed environments may be used (Panconesi & Rizzi, 2001). Problems may be considered as data streams (Zelke, 2009). Another important factor which should be considered is the nature of the data. This varies depending on the application for which they are acquired. For example, in blogs users tend to generate mostly text data. In the case of social tagging systems multimedia data such as images, videos, for example will dominate.
- **Assessing the relationships between users:** In most networks, the communities that are identified will have overlaps between their members due to the many different kinds of relationships that often connect the community members (Palla, Derényi, Farkas, and Vicsek, 2005). For example, users can belong to different groups depending on their: Level of Education, Working Environment, Hobbies and Family. Consequently, the overlap between the communities constitutes an important factor in the design of methods for community identification. A particular challenge is the definition of dynamic models that can reflect the features of community members in time, as well as their evolution time.
- **Requirement for metrics that will incorporate semantic features related to the content. This is in addition to the behavior and the psychology of the users.** Most methods of finding the communities in social networks are based on the study and exploitation of the properties of the graphs that illustrate the relationships among the members of a social network. The features that are often exploited are the power and the type of relationships between users, their profile, and the interaction between them. That is the comments to one another. That is the exchange of data for example. The data that are used in the network and their content. That is bookmarks, tags, reviews, pictures, video, for example can also contain important information which may contribute to the discovery of communities. Specifically, the semantic analysis of the data provides important information about, the users, their profiles, interests and preferences. Also it contains information about the respective social networks and their structure (Mika, 2005). Such analysis is done on different types of content such as tags, text, images. Therefore the derived multi-dimensional results can be utilized in the definition of metrics which will contribute to the discovery of unexpected groups of users with common interests and behaviors.

- **Effective clustering methods for the detection of communities:** The nature and vast sizes of data available from Web 2.0 applications necessitate the existence of more effective and scalable methods that can highlight the capacity and special characteristics of communities. For example the temporal and social trends that appear in these communities, and the relationships between the community members. Special attention should be given to the development of universal methods able to be applied to different types of networks. These may differ either on how the users access the data. That is whether it is fixed or mobile networks, or in the nature of the social or the Web 2.0 network. Dependant on whether it is a blog or a social recommendation system for example.

2 Capturing Groups of Data over the Web Graph

Recently there has been rising research interest in complex data networks obtained from various scientific fields which include biology, sociology and information technology and are primarily obtained from the Web. The networks derived from real-world data for example the Web, generally present a non-random organization. For example, the *degree* of the nodes in real-world networks often follows the *power law* distribution. This is according to whether there are many nodes which have large connectivity or a few nodes with little connectivity or degree. The uneven distribution of nodes is observed at the global level, but also locally (Fortunato & Castellano, 2008). There are groups of densely connected nodes creating groups or communities of data.

The existence of typical densely connected structures in many modern data networks has contributed to the emergence and evolution of a new research field. This is dedicated to the identification and capturing groups of data in complex networks. We will now give a summary of some of the most important methodologies that have so far been proposed or implemented. Given the large number of relevant methods in literature, only a selection of some of the most important methods will be included. These may be considered as indicative of some of the general categories. Typically, the approach is followed by finding groups of data and communities in complex networks. The implementation of the algorithms traditionally used in similar problems is considered. The algorithms used for graph partitioning and the algorithms used for node clustering are now considered:

Review of graph partitioning and node clustering methods

i. Graph partitioning

The first types of methods used for finding communities in complex networks were those which were related to the problem of graph partitioning. This required the segmentation of a graph into a number of subgraphs. This enabled the number of edges connecting nodes of different subgraphs to be minimized. Some of the important methods for graph partitioning are: the method used in the Kernighan-Li algorithm (Kernighan and Lin, 1970). The spectral partitioning method (Barnes, 1982) (Pothen, Simon, and Liou, 1990), and methods that do partitioning in levels. These are applied for example to the Multilevel Recursive-Bisection family of algorithms METIS (Karypis and Kumar, 1999). Other methods used to partition the graph are based on the minimization of measures such as *conductance* (Bollobás, 1998) (Šíma and Schaeffer, 2006) and *cut ratio* (Chan, Schlag, and Zien, 1993) (Wei and Cheng, 1989).

In general, graph partitioning methods require the prior definition of the number of subgraphs that occur after the graph is partitioned. In some cases even to define the size. Techniques that are applied to do graph bisection are recursively repeated for each resulting partition. These characteristics, combined with the lack of an appropriate metric to assess the quality of the final partitioning, render these methods inadequate for the identification of communities in complex networks (Newman, 2004).

ii. Clustering

Another early approach used to identify communities in networks adopted clustering techniques from the field of data mining. Namely, these are *hierarchical clustering* techniques were used for networks where the nodes appear to be organized hierarchically into groups (Hopcroft, Khan, Kulis, & Selman, 2004) (Jain & Dubes, 1988) (Scott, 2000). In addition as *partitional clustering* algorithms, with *k-means* (MacQueen, 1967) these are the most prominent algorithms representatives of this category. In hierarchical clustering algorithms, a node similarity metric is defined. This is used to compute the similarity for every pair of graph nodes to create a similarity matrix. Divisive clustering algorithms require the definition of a parameter *k* in advance, map each node to a point of a metric space and define a distance measure between points, =and then, the graph nodes are assigned to *k* groups so that a cost function based on the given metric distance is optimized. Despite the fact that the identification of groups of "similar" nodes is accomplished by either merging or divisive techniques, there are often cases when some nodes are not assigned to any group and they end up as isolated individuals.

More recent approaches to the problem of community identification have been based on the observation of the algorithms on these categories. Because of the types of restrictions they impose based on the prior definition of the number of groups, cannot usually be practically and efficiently applied to find communities from the complex networks. This observation highlights the need for algorithms designed exclusively to solve the problem of community identification in complex networks. Various approaches have been developed to solve this problem. For most of the categories a large number of representative algorithms have been developed. The most significant methodologies based on the classifications listed above are summarized in the following table. Some representative algorithmic approaches are also included.

Methodology	Emphasis/Main idea	Representative approaches	Characteristics
Divisive algorithms	Divisive algorithms follow the approach of repetitively removing edges that are considered to connect nodes belonging to the different communities, on the basis of some metric. Using this technique communities are gradually separated, while a community hierarchy is created.	Girvan & Newman, 2002 Girvan & Newman, 2004	• Edge removal based on the *edge betweenness* metric • Selection of the best partitioning from the derived hierarchy based on the *modularity* metric • High computational complexity
		Tyler, Wilkinson, & Huberman, 2003	• Extension to (Girvan and Newman, 2002) • Reduction of computational complexity
		Pinney and Westhead, 2006 Gregory, 2007	• Extension to (Girvan & Newman, 2002) • Support for overlapping communities
		Radicchi, Castellano, Cecconi, Loreto, and Parisi, 2004	• Edge removal based on the *edge clustering coefficient* metric
		Fortunato, Latora, and Marchiori, 2004	• Edge removal based on the *information centrality* metric
Modularity-based methods	The methods of this category use the *modularity* metric as a community evaluation function and attempt to optimize it.	Newman, 2004a	• Agglomerative hierarchical algorithm • Greedy technique
		Clauset, Newman, & Moore, 2004	• Extension to (Newman, 2004a) • Complexity reduction
		Blondel, Guillaume, Lambiotte, and Lefebvre, 2008	• Greedy technique • Low computational complexity
		Newman, 2006a	• *Modularity* optimization via *spectral bisection* • Not very accurate for networks with more than two communities
		White and Smith, 2005	• Approach the community identification task as a *spectral relaxation problem* • Fast algorithm for large sparse graphs
		Massen and Doye, 2005	• *Modularity* optimization via *simulated annealing* • Requirement for initial parameterization • Slow method applied to small graphs
		Duch and Arenas, 2005	• Uses the heuristic search technique *extremal optimization* • Relatively fast and accurate method
		Tasgin, Herdagdelen, and Bingol, 2007	• *Modularity* optimization using genetic algorithms

Spectral algorithms	Spectral algorithms exploit the algebraic properties of the matrices that can be derived from a graph, such as the Laplacian and Adjacency matrices,. Theaim is to cluster the nodes based on the similarity of the matrices' eigenvectors.	Donetti and Muñoz, 2004	• Exploitation of the eigenvectors of the Laplacian matrix • Mapping of the nodes to a metric space using the eigenvector components as coordinates • Calculation of similarity via the Euclidean or angle distance
		Capocci, Servedio, Caldarelli, and Colaiori, 2004	• Similarity calculation via the *Pearson* correlation coefficient of the eigenvectors of the *right stochastic matrix*
		Zarei and Samani, 2009	• Identification of "anti-community" structures in the complement graph (Its edges are the edges that the initial graph lacks in order to be complete) • Application of a spectral algorithm in the Laplacian matrix of the network • Ability to find small communities in small networks
Methods based on statistical inference	In general, *Statistical Inference* (Mackay, 2003) uses statistics and model hypothesis to infer the properties of a given population. That is the topology of a given graph. This approach to graph clustering is based on the assumption that the nodes are clustered in communities which can be identified by the graph linkage structure.	Hastings, 2006	• Techniques based on *Bayesian inference*, where the model is adjusted based on the maximization of a likelihood
		Newman & Leicht, 2007	
		Hofman & Wiggins, 2008	
		Reichardt & White, 2007	• Techniques based on *blockmodeling*
		Rosvall & Bergstrom, 2008	• Techniques based on *model selection*
		Ziv, Middendorf, & Wiggins, 2005	• Techniques based on *information theory*
Dynamic algorithms	Algorithms based on *spin models*	Reichardt and Bornholdt, 2004	• Usage of the *q-states Potts* model (Wu, 1982) which describes a system of spins that can be in one out of *q* different states • Definition of a spin state for each node and the study of the interactions among the spins of neighboring nodes • Evolution of the system via simulating annealing and the identification of communities of nodes characterized by similar spin states • Ability of identifying overlapping communities • Fast method, as most calculations require local information
		Son, Jeong, and Noh, 2006	• Uses the *Ferromagnetic Random Field Ising* (FRFI) model

	Algorithms based on *random walks* (Hughes, 1995). They utilize the fact that a random walker will be for longer time inside a community, as communities are characterized by dense structure of high connectivity.	Zhou, 2003	• Utilizes random walks to define a distance between pairs of nodes, and later a measure of dissimilarity
		Pons & Latapy, 2005	• Calculates the distance between nodes based on the possibility of moving from one node to the other within a predetermined number of steps
		van Dongen, 2000	• The Markov Cluster Algorithm (MCL) algorithm is used in many applications due to its simplicity • It applies *expansion* and *inflation* techniques at the graph's right stochastic matrix
	Algorithms based on the *synchronization* phenomenon which appears in every system with interactions (Pikovsky, Rosenblum, & Kurths, 2001) and is characterized by similar states of the system's members	Arenas, Díaz-Guilera, & Peréz-Vicente, 2006	• The existence of an oscillator with random phase at each node is assumed.,The synchronization of the oscillators of nodes belonging to the same community is expected to happen before full synchronization is observed
		Boccaletti, Ivanchenko, Latora, Pluchino, & Rapisarda, 2007	

3 Discovery of User Groups and Communities in Social Networks

The discovery of groups of users from Web 2.0 social networks has emerged as a particularly interesting field of application for community identification methods. These groups consist mainly of users who have common interests or goals, according to similar behavior patterns. They may be linked together by bonds of friendship. That is, if when supported by the requisite social network. The behavioral patterns of users refer to the way in which they participate in a Web 2.0 social network. This includes:

- creating new content
- commenting on existing content
- appending to content metadata (*tags*) so as to provide a description,
- creating bonds of friendship and providing cooperation with other users,
- Discussions between users,
- creating and / or participating in teams or groups that relate to specific topics.

Data that can be derived from a Web 2.0 social network to share the information between users, provide the relational information that indicates the way in which users connect to the content, and the interactions between users. It soon becomes apparent if this information can be utilized for discovery by other user communities. The resulting communities can prove particularly useful in applications such as *recommender systems*. Changing the composition of the recommendations according to the thematic group to which each user belongs in order to improve the accuracy. An important motivation for user community identification is the large number of users who choose to participate in social Web 2.0 applications. For example, according to statistical data, Facebook, a popular social network, has more than 400 million active registered users. Of these half connect to the network on any given day. The average Facebook user has 130 friends in the network. More than 25 billion pieces of information, such as web links, news stories, blog posts, notes, photo albums, for example, are shared between users monthly[3].

In the following paragraphs, a short review of the research that deals with the issue of finding communities in which users by means of on-line social networks share information.

One of the first published research works (Ohsawa, Soma, Matsuo, Matsumura, and Usui, 2002) dealt with the analysis of data from message boards. It depicts the relationships between users and between users and subjects as a graph structure using text co-occurrences. The topological structure of graph was then studied in relation to three factors: (i) how centralized is the graph's structure. Data on dominant users or topics and links, (ii) A measure of the coherence of the communication context, (iii) the orientation towards the formation of creative decisions. The defining of the respective metrics. The values of these metrics were associated with a classification of 'communities' using the following categories:

- topic-based,
- problem solving,
- focused on product / service evaluation,
- mutual supporting forums for the use of users,
- focus on the establishment of friendly relations between the participants,
- interested in taking part in discussions,

These have a degree of specific characteristics. The term "community" was used to represent the graph resulting from a discussion group. The existence of groups having a different orientation within a "community" was also considered. Special reference was made to user *roles*. This distinguished some of the users as "leaders" who can propose new ideas, drive discussion within a group, or circulate ideas of interest between the different groups. This research work is important from a theoretical standpoint. It does not suggest that an algorithm for the automatic extraction of groups has a different orientation from that of the "community'.

A subsequent research paper (Zhou, Manavoglu, Li, Giles, and Zha, 2006) aims at discovering user communities in social networks using an analysis of semantically rich text documents. This can include e-mails, instant messaging and message

[3] Facebook Statistics:
`http://www.facebook.com/press/info.php?statistics.`
Last-access date: 19/05/2010

boards. This method creates a Bayes network in order to model the formation of such documents in a social network. It uses the author or the recipient of the document and the topic as variables. The community is a latent variable. It should be noted that this method requires setting the number of communities and topics in advance. One of the proposed approaches leads to a polynomial distribution of user communities. There is also the possibility of finding a polynomial distribution of topics in the communities. The advantage of this method is that a semantic description is obtained for each community, through *topic tags* that define the topics. The proposed method was tested in the data/set that consists of the e-mails of the Enron company[4] and it produced results similar to those obtained by applying an algorithm based on Modularity Optimization (Clauset, Newman, and Moore, 2004).

Another research topic that relates to the problem of finding communities of users is the identification of users who have played the most active role in a community. The work of Zhang, Ackerman, and Adamic (Zhang, Ackerman, and Adamic, 2007) focused on systems for finding experts (*expertise finders*). This is used to find the users who have the necessary expertise on a topic. The on-line Java Forum community served as application field for this research work. The users pose and answer questions about the Java programming language. Applying appropriate ranking algorithms to the graph of the corresponding social network. Examples are ExpertiseRank,which is based on PageRank (Page, Brin, Motwani, & Winograd, 1999). A variant of HITS (Kleinberg, 1999), was used on simple metrics. It was proved that the graph structure can be used to evaluate the users of an on-line network of experts. The emergence of more experienced users can be observed. In addition, according to this study, the choice of an appropriate algorithm will result in a more accurate ranking also depends on the network structure.

A research paper dealing with a similar issue (Shin, Xu, and Kim, 2008) uses the term *POWER USERS* to characterize users of on-line communities. These are very popular and have a good reputation regarding their activities. It aims to design a method for identifying them. Several statistical features have been defined based on the comments and mail of the users. It also provides information on the features based on the relationships existing between users. After experimenting with the proposed features, *Cross Reference* (CR) emerges as the feature that provides the greatest precision in the identification of power users. This is provided that an appropriate threshold value has been defined. This feature is based on the number of comments exchanged between a given user and the other members. The research results were utilized in the development of a search engine for users. It was based on the cross reference feature.

Recent research (Kammergruber, Viermetz, and Ziegler, 2009) focuses on the problem of finding communities of users in *Social Tagging Systems*. The proposed method creates a vector for each user with the tags that He / She has used as components. Then, the *Cosine Similarity* Metric is used to express the similarity between the pairs of vectors. Thus the similarity of the behavior or interests of the respective users is noted. The program then applies the algorithm DBSCAN (Ester, Kriegel, Sander, and Xu, 1996) for the clustering of similar vectors which is then applied to the corresponding users. The resulting communities consist of

[4] `http://www.cs.cmu.edu/~enron/`

users who share common interests and the interests of the community of users are then represented by a group of tags. The authors propose the utilization of the resulting communities in the development of: friendship ties between the members of the community, the content of joint interest, relevant tags, and the development of a social tagging system of users.

This overview concludes with the use of current trends relating to the problem of identifying user communities in social networks. It is observed that there is a limited number of research works in this area. There are some studies about the utilization of user relationships in social networks. They model networks using graphs and focus on problems such as: The Analysis of the Structure and Characteristics of Small-Sized Social Networks (e.g. Ohsawa, Soma, Matsuo, Matsumura, and Usui, 2002). The recognition of the most active users in a network through the use of appropriate metrics (Zhang, Ackerman, and Adamic, 2007) (Shin, Xu, & Kim, 2008). The problem of user community identification considered in some research works. These focus and experiment using specific types of networks. They include the network that results from: (i) the modeling of all the e-mail messages exchanged during a time period between the members of a company (Zhou, Manavoglu, Li, Giles, & Zha, 2006), and (ii) a social tagging system (Kammergruber, Viermetz, and Ziegler, 2009). It is evident that there is a need to develop a methodology that can globally solve, with appropriate automatic adjustment, the problem of finding communities of users in social networks of most types.

Most research works link two users in the graph model of the given network. When their relationships are not clearly defined (e.g. through friendship links), on the basis of Text Analysis and Word / Expression co-occurrence, The use of Traditional Similarity Measures, such as the *cosine similarity and also used.* An important issue that is not fully addressed in the literature is the possibility of extracting an automatic semantic description for the communities and their evaluation. Based on this, it is evident that there is a need to develop additional user similarity and community evaluation measures. These will take into account features of the behavior of the users. Examples are the content such as Text, Image, for example. In addition the relevance and cohesion of the communities, must respectively be assessed. It is important to include the simple application of traditional techniques when identifying communities in graphs. It is also important to develop a methodology that will take into account the special characteristics concerning the behavior of users in social networks. This would include the execution time required for a given activity, the emotional state, the special roles of users, and the degree of privacy necessary. This is to name but a few of the possible variables.

4 Motivation for Community Identification and Indicative Application Areas

The analysis of data concerning the users of social networks and the Web 2.0 applications can be utilized by various applications in order to provide new opportunities for ; promotion, recommendation and projection of content. Other factors are the improvement and the quality of the services and thereby increase user satisfaction. Some indicative application types follow in which social network analysis methods, and more specifically methods for finding communities, which can be integrated and used. Investigation must be made as to whether there are any imminent relevant applications, including related examples.

- **Personalization:** The Social Web represents a new philosophy in which the users themselves are the main producers of the content (User Contributed Content). The users create, manage and spread the content and the information (Coppola, Lomuscio, Mizzaro, and Nazzi, 2008). This is done by the use of Wikies, Blogs, Social Tagging and multimedia sharing systems. For example YouTube, flickr, Twitter, Jaiku[5], Social Networking Applications such as Facebook are used. Semantic Web technologies provide the opportunity of developing better personalized applications and services. These may provide new capabilities, which allow users to describe aspects such as social networks, environment data in a standardized way (Heath and Motta, 2006). Semantic tools such as the RDF model, the SPARQL language, ontologies, the OWL language, and the microformat approach, allow the analysis and representation of social networks. The resulting data is presented in a structured and consistent manner. Consequently, it is possible to combine data from different sources and facilitate the utilization of "social" data (Ereteo, Buffa, Gandon, Grohan, Leitzelman, and Sander, 2008).

- **Content Recommendation:** There has been an explosive like growth of information in social networking applications in the last few years. The role of content recommendation applications has consequently become very important (Adomavicius and Tuzhilin, 2005). Existing content recommendation programs are mainly based on the content required by users (content-based) (Pazzani & Billsus, 2007). The relationships between the users is through their common requirements. See Collaborative Filtering (Herlocker, 2004). Hybrid techniques of content recommendation have also been developed (Burke, 2002). In the context of Web 2.0 applications to new problems related to content recommendation have appeared., Examples are recommendations for tags (Sigurbjörnsson and Zwol, 2008). The detection of user communities has had a direct impact on the recommendation techniques. The information retrieval model for a user varies depending on the community in which they belong (Almeida and Almeida, 2004). Despite the growing importance of content recommendation engines arising from web applications (such as Amazon[6] and Digg[7]), the existing mechanisms have not benefited from community detection technologies.

---

[5] `http://www.jaiku.com/`

[6] *Amazon.com. Recommendations Item-to-Item Collaborative Filtering* `http://www.win.tue.nl/~laroyo/2L340/resources/Amazon-Recommendations.pdf`

[7] *Digg's new recommendation system relies on the wisdom of crowds* `http://www.technologyreview.com/Infotech/21045/page1/`

- **Trend Identification:** An analysis of the communities can contribute to the recognition of recent activity which uncovers emerging trends within a social tagging system. In (Sun and others, 2008) a statistical method used to identify interesting topics from a dataset from Delicious. In (Hotho and others, 2006) a trend identification measure is proposed that "catches" trends related to a subject based on a variant of the PageRank algorithm. Finally, in (Wetzker and others, 2008) a statistical model that identifies topics and indicates trends in a dataset of Delicious is shown and apply a smoothing process that leads to a better calibration of the model.

5 The Aims of This Book

This book has assembled major topics which embed new trends and practices in terms of Web Data Management with a focus on modern technologies and Web 2.0 reality. More specifically, the book is organized into three major sections which include 10 chapters in a step-by-step manner. It starts with structures, proceeds to technologies and searching and concludes with an emphasis on the social media of the Web 2.0. More specifically, the major themes of this book are as follows:

- Structures and Metrics

  Massive Graph Management for the Web and Web 2.0
  Web Engineering and Metrics
  Modern Web Technologies

- Management of Searching and Querying

  Federated Data Management and Query
  Optimization for Linked Open Data
  Queries over Web Services
  Towards an Adaptively Approximated Search in Distributed Architectures

- Social Media and Semantics

  Online Social Networks: Status and Trends
  Enhancing Computer Vision using the Collective Intelligence of Social Media
  From XML to AXML
  Migrating Legacy Assets through SOA to Realize Network Enabled Capability

The editors hope that this book will be easy to follow and be of interest to engineers and computer scientists who already have a basic background on topics such as data mining, databases and Web technologies. The book will be useful to the researchers as well as graduate students or (advanced) undergraduate students in engineering or computer science academic programs.

References

[1] Adomavicius, G., Tuzhilin, A.: Towards the Next Generation of Recommender Systems: A Survey of the State of the Art and Possible Extensions (2005)

[2] Alba, R.D.: A Graph-Theoretic Definition of a Sociometric Clique. J. Math. Soc. 3, 113–126 (1973)

[3] Almeida, R.B., Almeida, V.A.: A community-aware search engine. In: Proceedings of the 13th international Conference on World Wide Web, WWW 2004, May 17 - 20, pp. 413–421. ACM, New York (2004)

[4] Arenas, A., Díaz-Guilera, A., Peréz-Vicente, C.: Synchronization reveals topological scales in complex networks. Physical Review Letter 96, 114102 (2006)

[5] Barnes, E.R.: An algorithm for partitioning the nodes of a graph. SIAM J. Algebraic Discrete Methods 3(4), 541–550 (1982)

[6] Batagelj, V., Zaveršnik, M.: Generalized cores. Eprint arXiv:cs/0202039 (2002), `http://www.arxiv.org`

[7] Baumes, J., Goldberg, M.K., Magdon-Ismail, M.: Efficient Identification of Overlapping Communities. In: Kantor, P., Muresan, G., Roberts, F., Zeng, D.D., Wang, F.-Y., Chen, H., Merkle, R.C. (eds.) ISI 2005. LNCS, vol. 3495, pp. 27–36. Springer, Heidelberg (2005)

[8] Boccaletti, S., Ivanchenko, M., Latora, V., Pluchino, A., Rapisarda, A.: Detecting complex network modularity by dynamical clustering. Phys. Rev. E 75(4), 45102 (2007)

[9] Bollobás, B.: Modern Graph Theory. Springer, New York (1998)

[10] Burke, R.: Hybrid Recommender Systems: Survey and Experiments (2002)

[11] Capocci, A., Servedio, V., Caldarelli, G., Colaiori, F.: Detecting communities in large networks. Physica A 352, 669–676 (2004)

[12] Cattuto, C., Baldassarri, A., Servedio, V.D.P., Loreto, V.: Emergent Community Structure in Social Tagging Systems. Advances in Complex Systems 11(4), 597–608 (2008)

[13] Chan, P.K., Schlag, M.D., Zien, J.Y.: Spectral K-way ratio-cut partitioning and clustering. In: Proceedings of the 30th International Conference on Design Automation, pp. 749–754. ACM Press, Dallas (1993)

[14] Chiang, Y.-J., Goodrich, M.T., Grove, E.F., Tamassia, R., Vengroff, D.E., Vitter, J.S.: External Memory Graph Algorithms. In: Proceedings of the 6th Annual ACM-SIAM Symposium on Discrete Algorithms San Francisco, California, United States, Symposium on Discrete Algorithms, January 22 - 24, pp. 139–149. Society for Industrial and Applied Mathematics, Philadelphia (1995)

[15] Clauset, A.: Finding local community structure in networks. Physical Review E 72 (2005)

[16] Clauset, A., Newman, M.E., Moore, C.: Finding community structure in very large networks. Physical Review E 70, 66111 (2004)

[17] Coppola, P., Lomuscio, R., Mizzaro, S., Nazzi, E.: m-Dvara 2.0: Mobile & Web 2.0 Services Integration for Cultural Heritage, Social Web and Knowledge Management. In: Social Web 2008 Workshop at the 17th World Wide Web Conference (WWW 2008), Beijing, China, April 22 (2008)

[18] Donetti, L., Muñoz, M.A.: Detecting network communities: a new systematic and efficient algorithm. Journal of Statistical Mechanics: Theory, P10012 (2004)

[19] Duch, J., Arenas, A.: Community detection in complex networks using Extremal Optimization. Phys. Rev. E 72(2), 27104 (2005)
[20] Ereteo, G., Buffa, M., Gandon, F., Grohan, P., Leitzelman, M., Sander, P.: A state of the Art on Social Network Analysis and its Applications on a Semantic Web. In: Proceedings of the ISWC 2008 Workshop on Social Data on the Web (SDoW 2008), Karlsruhe, Germany, October 27 (2009)
[21] Ester, M., Kriegel, H.-P., Sander, J., Xu, X.: A Density-Based Algorithm for Discovering Clusters in Large Spatial Databases with Noise. In: Proceedings of the 2nd International Conference on Knowledge Discovery and Data Mining (KDD 1996), pp. 226–231 (1996)
[22] Fortunato, S., Barthelemy, M.: Resolution limit in community detection. Proceedings of the National Academy of Sciences USA 104(1), 36–41 (2007)
[23] Fortunato, S., Castellano, C.: Community structure in graphs. In: Encyclopedia of Complexity and System Science. Springer, Heidelberg (2008)
[24] Fortunato, S., Latora, V., Marchiori, M.: A method to find community structures based on information centrality. Physical Review E 70, 56104 (2004)
[25] Giannakidou, E., Kompatsiaris, I., Vakali, A.: SEMSOC: Semantics Mining on Multimedia Social Data Sources. In: Proceedings of the 2nd IEEE International Conference on Semantic Computing, Santa Clara, CA, USA (2008a)
[26] Girvan, M., Newman, M.E.: Community structure in social and biological networks 99, 7821–7826 (2002)
[27] Gregory, S.: An Algorithm to Find Overlapping Community Structure in Networks. In: Kok, J.N., Koronacki, J., Lopez de Mantaras, R., Matwin, S., Mladenič, D., Skowron, A. (eds.) PKDD 2007. LNCS (LNAI), vol. 4702, pp. 91–102. Springer, Heidelberg (2007)
[28] Hastings, M.B.: Community Detection as an Inference Problem. Phys. Rev. E 74(3), 35102 (2006)
[29] Heath, T., Motta, E.: Personalizing Relevance on the Semantic Web through Trusted Recommendations from a Social Network. In: International Workshop on Semantic Web Personalization, Montenegro, June 12 (2006)
[30] Herlocker, J.L., Konstan, J.A., Terveen, L.G., Riedl, J.T.: Evaluating collaborative filtering recommender systems (2004)
[31] Hofman, J., Wiggins, C.: A Bayesian approach to network modularity. Physical Review Letters 100, 258701 (2008)
[32] Hopcroft, J., Khan, O., Kulis, B., Selman, B.: Tracking Evolving Communities in Large Linked Networks. Proceedings of the National Academy of Sciences 101, 5249–5253 (2004)
[33] Hotho, A., Jaschke, R., Schmitz, C., Stumme, G.: Trend Detection in Folksonomies. In: Proceedings of SAMT, pp. 56–70 (2006)
[34] Jain, A.K., Dubes, R.C.: Algorithms for Clustering Data. Prentice-Hall, Upper Saddle River, NJ (1988)
[35] Kammergruber, W., Viermetz, M., Ziegler, C.: Discovering Communities of Interest in a Tagged On-Line Environment. In: CASON 2009: Proceedings of the 2009 International Conference on Computational Aspects of Social Networks, pp. 143–148. IEEE Computer Society, Washington (2009)
[36] Karypis, G., Kumar, V.: A fast and high quality multilevel scheme for partitioning irregular graphs. SIAM Journal on Scientific Computing 20(1), 359–392 (1999)
[37] Kernighan, B., Lin, S.: An efficient heuristic procedure for partitioning graphs. The Bell System Technical Journal 49, 291–307 (1970)

[38] Kleinberg, J.M.: Authoritative sources in a hyperlinked environment. J. ACM 46(5), 604–632 (1999), DOI= http://doi.acm.org/10.1145/324133.324140
[39] Koutsonikola, V., Vakali, A., Giannakidou, E., Kompatsiaris, I.: Clustering of Social Tagging System Users: A Topic and Time Based Approach. In: Vossen, G., Long, D.D.E., Yu, J.X. (eds.) WISE 2009. LNCS, vol. 5802, pp. 75–86. Springer, Heidelberg (2009)
[40] Koutsonikola, V.A., Petridou, S.G., Vakali, A.I., Hacid, H., Benatallah, B.: Correlating Time-related Data Sources with Co-clustering. In: Bailey, J., Maier, D., Schewe, K.-D., Thalheim, B., Wang, X.S. (eds.) WISE 2008. LNCS, vol. 5175, pp. 264–279. Springer, Heidelberg (2008)
[41] Li, X., Snoek, C.G.M., Worring, M.: Learning Social Tag Relevance by Neighbor Voting. IEEE Transactions on Multimedia 11(7), 1310–1322 (2009)
[42] MacQueen, J.B.: Some methods for classification and analysis of multivariate observations. In: 5th Berkeley Symposium on Mathematical Statistics and Probability, pp. 281–297. University of California Press, Berkeley (1967)
[43] Massen, C.P., Doye, J.P.: Identifying "communities" within energy landscapes. Phys. Rev. E 71(4), 46101 (2005)
[44] Mika, P., Flink, J.: Semantic Web technology for the extraction and analysis of social networks. Web Semantics 3(2), 211–223 (2005)
[45] Milo, R., Itzkovitz, S.: Network motifs: Simple building blocks of complex networks. Science 298, 824–827 (2002)
[46] Nepusz, T., Petróczi, A., Négyessy, L., Bazsó, F.: Fuzzy communities and the concept of bridgeness in complex networks. Phys. Rev. E 77(1), 16107 (2008)
[47] Newman, M.E.J.: Analysis of weighted networks. Phys. Rev. E 70(5), 56131 (2004)
[48] Newman, M.E.J.: Fast algorithm for detecting community structure in networks. Phys. Rev. E 69, 66133 (2004a)
[49] Newman, M.E.J.: Finding community structure in networks using the eigenvectors of matrices. Physical Review E 74 (2006)
[50] Newman, M.E.J.: Modularity and community structure in networks. Proceedings of the National Academy of Sciences of USA 103, 8577–8582 (2006a)
[51] Newman, M.E., Girvan, M.: Finding and evaluating community structure in networks. Phys. Rev. E 69(2), 26113 (2004)
[52] Newman, M.E., Leicht, E.A.: Mixture models and exploratory analysis in networks. Proc. Natl. Acad. Sci. USA 104, 9564–9569 (2007)
[53] Ohsawa, Y., Soma, H., Matsuo, Y., Matsumura, N., Usui, M.: Featuring web communities based on word co-occurrence structure of communications: 736. In: WWW 2002: Proceedings of the 11th international conference on World Wide Web, p. 742. ACM, New York (2002)
[54] Page, L., Brin, S., Motwani, R., Winograd, T.: The PageRank Citation Ranking: Bringing Order to the Web. Technical Report. Stanford InfoLab (1998)
[55] Palla, G., Derényi, I., Farkas, I., Vicsek, T.: Uncovering the overlapping community structure of complex networks in nature and society. Nature 435, 814–818 (2005)
[56] Panconesi, A., Rizzi, R.: Some Simple Distributed Algorithms for Sparse Networks. Distributed Computing 14(2), 97–100 (2001), DOI= http://dx.doi.org/10.1007/PL00008932
[57] Pazzani, M.J., Billsus, D.: Content-Based Recommendation Systems (2007)

[58] Pinney, J., Westhead, D.: Betweenness-based decomposition methods for social and biological networks. In: Barber, P.B.S., Barber, S., Baxter, P., Mardia, K., Walls, R. (eds.) Interdisciplinary Statistics and Bioinformatics: Proceedings. Leeds University Press, Leeds (2006)
[59] Pons, P., Latapy, M.: Computing Communities in Large Networks Using Random Walks. In: Yolum, p., Güngör, T., Gürgen, F., Özturan, C. (eds.) ISCIS 2005. LNCS, vol. 3733, pp. 284–293. Springer, Heidelberg (2005)
[60] Pothen, A., Simon, H., Liou, K.P.: Partitioning sparse matrices with eigenvectors of graphs. SIAM journal of Matrix Analysis and Application 11, 430–452 (1990)
[61] Quack, T., Leibe, B., Van Gool, L.: World-scale mining of objects and events from community photo collections. In: Proceedings of the 2008 international Conference on Content-Based Image and Video Retrieval, Niagara Falls, Canada (2008)
[62] Radicchi, F., Castellano, C., Cecconi, F., Loreto, V., Parisi, D.: Defining and identifying communities in networks. Proceedings of the National Academy of Science of the United States of America 101, 2658–2663 (2004)
[63] Reichardt, J., Bornholdt, S.: Detecting Fuzzy Community Structures in Complex Networks with a q-state Potts Model. Phys. Rev. Lett. 93, 218701 (2004)
[64] Reichardt, J., White, D.: Role Models for Complex Networks. European Physical Journal B 60, 217–224 (2007)
[65] Rosvall, M., Bergstrom, C.T.: Maps of random walks on complex networks reveal community structure. Proceedings of the National Academy of Sciences of USA 105, 1118 (2008)
[66] Schaeffer, S.E.: Graph Clustering. Computer Science Review 1(1), 27–64 (2007)
[67] Scott, J.: Social Network Analysis, a handboook. SAGE publications, London (2000)
[68] Seidman, S.B., Foster, B.L.: A graph theoretic generalization of the clique concept. Journal of Mathematical Sociology 6, 139–154 (1978)
[69] Shepitsen, A., Gemmell, J., Mobasher, B., Burke, R.: Personalized recommendation in social tagging systems using hierarchical clustering. In: RecSys 2008: Proceedings of the 2008 ACM conference on Recommender systems, pp. 259–266. ACM, New York (2008)
[70] Shi, J., Malik, J.: Normalized cuts and image segmentation. IEEE Transactions on Pattern Analysis and Machine Intelligence 22(8), 888–905 (2000)
[71] Shin, H., Xu, Z., Kim, E.: Discovering and Browsing of Power Users by Social Relationship Analysis in Large-Scale Online Communities. In: WI-IAT 2008: Proceedings of the 2008 IEEE/WIC/ACM International Conference on Web Intelligence and Intelligent Agent Technology, pp. 105–111. IEEE Computer Society, Washington (2008)
[72] Sigurbjörnsson, B., van Zwol, R.: Flickr tag recommendation based on collective knowledge. In: Proceeding of the 17th international Conference on World Wide Web, WWW 2008, Beijing, China, April 21 - 25, pp. 327–336. ACM, New York (2008)
[73] Šíma, J., Schaeffer, S.E.: On the NP-Completeness of Some Graph Cluster Measures. In: Thirty-second International Conference on Current Trends in Theory and Practice of Computer Science (Sofsem 2006), pp. 530–537. Springer, Berlin (2006)
[74] Son, S.-W., Jeong, H., Noh, J.-D.: Random field Ising model and community structure in complex networks. Eur. Phys. J. B 50(431) (2006)
[75] Tang, L., Liu, H.: Graph Mining Applications to Social Network Analysis. In: Managing and Mining Graph Data. Springer, Heidelberg (2009) (in press)
[76] Tasgin, M., Herdagdelen, A., Bingol, H.: Community Detection in Complex Networks Using Genetic Algorithms (2007) (preprint)

[77] Tyler, J.R., Wilkinson, D.M., Huberman, B.A.: Email as spectroscopy: automated discovery of community structure within organizations. In: Communities and technologies, pp. 81–96. Kluwer, B.V, Deventer, The Netherlands (2003)
[78] van Dongen, S.: Graph Clustering by Flow Simulation, Ph.D. thesis. Ph.D. thesis, Dutch National Research Institute for Mathematics and Computer Science, University of Utrecht, Netherlands (2000)
[79] Vakali, A., Kompatsiaris, Y.: Detecting and Understanding Web communities. In: Proceedings of the WebSci 2009: Society On-Line, Athens, Greece, March 18-20 (2009)
[80] Vázquez, A., Dobrin, R., Sergi, S., Eckmann, J.–P., Oltvai, Z.N., Barabási, A.–L.: The topological relationship between the large-scale attributes and local interaction patterns of complex networks. Proceedings of the National Academy of Sciences of USA 101(52), 17940–17945 (2004)
[81] Wetzker, R., Plumbaum, T., Korth, A., Bauckhage, C., Alpcan, T., Metze, F.: Detecting Trends in Social Bookmarking Systems using a Probabilistic Generative Model and Smoothing. In: Proceedings of the International Conference on Pattern Recognition (ICPR). IEEE, Los Alamitos (2008)
[82] Wetzker, R., Zimmermann, C., Bauckhage, C.: Analyzing social bookmarking Systems: A del.icio.us cookbook. In: Proceedings of ECAI 2008 Workshop on Mining Social Data (MSoDa), Patras, Greece, July 2008, p. 2630 (2008)
[83] White, S., Smith, P.: A Spectral Clustering Approach to Finding Communities in Graphs. In: Proceedings of the SIAM Data Mining Conference (SDM), Newport Beach, California, pp. 76–84 (2005)
[84] Zanardi, V., Capra, L.: Social Ranking: Uncovering Relevant Content Using Tag-based Recommender Systems. In: RecSys 2008: Proceedings of the 2008 ACM conference on Recommender systems, pp. 51–58. ACM, New York (2008)
[85] Zarei, M., Samani, K.A.: Eigenvectors of network complement reveal community structure more accurately. Physica A: Statistical Mechanics and its Applications 388(8), 1721–1730 (2009)
[86] Zelke, M.: Algorithms for Streaming Graphs: Approaching Graph Problems with Limited Memory and without Random. Suedwestdeutscher Verlag fuer Hochschulschriften (2009)
[87] Zhang, J., Ackerman, M.S., Adamic, L.: Expertise networks in online communities: structure and algorithms. In: WWW 2007: Proceedings of the 16th international Conference on World Wide Web, Banff, Alberta, Canada, pp. 221–230. ACM, New York (2007)
[88] Zhang, Y., Xu Yu, J., Hou, J.: Web Communities: Analysis and Construction. Springer, Heidelberg (2006)
[89] Zhou, D., Manavoglu, E., Li, J., Giles, C.L., Zha, H.: Probabilistic models for discovering e-communities. In: WWW 2006: Proceedings of the 15th international conference on World Wide Web, pp. 173–182. ACM, Edinburgh (2006)
[90] Zhou, H.: Network landscape from a Brownian particle's perspective. Phys. Rev. E 67, 41908 (2003)
[91] Ziv, E., Middendorf, M., Wiggins, C.H.: Information-theoretic approach to network modularity. Phys. Rev. E 71(4), 46117 (2005)

Chapter 2
Massive Graph Management for the Web and Web 2.0

Maria Giatsoglou[1], Symeon Papadopoulos[1,2], and Athena Vakali[1]

[1] Aristotle University, 54124, Thessaloniki, Greece
mgiatsog@csd.auth.gr,
avakali@csd.auth.gr
[2] Informatics and Telematics Institute, CERTH, 57001, Thermi, Greece
papadop@iti.gr

Abstract. The problem of efficiently managing massive datasets has gained increasing attention due to the availability of a plethora of data from various sources, such as the Web. Moreover, Web 2.0 applications seem to be one of the most fruitful sources of information as they have attracted the interest of a large number of users that are eager to contribute to the creation of new data, available online. Several Web 2.0 applications incorporate *Social Tagging* features, allowing users to upload and tag sets of online resources. This activity produces massive amounts of data on a daily basis, which can be represented by a *tripartite graph* structure that connects users, resources and tags. The analysis of *Social Tagging Systems* (STS) emerges as a promising research field, enabling the identification of common patterns in the behavior of users, or the identification of communities of semantically related tags and resources, and much more. The massive size of STS datasets dictates the necessity for a robust underlying infrastructure to be used for their storage and access.

This chapter contains a survey of existing solutions to the problem of storing and managing massive graph data focusing particularly on the implications that the underlying technologies of such frameworks have on the support/operation of Web 2.0 applications using them as back-end storage solutions, as well as on the efficient execution of web mining tasks. Considering the category of STS as an example of Web 2.0 applications, the requirements that are posed for the management of STS data are thoroughly discussed. On the basis of these requirements three frameworks have been developed, using state-of-the-art technologies as backbones. The results of benchmarks conducted on the developed frameworks are presented and discussed.

1 Introduction

The widespread adoption of Web 2.0 tools and technologies that took place during the last years has fundamentally changed the way information is published on the Web. A plethora of Web 2.0 applications, including Social Tagging Systems, Wikis, and Blogs, have emerged, amongst which there are some that recently gained

A. Vakali & L.C. Jain (Eds.): New Directions in Web Data Management 1, SCI 331, pp. 19–58.
springerlink.com

profound success. Some of the most well-known examples of successful Web 2.0 applications are: *Facebook*[1], a social networking website counting hundreds of millions of users, *Flickr*[2], a photo management and sharing application that allows users to tag pictures and form communities, and *delicious*[3], a social bookmarking web service where users can store, share and retrieve bookmarks. What is common between all Web 2.0 applications is that the activity of users results in data that are interconnected through associations, thus forming a network.

The breakthrough of Web 2.0 applications was accompanied by the eagerness of a large proportion of people to join them and to actively contribute to the generation and publishing of Web content. This type of user activity produces massive amounts of data on a daily basis, regarding the uploaded content itself, as well as the relations formed between (a) users, (b) users and shared/uploaded content, and (c) content and metadata (such as *tags*) associated to it by users. A rough idea of the amount of these data can be drawn taking Facebook as an example, where each week more than 3.5 billion pieces of content (such as web links, news stories, blog posts, notes, photos) are shared [67]. The data magnitude, the need to model their relation structure, as well as to efficiently store and retrieve them, have created new challenges in the field of data management. Classic data management solutions, such as data warehouses, seem to be inadequate to store efficiently massive sets of relational data. Moreover, emphasis has been moved from traditional entry-based data access, e.g. customer records, to *navigational access* that allows reaching e.g., the references of an article, the friends of a user via friendship links, etc. The design and implementation of a robust data management framework that manages to maintain a stable performance as the size of data increases, and support navigational queries in a optimal way is still a challenging task for web-scale retrieval systems.

The existence of such massive amounts of data containing complex and emerging structures has also given new impetus to the field of data mining. The information of how users or online resources relate to each other, as well as how users react to resources has captured the interest of researchers, as it was soon realized that it could be exploited to deduce interesting conclusions about how groups of people characterize resources and interpret content, or even what pieces of information tend to be more popular among them. The analysis of Web 2.0 data is further motivated by the notion that the collaboration and contribution of many individuals results in the "formation" of a shared or group intelligence, characterized as *collective intelligence*. Collective intelligence is a new source of information that can be utilized in a variety of applications, as it is produced by the contribution of multiple people representing different views and ideas. For example, it can be exploited in order to uncover groups of either users that share common interests, resources that seem to belong to the same thematic region, or tags (usually referred to as *communities* of users, resources, or tags, respectively). The discovery of such meaningful communities can be utilized in applications such as recommender systems, in order to

[1] `http://www.facebook.com/`
[2] `http://www.flickr.com/`
[3] `http://delicious.com/`

increase their efficiency. For example, *tag communities* can be used in a system that recommends tags to users that they would possibly find relevant to a given resource.

The analysis of relational data produced by Web 2.0 applications, however, requires the use of special methods and poses a question on the data structure that should be used for storing and accessing them. A natural way to model Web 2.0 data seems to be the *network* or *graph* model where *nodes* represent entities/objects and *edges* represent the relations that exist between them. In order to enable the progress of research on such relational datasets continuously increasing in size, a prerequisite is the availability of a robust framework, appropriate for storing and accessing graph-based data. Some of the most challenging issues that should be carefully taken into consideration are: the storage of large graphs (e.g. of 10^9 nodes and 10^{10} edges) in a form that will be as compact as possible, the support for reasonably fast graph traversals and updates, and the design of a framework that will be easy to use and adaptable to the specifications of individual applications.

This chapter provides a review of several solutions and infrastructures used for the storage and analysis of very large graphs, and also discusses and compares their individual characteristics and limitations. Moreover, the special case of using a Social Tagging Systems (STS) as a source of data that can be modeled as a tripartite graph is thoroughly discussed, as an interesting application area. After considering and summarizing the requirements for the storage and analysis of data from STS, we present the results of a set of benchmark experiments that have been designed to compare the performance of three STS data management frameworks built upon different graph persistence technologies, with respect to the storage and management of graph-based data derived from social tagging applications.

The rest of the report is structured as follows. Section 2 discusses the challenges presented by the analysis of massive graphs and includes a categorization of the different available solutions for the management of graph-structured data. Section 3 and Section 4 provide an overview of some of the most recent graph management solutions that belong to the category of transaction graph databases and data mining-oriented solutions, respectively. Section 5 presents STS as an application setting, describing some of the state-of-the-art data mining tasks that are currently being applied in the area, and also summarizes the requirements that these tasks impose on the underlying framework used. Section 6 describes the architecture of three frameworks that have been developed for the management of STS data, presents a set of benchmarks experiments designed to test and compare their performance, and discusses the benchmark results. Finally, Section 7 concludes the chapter.

2 Handling Massive Graphs on the Web

The study of the Web has recently emerged as a new research field. Researchers started to model the Web as a network consisting of nodes representing web pages and edges representing the hyperlinks that connect them, forming the so-called *Web Graph*. The edges in such a model can be: directed (e.g. a hyperlink leading from web page x to web page y), or undirected (e.g. a hyperlink leading from web page x

to web page y and vice versa). One of the earliest application domains that exploited the graph model of the Web to extract knowledge was the domain of the Web search engines [10,37]. However, as the Web gained more and more popularity, the number of web pages made available for analysis, acquired usually with the help of web crawlers, was rapidly increasing. While the Web started to reach the size of billions of web pages with ten or hundred times more edges, technological advances made it possible to collect for analysis datasets of sizes proportional to the aforementioned numbers. The availability of such large datasets posed new questions on what techniques and algorithms should be employed to analyze the data.

The obvious problem is that as sizes are getting bigger, the main memory of an average personal computer does not suffice anymore in order to load and manipulate all of the data at once. This has created the need for the development and employment of alternative storage and analysis techniques (Figure 1). Some of the most straightforward approaches are:

- to compress the data so as to reach a size small enough in order to fit in an average computer's RAM and then analyze them,
- to store the data in an external memory repository, and fetch them in batches when required by the analysis algorithm, combining possibly a caching schema to increase performance,
- to use a cluster or a grid of computer nodes in order to distribute the data so as to fit into each node's RAM for faster analysis, and then aggregate the result.

A prerequisite for efficient access to Web and Web 2.0 data within information retrieval scenarios or during the execution of demanding analysis operations is the existence of a robust underlying graph management framework. Frameworks for large graphs' management are usually disk-based, enabling the persistent storage of the large amounts of graph data. There are numerous approaches as to how to store and provide access to such data, that make use of existing infrastructures. Figure 2 depicts a categorization of existing persistent graph frameworks. Existing solutions can be distinguished in two generic categories depending on the reason why the

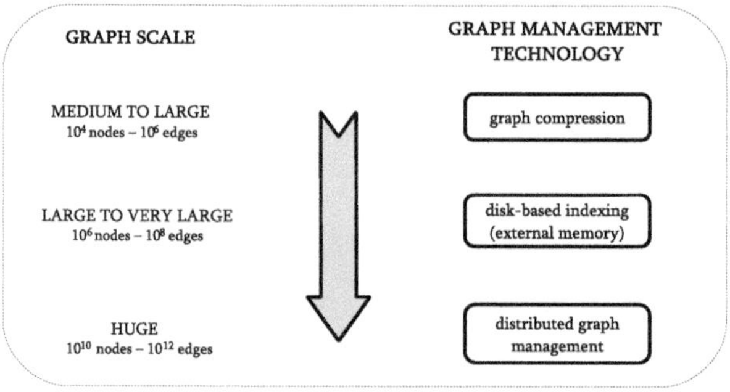

Fig. 1. Techniques to store and analyze graph data depending on graph scale

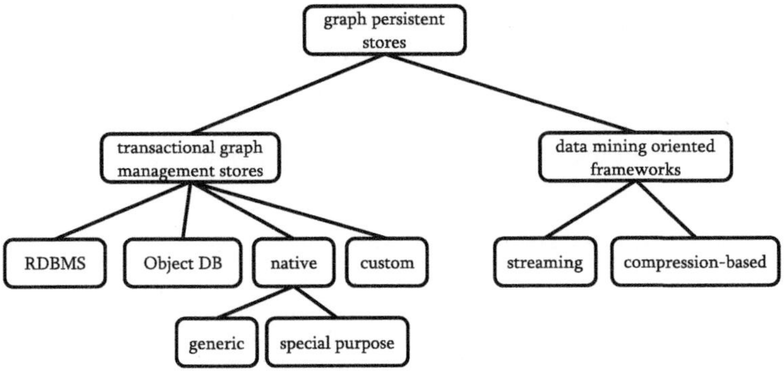

Fig. 2. Categories of graph management frameworks

storage and availability of the data is required: (a) transactional graph databases, and (b) data-mining oriented solutions.

Transactional graph databases can be used for the management of graphs where data (modeled as graph nodes or edges) can be inserted, deleted or updated on demand. This type of frameworks support ACID transactions to ensure reliable processing of database operations. The underlying infrastructures are disk-based enabling the persistent storage of large graphs. The infrastructures that can be used in a transactional graph database can be classified as follows:

- Frameworks based on Relational Database Management Systems (RDBMS),
- Frameworks based on Object Databases,
- Native graph stores, characterized as either: (i) generic, or (ii) special-purpose,
- Custom solutions.

In addition, as mentioned above, there is a requirement for frameworks to store and allow access to web graphs for data mining purposes. The most usual case in graph data mining (or graph mining) is to examine static datasets, and analyze data with algorithms that involve random navigational access to graph nodes and edges. Graph mining operations therefore pose different requirements to the respective data management framework, e.g. there is no need for graph updates, and also the graph accessing mechanisms should be as fast as possible in order for the algorithms to execute in a reasonable time. In general, data mining-oriented solutions can be distinguished in two subcategories:

- Streaming,
- Compression-based.

The categorizarion of graph frameworks depicted in Figure 2 is based on their suitability for a particular application setting. However, frameworks belonging to these categories may address the problem of scalability of the graph data in a different way. In particular, when the size of data is very large, persistent graph frameworks based on distributed computing infrastructures can be used in order to exploit the storage capacity of multiple computer nodes. In Sections 3 and 4 the categories of: (i) transactional graph databases and (ii) data mining-oriented graph management solutions, respectively, are thoroughly discussed. Each section presents representative examples of frameworks and methods belonging to the respective category, including examples of the special case of distributed graph management solutions.

3 Transactional Graph Databases

Transactional graph databases are disk-based dynamic graph management solutions that operate on the basis of *transactions*. The following subsections intend to provide a thorough insight in the different types of back-end infrastructures that can be used in a transactional graph database.

3.1 RDBMS-Based Frameworks

One early approach for storing networks has been the use of RDBMS, such as *MySQL*. The obvious reason is that RDBMS have been established as the dominant choice for storing data due to their simplicity, robustness, and flexibility as a generic data storage and manipulation mechanism, compared to their alternatives. Moreover, they provide native support for integrity constraint checking, removing this burden from the application side. However, nowadays relational databases receive criticism based on the argument that they are not efficient for managing relational data.

Critics claim that the RDBMS structure is too rigid for storing networks of data, considering that they store both data and their relationships in the form of tables. In particular, the use of tables makes it difficult to fit new kind of data, as their structure should be strictly defined from the beginning and cannot be altered later. Moreover, their most serious limitation is that relational databases are not scalable enough for graph access operations, especially when the size of data is continuously increasing.

However, some people support the use of traditional RDBMS for storing and analyzing graphs. For example, one recent approach [57] proposes:

- the storage of the graph nodes in an SQL table, using an integer identifier for each node as the primary key of the respective record, and also
- the storage of the graph edges in a separate table, using the source and destination nodes for each edge as foreign keys to the nodes table.

Requirements such as the uniqueness of an edge or the prevention of self-loops are ensured with the use of SQL *CHECK* constraints. The graph can then be traversed by either SQL querying, SQL standards *Common Table Expressions* (CTEs) that enable recursions though the nodes, or by using temporary tables [28]. In addition, the construction of the graph's *transitive closure*[4] with the use of CTEs is proposed. A graph's closure can be used to answer queries related to social networks, such as the degree of separation or the possible paths between two nodes. Nevertheless, the

[4] The transitive closure of a graph is a graph which contains an edge (u,v) whenever there is a directed path from u to v.

proposed methods might be too slow depending on the size of the graph and the application performance requirements, so there may be a need for employing caching schemes on top of such a framework. In the following paragraphs two RDBMS are presented, namely (i) H2 and (ii) Oracle DB, which are considered as a suitable basis for graph management frameworks.

H2 database. Using a fast database engine can partially mitigate the performance shortcomings of RDBMS-based graph frameworks. The H2 database engine [69] is a native Java RDBMS that appears a promising choice. Benchmark results show that not only the memory usage of H2 database is smaller, but also its query optimizer results in query times shorter than the times achieved by most competing RDBMS. Moreover, H2 is considered to be scalable as it creates both in-memory and disk-based tables, using hash table and tree indexing or B-tree indexing, respectively. Another important asset is that with H2 there is no limit on the size of the result set of a query, as it buffers the results to disk after a certain size of data is exceeded.

Oracle Database. This database supports modeling networks of data as graphs and analyzing them (since the 10*g* version). These functionalities are included in Oracle's *Network Data Model* (NDM)[5] [49]. NDM enables the storage of the network nodes, links (directed or undirected), as well as ordered lists of links that contain no repeating links or nodes, and are referred to as *paths*. Graphs are represented in object-relational form in the database, using separate tables, whereas queries and updates are performed via PL/SQL. NDM allows posing certain network constraints such as minimum bounding rectangle, path cost, and path depth, and also supports graph operations including shortest path between nodes, minimum cost spanning tree, k-nearest neighbors, k-shortest paths, as well as node and link buffering.

NDM analyzes networks after loading them entirely in memory, thus posing boundaries on the size of network that it can support. Its network analysis capabilities were enhanced in the 11*g* version of Oracle, with the introduction of the *load-on demand* (LOD) approach that made the analysis of larger networks possible [62]. With LOD, the network is not loaded in memory from the beginning, but is partitioned and after that, only the partitions that are required for analysis are loaded in memory automatically. Moreover, partition loading can be accelerated by generating and using BLOB representations.

3.2 Object Database-Based Frameworks

Object or *Object-oriented* (OO) *databases* constitute an alternative solution to RDBMS, combining object-oriented programming language capabilities with traditional persistent data storage and management features. Their use enables developers to model and store complex data as objects, without the need of defining and abiding to a specific relational schema, and simplifies the modification process that is required in case the data model changes. Another argument in favor of object databases with respect to RDBMS is their support for an object schema for data representation both within the application as well as for persistent storage, without

[5] Part of the Oracle Spatial component.

the need for an Object-Relational mapping [60], which is usually a rather cumbersome task. However, the use of *object* instead of *relational* databases results in bigger files for the same data, as they do not separate the structure from the data themselves. Regarding relationships between data, relational and object databases follow two different approaches; (i) relational relationships are usually based on set theory idioms, while (ii) object relationships are mainly based on idioms adopted from graph theory, such as trees, thus depending on the approach, information is accessed in different ways [74]. Moreover, object databases are in general considered to be faster than relational databases for specific access patterns such as navigational access, whereas this is not the case for direct queries to objects.

Object databases can be readily used for storing graph data, mapping the graph structure on an object schema. With such a mapping, e.g. each graph node can be represented by an object of the class *node* with the edges being represented as relationships between the appropriate *node* objects. This constitutes a simpler and more natural way of storing graph data than using a relational database, and is expected to be a faster solution due to the navigational nature of the graph access patterns. One shortcoming of using object databases is the bigger size of the database files.

Although not so widely used as RDBMS, there is a variety of object databases available. In the following paragraphs three popular open-source object databases are presented: (i) Oracle Berkeley DB, (ii) db4o, and (iii) Neodatis ODB.

Oracle Berkeley DB or *Berkeley DB* is an open-source object database, that can be embedded in applications developed in various programming languages, such as Java, C++, Perl, and Python. The use of the Berkeley DB library allows developers to freely decide how data will be stored in a record, without enforcing any constraints on the data. The database comes in three different editions that are also configurable to fit any application's special requirements, with some editions/configurations supporting traditional database features such as ACID transactions, locking, concurrency management, and replication [75].

Berkeley DB stores data as key/data pairs and supports B-tree, hash table, record and queue access methods. It does not support SQL queries, whereas queries can be performed with the use of indexes to each record. According to its developers, Berkeley DB is very scalable, supporting small databases that fit entirely in memory, as well as extremely large disk-resident databases of sizes up to 256 terabytes of data. In order to speed up access to data that are frequently accessed, Berkeley DB offers an in-memory cache [68].

db4o is another open-source object database library, that can be embedded in Java and .NET applications. Similar to Berkeley DB, db4o combines traditional database features, such as robustness, reliability, replication, concurrency support, with simplification of the data storage procedure. An interesting feature is that db4o not only creates automatically the data model that is required to store data objects during a transaction, but also updates the models on-demand [66]. db4o supports *Native Queries* (NQ) instead of string-based APIs, such as SQL, in order to enable database access using the programming language that has been used for the development of the application. Moreover, it supports the *Query by Example* (QbE) API to enable easy searching for matching objects, as well as the LINQ extensions for .NET. db4o uses B-trees for indexing, supports caching for efficient access to objects, and also provides an in-memory mode. As far as scalability is concerned, db4o can create database files of up to 254 GB.

After conducting the Poleposition database benchmark [6], between db4o and other relational databases, such as MySQL, JavaBD and SQLite, combined with object-relational mappers (JDBC or *Hibernate*), db4o was found to perform better than its competitors for read, write, query, and delete operations, when they involve accessing complex object structures or deep hierarchies. Moreover, its performance was acceptable, although worse than one competitor, for simple flat objects [65].

NeoDatis ODB is also an open-source object database library, embeddable in Java and .NET applications, that supports ACID transactions and can be used in a multi-threaded environment. In ODB every entity (class or object) is characterized by an Object Identifier (OID), which is associated with the respective physical position of the entity in the database file. OIDs are used by pointers in the database for accessing directly a specific object, or for storing relations between objects. They are grouped in blocks that contains the OIDs of the objects that are instances of a given class, in order to enable quick access to them. ODB has also a caching mechanism for mappings from OIDs to objects and reversely, and supports B-tree indexing. ODB provides the following query possibilities for data retrieval: (i) all objects of a specific class, (ii) a subset of objects of a specific class via CriteriaQuery, (iii) a subset of objects of a specific class via NativeQuery, (iv) direct id-based object retrieval, or (v) specific object value retrieval [70].

Based on the results on the Poleposition benchmark, it appears that ODB performs on average better than db4o on most circuits, although there are also some results that indicate that *db4o* is slightly faster than ODB for some circuits [77].

3.3 Native Graph Stores

A natural way to store large graph-shaped datasets seems to be through the use of a persistence engine that directly encodes the graph structure. This type of graph store can be characterized as *native* and should in general support the representation and storage of both nodes including node-related properties, as well as attributed links connecting pairs of data nodes. In the following sections some examples of existing native graph stores will be given, including stores that are generic, i.e. designed to enable the storage of various types of graphs, or are intended for the storage of special graph types, e.g. RDF or XML.

Generic graph stores: *Graph databases* have been recently presented as an efficient way to handle networks of data. Unlike RDBMS, graph databases are designed with inner support for entities that represent *nodes* and *relationships* (or edges), thus making it possible to store and access data in a more efficient and simple way. They aim to provide a complete environment that will make the storage, indexing and quick retrieval of graph data easy, and at the same time retaining traditional

[6] `http://www.polepos.org/`

database properties such as: transactions, durable persistence, concurrency control, and transaction recovery. Graph databases have also been designed taking seriously into consideration the matter of scalability.

One of the first and more complete efforts towards the direction of a generic native graph store has been the development of Neo4j [73] and its release as an open-source graph database. Neo4j is an embedded, disk-based, transactional graph persistence engine that stores data in the form of graphs. Apart from the capabilities of storing nodes and edges and also properties related to them (they are collectively referred to as *primitives*), Neo4j has an easy-to-use, rather straightforward API and provides a variety of extra graph manipulation facilities, such as checks for possible inconsistencies and support for both directed and undirected edges. Moreover, it requires constant time for adding, removing, or accessing a property and creating, deleting, or accessing a node or relationship, whereas it requires linear time for accessing the relationships that involve a given node. However, although in general Neo4j can be considered as fast when concurrent reads take place, it is slower with concurrent updates. This requires careful consideration of the number of operations that will be packed in a Neo4j transaction, which is also affected by the available size of RAM. Moreover, transactions may be useful for ensuring data integrity, but sometimes they can seriously decrease the speed of operations.

It is claimed that Neo4j can scale up to billions of nodes, relationships and properties, but this is a maximum capability relevant only for servers with more than 16 GB of RAM. In general, the scalability of Neo4j is greatly affected by the hardware specifications of the computer station hosting it. For example, it is claimed that an average laptop with 1-2 GB RAM handles tens of millions of primitives, whereas a standard server of 4-8 GB RAM handles hundreds of millions of primitives. However, our experiments with Neo4j (see Section 6) did not give proof for such scalability.

Although Neo4j does not provide a native indexing mechanism yet, it supports indexing facilities by use of the *Apache Lucene* text indexing library. This utility allows indexing nodes with key-value pairs, just like properties, so that they can be queried and retrieved using a given key. The querying process can be accelerated via a LRU cache that holds the most recently accessed results. A limitation of this indexing scheme, however, is that it does not allows indexing relationships.

Another example of native graph store is grDB [30], In grDB graph data are stored grouped in *blocks*, with the block being the smallest amount of information inserted or extracted from the database. The information that grDB stores for a graph is structured in the form of adjacency lists for each node using an integer identifier per node. A grDB instance consists of the *storage component*, that stores the blocks containing parts of the adjacency lists of one or more nodes, and the *block cache component*, that caches some storage blocks in order to improve performance. It also supports multiple levels of storage files.

Special-purpose graph stores: Data encoded in the XML format exhibit an innate tree-like structure that could be used for modeling certain relations that exist in web graphs. More specifically, since a tree is by definition a connected graph that does not contain any cycles, XML could be possibly used for modeling data nodes with relations that conform to these limitations, or at least can be normalized in more than one trees. For the efficient storage and management of XML data, special databases have been developed. Although the design and functionality of these special-purpose graph stores have been optimized for the storage and retrieval of XML data, they could provide a framework for the storage of graph data (with the aforementioned limitations). Native XML databases, such as *Apache Xindice* and *Tamino XML Server* [79], constitute an interesting alternative to RDBMS, as they do not require the definition of a schema (*schema-free*), thus allowing storing records (XML documents) including semi-structured data that do not necessarily follow a strict predetermined structure. In such databases the storage and retrieval of XML documents takes place according to a (logical) model, such as the XPath model, whereas data retrieval is usually performed by means of the XQuery language. On most occasions, indexing is used to accelerate the querying process [61]. XML databases have, however, received criticism about not being very scalable, as in general XML queries and other mechanisms result in very slow retrievals across large document repositories [58,59]. It also should be mentioned that XML databases are not required to have any particular underlying physical storage model, as they can be built on top or other data storage infrastructures.

Apart from XML databases, structured data can also be stored in RDF[7] or OWL[8] repositories. In general, RDF is a semantically richer way to represent graph-based data, in the form of RDF statements, i.e.subject-predicate-object expressions, known as *triples*, that connect with a specific relationship the subject to the object of the statement. OWL is an extension of RDF that exhibits more expressive power than RDF and enables efficient reasoning. RDF repositories are frameworks dedicated to the management of RDF data in general, that could also be used for the management of web graph data. OWL repositories could also be used for the same purpose, however they are considered to be more specialized than RDF repositories, with the expressive power of OWL being rather needless for the modeling of simple web graph data. Some of the most efficient repositories that support the storage of graph-shaped data either in RDF, OWL, or both, as well as SPARQL queries are Jena [71], Sesame [78], AllegroGraph [64], Virtuoso [80] and OWLIM [76].

3.4 Custom

Apart from the previous infrastructures, custom disk-based solutions that do not belong to a specific category can be employed for the management of web graph data. For instance, the use of a framework based on *Lucene* is proposed. Lucene is text search engine library, that can be easily incorporated in any application that requires text indexing and searching. Indexing with Lucene offers high scalability, cross-platform support, rather small memory requirements, and also fielded search capabilities. Apart from indexing and searching data from other sources, Lucene

[7] `http://www.w3.org/RDF/`
[8] `http://www.w3.org/TR/owl-features/`

also provides the possibility of storing the data in their original form. In general, data are indexed in Lucene as *documents* that contain *fields* of text.

The generic nature of Lucene in combination with its scalability renders it a promising candidate back-end infrastructure for a graph storage framework. A possible implementation would index and store nodes and edges with Lucene, creating a separate document for each entity, and using *terms* to store the properties of each entity. Depending on the application and type of data, for each document, the terms that store the properties that are intended to be used as keywords for querying, will be indexed. In Section 6.1, an implementation of a framework for managing STS data based on Lucene is described in more detail.

3.5 Distributed Transactional Databases

Distributed graph management frameworks are recent solutions that try to solve the problem of limited memory, by distributing the graph in more than one computer nodes that form a cluster. In order to achieve this and for the resulting framework to be efficient, distributed frameworks should employ an appropriate graph partitioning policy and also a query mechanism that will seek and retrieve data from the appropriate computer nodes, minimizing needless queries to irrelevant nodes.

An early research work in distributed transactional graph management, namely MSSG [30] presents a middleware framework for storing, accessing and analyzing massive-scale semantic graphs with update capabilities. The development of MSSG aims to support the storage and analysis of very large graphs reaching trillions of vertices and edges. In order to handle such massive datasets, MSSG has been designed as a distributed database, that supports a large cluster architecture of computer nodes for storing data. Moreover, the framework utilizes the grDB graph database (described in subsection 3.3). The framework, combined with a new parallel external memory breadth-first search algorithm enables fast query responses to the database. The way that MSSG functions is described in brief in the following paragraphs, however it should be stressed that little information has been made available as to how MSSG partitions the graph in order to enable distributed storage.

MSSG was designed using *DataCutter* [5], a development and deployment framework for establishing "filter" services that operate on data "streams" between storage systems and user applications, as a base infrastructure, with the *Ingestion Service*, the *Query Service*, and the *GraphDB Service* modules having been added as integrated components and interfaces. In brief:

- the Ingestion Service is used for entering graph data that are stored to the back-end storage nodes after having been clustered,
- the Query Service allows the analysis of the stored graph,
- the GraphDB Service provides an interface for the available methods implemented for storing and accessing graph data.

The adjacency list of a node can be stored in either a single computer of the cluster, or it can be distributed in more than one computers. Experimental results indicate that the MSSG framework can handle large graph datasets, managing to store and query a graph of 100,000,000 nodes and 1,999,999,640 directed edges, even though a query with length 5 between the source and destination node is answered in about 12 minutes, which is relatively slow. Experiments also showed that grDB outperforms BerkeleyDB and MySQL in storage and retrieval, considering the tested graphs. Moreover, the performance of grDB on a search query is relatively close to the performance of the implemented in-memory methods under test.

4 Data Mining-Oriented Solutions

In situations where the storage and analysis of static graphs is required, database transactions can be omitted for the sake of performance, and alternative solutions are usually employed. The most efficient solution seems to be to manage to fit the graph structure in main memory by means of graph compression techniques. Another more scalable possibility is to encode the graph's structure in human-readable text files stored in the computer's filesystem and stream the data into memory for analysis. However, this approach requires the adaptation of data mining algorithms to the streaming or semi-streaming model.

In the following subsections both compression-based as well as streaming solutions for the analysis of graph data will be discussed. Moreover, some recent distributed solutions for the management of massive graphs will be discussed.

4.1 Compression-Based Databases

When the available main memory does not suffice to load the whole graph dataset, but fast access to data is required, an efficient graph data compression method is necessary. In the following paragraphs we will present some examples of compression-based graph databases.

WebGraph. One of the earliest and more successful efforts in the compression of web data has been the *WebGraph* framework [6], a suite of codes, algorithms and tools for storing and manipulating large web graphs. The algorithms of WebGraph were based on the *Link* Database [33], an earlier work employing compression techniques to store web graphs that can fit in main memory. Both *Link* and WebGraph perform well in compressing large graphs, combining a number of techniques, such as *referentiation* and *intervalization*. However, WebGraph outperformed its predecessor achieving compression rates of e.g. 3.08 bits per link for a graph consisting of 118 million nodes and 1 billion links.

The success of the WebGraph compression approach is justified considering that the properties of *locality*, *similarity* and *consecutivity* that are typical on the Web were seriously taken into consideration during its development. The property of locality describes the fact that pages belonging to the same host often point to each other via navigational links. Therefore, if we consider a lexicographical ordering of URLs, the source and destination URLs of a link are "close". The property of similarity expresses the observation that pages whose URLs are lexicographically "close", tend to have links to common destination pages (*successors*). Consecutivity means that the successors of a web page also tend to be lexicographically "close", as they usually belong to the same level of site hierarchy. In order to exploit the aforementioned properties of the Web, WebGraph applies the following technique.

- Given a set of URLs and the information that some of them are linked, URLs are sorted lexicographically and assigned integer identifiers.
- The successor lists of each node are created and sorted by the node identifier.
- The successor list of a node x is expressed with respect to the successor list of a node y with smaller identifier via a *reference list* comprising (a) the *copy list*, i.e. a list of the two nodes' common successors and (b) the *list of extra nodes*, i.e. the set of the successors of x not present in the successor list of y.
- Applying the technique of *differential compression* with encoding methods such as γ coding to the copy list, WebGraph manages to code a link in less that one bit. The list of extra nodes is also compressed using *integer intervalization* and *gap encoding*.

After the compressed graph has been created, WebGraph provides methods for accessing the graph either *randomly* (selecting to access random nodes) or *sequentially* (iterating over all nodes defining the sequence by increasing number identifier). The provided access algorithms are very efficient as they employ lazy techniques to access the compressed graph, thus delaying decompression until it is actually needed. Moreover, WebGraph offers a number of parameterization options in order to allow a trade-off between the *compression ratio* and the time needed to compress the graph, as well as between the decompression speed and the size of the offset array. The compressed graph can either be loaded to RAM, or accessed offline.

The developers of WebGraph also investigated and experimented with different codes to encode the *gaps* that exist between nodes belonging to an ordered successor list [7], after proving empirically that they follow a power-law distribution. They introduced a new set of flat codes for integers, the ζ codes, and proved experimentally that on most cases they are superior to traditional coding methods such as, *Elias* γ, *Elias* δ and *variable-length nibbling*, when they are applied to integers that follow a power-law distribution similar to the distribution of the successor list gaps.

Extensions of WebGraph. A recent work [8], focuses on determining whether WebGraph could be used for efficiently compressing graphs created by data from sources other than web graphs, such as a Social Tagging System. Based on the notion that the compression rate achieved when compressing a web graph depends greatly on the ordering of the nodes, several ordering methods either: *extrinsic* (using information other than that conveyed in the graph itself), or *intrinsic* (using only the information conveyed by the graph structure), have been investigated to determine their effect on the compression rate. As the efficiency of an extrinsic method, such as URL ordering, is doubtful for the case of a network other than a web graph, finding an intrinsic ordering that yields good compression rates is generally considered a challenging problem. In [8], the proposed method is to:

1. order the nodes of the graph randomly,
2. create the adjacency matrix considering each row as a sequence of bits, with 0 denoting the absence, and 1 the existence of a link between two nodes,
3. permute the rows and columns of the matrix so that in the resulting matrix two rows are similar only if they appear consecutively, or almost consecutively.

The two methods that were applied were to either find a permutation that sorts the rows of the adjacency matrix based on the lexicographic ordering, or find a permutation based on the Gray ordering of the row bit vectors. Moreover, two mixed methods combining extrinsic as well as intrinsic characteristics were tested. Both methods use the Gray ordering but limit its application based on the information of the distribution of the nodes within hosts. Experiments with the aforementioned methods using URL ordering showed that (a) the efficiency of each method depends on the structure of the graph itself, (b) intrinsic methods perform very well for the inverse graph, and (c) mixed methods yield better performance on every tested graph.

A recent study [17] focuses on determining whether social networks can be efficiently compressed. This work was motivated by the approach followed in WebGraph considering the properties of locality and similarity that exist for web graphs in order to improve compression ratios. The question posed in this study is whether social networks in general can be effectively compressed by a method similar to the one followed in WebGraph. An easy observation is that the URL lexicographic ordering of the nodes that is a part of the WebGraph compression technique and also a reason for its success cannot be applied to generic social networks. Thus, a new node ordering technique is proposed that uses a simple heuristic based on *shingles*. If we consider two sets A and B, and σ as a random permutation of the elements in $A \cup B$, then $M_\sigma(A) = \sigma^{-1}(min_{a \in A}\{\sigma(a)\})$ is the smallest element in A according to σ, and is called a shingle. The probability that the shingles of set A and set B are identical equals to the *Jaccard coefficient* of the two sets, which is a measure of their similarity. The proposed method regards the out-neighbors of each graph node as separate sets, computes their shingles for an appropriate permutation and then orders the graph nodes according to their corresponding shingles. As a result, nodes with many out-neighbors in common will end up to be close to each other. An alternative technique, *double shingle ordering*, has also been proposed that uses a second shingle for breaking ties produced by the first one.

After the nodes have been ordered, their adjacency lists are compressed using a technique similar to the one employed in WebGraph. Apart from referential and gap encoding, the technique introduces an alternative method for encoding the links that are *reciprocal*, that is the links that are undirected. In particular, this method encodes the reciprocal links in the adjacency list of the node with the smallest integer identifier, and also adds a bit flag for each neighbor encoded in the adjacency list, that declares whether the link is reciprocal or not. With this approach, reciprocal links are encoded only once, thus improving the compression ratio, but this also causes slower queries in the compressed graph.

Experimental results on various datasets indicate that the proposed compression method yields better compression ratios than WebGraph when applied to social networks that are highly reciprocal in structure. Moreover, after experimenting with various ordering techniques such as, *Gray*, *natural*, and *random* orderings, the *double shingle ordering* managed to achieve the best compression performance. The success of *shingle ordering* with respect to the other methods is attributed to: (a) the reduction of the lengths of the gaps that exist between the neighbors of the adjacency lists, and (b) the exploitation of the properties of locality and similarity. Finally, experimental results indicated that social networks appear to be less compressible than web networks, mainly due to the presence of nodes with low degree.

Taking the above into consideration, WebGraph seems to be a very effective solution as it manages to store a graph in limited disk space and also fetches the neighbors of a node when requested in little time. However, one drawback of the WebGraph framework is that it represents each node with a number without giving the possibility of compressing more information related to a given node. Moreover, it does not provide edge indexing capabilities.

Re-Pair. Several researchers, motivated by the WebGraph compression approach and using the WebGraph framework as a basis for comparisons, tried to find methods with improved performance in terms of compression rates or graph access speed. One such effort [18] proposed a method based on the *Re-Pair* compression technique [36] in order to store a representation of a given graph. Re-Pair is a phrase-based compressor that receives as an input a sequence of symbols, finds the most frequent pair of symbols in it and replaces it with a new symbol, storing the corresponding mapping in a dictionary. This procedure is repeated until every pair in the sequence is unique. Although it is a rather fast (linear-time) technique, it requires a large amount of memory, especially when the initial sequence is long. Therefore, an approximate technique is proposed [18] that can be applied in external memory. In any case, an in-memory hash-table is required to hold the unique pairs of symbols occurring in the sequence (represented by their position in it) along with their frequency. After the hash-table is filled up to a load threshold, no new pairs are inserted, although the traversal is completed so as to calculate the frequencies of the pairs that have already been inserted. Afterwards, the k most frequent pairs are selected and replaced in the sequence with new symbols with a new traversal. The process is repeated traversing the sequence from the position where the insertion of new pairs in the hash-table had previously stopped. When the sequence of symbols resides on secondary memory, the hash-table can store more pairs of symbols as it can occupy all the main memory that is available and also special techniques have been employed so as to avoid unnecessary random access to disk.

In order to apply the Re-Pair technique on a graph representation, the graph is modeled as a sequence of integers representing the graph nodes, each one followed by its adjacency list. However, each node maps to two different distinct integers; one that is used when the integer is placed in the sequence before its adjacency list and one that is used when it is included in the adjacency list of other nodes, thus preventing the integers that mark the start of an adjacency list from being replaced. These alternative representations are removed from the sequence after Re-Pair has been applied, and they are stored in main memory along with pointers to the beginning of their adjacency list. In general, the proposed method takes advantage of the similarity of the adjacency lists. Moreover, in order to achieve better compression rates, differential compression can also be applied to the lists.

Experiments showed that the proposed method yields compression rates comparable to WebGraph providing faster graph navigation. For example, a graph with 22,744,080 nodes and 639,999,458 edges was compressed into 420 MB (a plain representation would require around 2.4 GB of RAM), achieving two times faster navigation to the compressed graph than when using the WebGraph compression. When differential compression is also applied, slightly better compression rates are achieved, but graph navigation is somewhat delayed.

Virtual Node Miner. Another approach, the *Virtual Node Miner* [14], provides a solution for web graphs that need to be updated after having been compressed, and that also performs well without requiring URL sorting, which is a relatively time-consuming process. The main innovation is that it employs a *pattern mining* approach in order to compress a web graph, using an effective itemset mining algorithm that finds directed bipartite cliques. Moreover, the fact that this method is not based on URL encoding, indicates that it may possibly be used in application domains other than web graphs, such as social networks. The proposed algorithm considers the outlinks (or inlinks) of each graph node as an itemset and aims to identify frequent subsets that in fact represent common links between the graph nodes. The algorithmic steps are described roughly below:

- The graph nodes are clustered on basis of the similarity of their adjacency lists. This step uses *k min-wise independent hash functions* to sample the adjacency lists of each node and then sort the rows of the resulting adjacency matrix lexicographically, in order to bring closer similar adjacency lists and form clusters.
- For every cluster, the algorithm searches for frequent recurring patterns of neighbors, which are actually directed bipartite cliques.
- Every pattern is replaced by a new node, called a *Virtual Node*, that has outlinks to the nodes that formed each specific pattern. After that, the nodes that demonstrated the pattern in their adjacency lists, replace all the outlinks to the nodes that belong to the pattern with just a single outlink to the Virtual Node.
- The process is repeated allowing Virtual Nodes to be reused as actual graph nodes.
- The remaining edges are compressed with an appropriate compression method, such as: ζ or *Huffman* coding.

The resulting graph has a number of extra nodes, the Virtual Nodes, but significantly less links, therefore it is considered to be *compressed*. Experiments indicated that the Virtual Nodes added are about 20% of the original number of nodes, and therefore a moderate overhead to the offset array. Moreover, when compared with WebGraph, experiments showed that the above methods are comparable regarding the compression they achieve. The proposed method has been proven to be rather scalable, as it manages to compress a graph with 3 billions of edges on a computer with 16 GB of RAM in about 2.5 hours. The time required for compression also scales linearly with the graph's size. It is also of interest that if the available memory does not suffice, Virtual Node Miner can run in batches, thus enabling incremental updating of the compressed graph.

Research with the Virtual Node Miner continued with an effort that used this compression technique to adjust several web graph algorithms so that they could run directly on the compressed graphs and thus demonstrate reduced time complexity [32]. The basis for these algorithms was the invention of a method operating on the compressed graph, that speeds up the multiplication of a graph's adjacency matrix. This multiplication routine was used for computing random walk distributions, finding top singular vectors, estimating the size of neighborhoods, and others, and the resulting methods were used to speed up the implementation of well-known link analysis algorithms such as PageRank [10,37], SALSA [38] and HITS [35]. Experiments showed that the performance of the proposed algorithms was better than the traditional implementations, increasing speed almost up to a ratio close to the respective compression ratios.

4.2 Streaming Solutions

Probably the most intuitive way to encode a graph in a human readable file is either in the form of adjacency lists [9], or in the form of an edge list (e.g. in its simpler form an edge can be expressed as a pair of nodes that are related). The existence of disk-resident files satisfies the persistence requirement for the storage of the graph data, with their sizes being limited only in terms of the available size of external memory. However in order to perform graph analysis tasks, data should be loaded in main memory in an efficient way. In the *streaming model* graph data are streamed from the disk into memory as a sequence of edges. However, the streaming model poses some constraints on the graph mining algorithms, which should be designed so that they can process the edges of a graph in an arbitrary order given only a limited RAM space and desirably making only one pass over them [24]. In order to achieve this, algorithms should be able to make space-efficient data summaries in RAM as data are streamed. This is a considerable challenge, since in general the streaming model poses the constraint of using $O\{polylogN\}$ space and per-item processing time for a given graph with N nodes [41].

There have been some efforts in trying to solve simple graph problems in the stream model, such as the problem of counting triangles in a graph [15].The triangle counting problem in the streaming model is defined as finding the ε-approximation of the number of triangles in a graph with probability at least 1-δ, making one pass over the data stream. The method proposed in [15] assumes that the set of the graph's nodes is known in advance and the graph's edges appear as a stream. It manages to calculate approximately the number of triangles via a technique that uses reservoir sampling, requiring $O(\frac{1}{\varepsilon^2} \times \log\frac{1}{\delta} \times (1+\frac{|T_1|+|T_2|}{|T_3|}))$ memory cells, where T_i stands for the number of triples of nodes in the graph which have i edges between them. In [13] authors try to provide lower bounds for the more complex problem of finding pairs of vertices that share c neighboring nodes. They give proof that any

[9] The adjacency list of a node is a sequence of its neighboring nodes.

one-pass, randomized data-stream algorithm that determines if a pair of nodes in a directed graph with N nodes shares more than c neighbors requires $O(\sqrt{c} \times n^{\frac{3}{2}})$ bits of space. The large memory bound of the aforementioned method indicates that the application of the streaming model for general graph problems seems to be difficult due to the strict space constraint it imposes [23].

A more relaxed model is the *semi-streaming model* that was initially suggested in [41], as a solution for graph problems where the available main memory suffices for the storage of the graph's nodes, but not for the storage of the graph's edges. This model bounds the storage space for an algorithm that operates on a graph stream by $O(N \times polylogN)$. In [23] the semi-streaming model was further elaborated, allowing also a small number of sequential passes over the graph data. Authors in [23] dealt with various graph problems in the semi-stream model such as the computation of the shortest-path distances between the graph's nodes, as well as the diameter and girth of a graph. They showed that these problems can be approximately solved even with one-pass over the data, via an approximation technique that uses *graph spanners*[10] to calculate shortest distances. In addition to the aforementioned problems, algorithms for the problem of graph matching in the semi-streaming model were also presented in the same research paper. Feigenbaum et al. [24] later improved their method for constructing graph spanners decreasing the processing time per edge from $O(N)$ to $O(polylogN)$. Moreover, they proved that the computation of Breadth-First-Search (BFS) trees is not efficiently executed in the semi-streaming model.

Another recent work [2], studies the problem of *graph sparsification* in the one-pass semi-streaming model. Graph sparsification involves the construction of a compact representation of a given graph through which the size of any cut can be estimated. This problem is therefore connected to the problem of finding an approximate min-cut in a graph. The method proposed in [2] constructs and stores a summary of the graph in main memory, that is updated based on the newly-arrived edges. The original algorithm for finding the sparsification of a graph involves the calculation of the connectivity of every new edge which is impossible unless all the graph's edges are available. However, the proposed method calculates the connectivity of each new edge on the current sparsification, achieving an $1 \pm \varepsilon$ approximation of the cut values of a graph with N nodes and M edges, while requiring $O(N(\log N + \log M) \times \log \frac{M}{N} \times (1+\varepsilon)^2/\varepsilon^2)$ edges in main memory.The semi-streaming model has also been applied to the problem of local triangle counting in graphs [4] (i.e. given a node u, count the number of triangles that are incident to node u). In this research work, apart from the space constraint of the semi-streaming model, algorithms are allowed to $O(logN)$ passes over the data that reside in the external memory. Two algorithms are proposed: one that requires the storage of some intermediate counters in external memory and another that maintains all information in main memory. Given a newly-arrived edge (u, v), both algorithms are based on the approximate calculation of the Jackard coefficient between the two sets of nodes that are adjacent to nodes u and v, respectively.

---

[10] A subgraph $G'(V,E')$ is a t-spanner of graph $G(V,E)$ if the distance between any pair of nodes in G' is at most t times the distance in G.

A similar graph access approach that is mentioned here, but not presented in detail, is the semi-external model [1]. This model allows for enough main memory to store the graph nodes, but not the graph's edges as well. On the contrary, the graph's edges are stored in external memory, with the model allowing random access to them. However, random access to the disk-residing edges can make the whole process seriously slow.

4.3 Distributed Data Mining-Oriented Solutions

The requirement to perform data mining on massive graphs in a relatively short time has also motivated research in the field of distributed data mining-oriented solutions. Bader and Madduri have recently presented a study including combinatorial techniques for the analysis of large-scale dynamic networks [39]. Their innovation is that they have designed and implemented efficient graph data structures and kernels for modeling temporal graphs of massive sizes that are processed on parallel systems. Temporal information related to e.g. the update or insertion of a node or an edge, are handled by assigning time-stamps to the respective nodes or edges. After experimenting with a number of structures, they proposed a hybrid data structure combining dynamic resizable adjacency arrays for low-degree vertices, with simple self-balancing binary trees, referred to as "treaps" [46], for high-degree vertices. This structure was found to achieve good performance for both insertions as well as deletions, that may be batched or streaming. The data models as well as the algorithms have been designed for multithreaded servers, with multiple cores and a significant amount of both shared cache and main memory. These architectures have been proven to perform much faster in graph analysis algorithms than optimized external memory based architectures [3].

In order to solve or avoid conflicts when e.g. multiple threads try to add data to the adjacency list of the same node, various methods are proposed, such as: following the simple lock-based approach, or allowing each processor to have access to the adjacency lists of only a subset of the graph nodes. In addition, several algorithms have been designed and implemented to execute efficiently graph operations such as: connectivity, path-related and centrality queries. Experiments show that the proposed algorithms scale well on parallel architectures, with e.g. an algorithm based on an implementation of the link-cut tree being able to process queries in time proportional to the diameter of the network. It is also important that the proposed implementation can answer queries related to the evolution of the graph during time.

MapReduce: MapReduce [20] is a programming model with an associated implementation for processing large data sets that may be stored in a distributed filesystem or database. The proposed model, introduced by Google, is applicable for computational problems that can be formed as a set of key-value pairs, e.g. web page indexing based on keywords. The computational process is in general divided into two steps: *map* and *reduce*. Programmers are responsible for creating an application-specific map function that processes the input key/value pair to generate a set of intermediate key/value pairs, and also implement a reduce function that merges all intermediate values associated with the same intermediate key. Each operation initializes with the splitting of the input files and continues with the assignment of different map and reduce tasks to worker nodes by a special master node. The master node is also responsible for the final aggregation of all results and the production of the output to the original computational problem.

This model has proven to be very efficient for problems that involve accessing large sets of data, however it is disputable whether it can be applied for graph related problems. Some graph related problems can be successfully solved by use of MapReduce. For instance, the computation of PageRank over the Web can be implemented as a chained MapReduce application. However, the main difficulty with solving graph-related problems with MapReduce is that it is is very inefficient for graph traversals, as map workers have access to only a part of the graph. A recent research work [19] investigates the possibility of decomposing graph operations, such as graph simplification, triangle and rectangle enumeration, finding trusses and components, and performing Barycentric clustering, into a sequence of MapReduce processes. In order to overcome the problem that exists with graph traversals, techniques such as the use of multiple map and reduce iterations, or the use of custom optimized graph representations, such as sparse adjacency matrices are proposed.

5 A Case for Web 2.0 Graph Stores: Social Tagging Systems

In this section we focus on a recently evolved research area: the analysis of *Social Tagging Systems*. An introduction to Social Tagging Systems is provided, along with a short review of the most current progress made in several related analysis tasks, and we discuss their special characteristics. Social Tagging Systems are presented as an application setting for massive graph data management frameworks, due to the special requirements that their analysis imposes on the underlying infrastructure.

5.1 Introduction to Social Tagging Systems

An important functionality that has been embraced by many on-line applications is *Social Tagging*. Social Tagging Systems (STS) enable their users to upload content, and to annotate it by means of freely chosen keywords, called *tags*. By relating resources with tags, users enrich them with a semantic meaning that can be of use to other people that come across it. Moreover, the information from STS can be exploited by use of data mining in order to provide enhanced services to users, e.g. recommendations, sophisticated content navigation (e.g. by means of a concept hierarchy representing a resource collection). The study of STS has led to the formalization of *folksonomies*, i.e. lightweight knowledge structures that emerge from the use of a shared vocabulary to characterize resources (*emergent semantics*) [31,40]. The folksonomy model has been established as the most widely-used means to represent and analyze STS-related information, thus its definition is given below.

Definition 1. A folksonomy is defined as the tuple $\mathbb{F} = (U,T,R,Y)$, where U, R and T are the disjoint sets of users, resources and tags, respectively, and $Y \subseteq U \times R \times T$ is a triadic relation between them, representing the annotation of a resource with a tag by a user. Another way to represent the folksonomy is as an undirected hypergraph $G = \{V, E\}$ consisting of a set of nodes $V = U \cup T \cup R$ that are connected by hyperedges that formulate the set $E = \{\{u, r, t\} | (u, r, t) \in Y\}$.

Rather than working on a hypergraph, on many occasions and depending on the corresponding analysis task, a simplified bipartite graph is produced representing the associations between either: (a) users and resources, (b) users and tags, or (c) resources and tags.

This technique makes the graph analysis easier, as it transforms the hyperedges of the tripartite hypergraph into simple edges. The resulting edges are usually weighted, e.g. in the user-tag bipartite graph, an edge exists between a user and a tag if the user has used this tag to annotate at least one resource, and is weighted by the number of resources that have been annotated with this tag. This graph can be symbolized as: $UT = \{U \times T, E_{ut}\}, E_{ut} = \{(u,t) | \exists u \in U : (u,r,t) \in E\}, w : E_{ut} \rightarrow \mathbb{N}, \forall e : (u,t) \in E_{ut}, w(e) := |\{r : (u,r,t) \in E\}|$ [31]. Relevant expressions can be formulated for the bipartite graph between resources and tags (RT), as well as for the graph between users and resources (UR). A bipartite graph can be represented with a model, such as an adjacency matrix, with each row relating to a member of the first entity type and each column to a member of the other, whereas the value of a cell stores the number of co-occurrences of the respective entity members.

A further simplification can take place, resulting in a graph that represents the co-occurrences between members of the same entity type only. For example, considering the user-tag bipartite graph, two graphs can be produced; one that comprises tags as vertices and edges that represent the annotation of some resource with two tags by a common user, and another that comprises users as vertices and edges that represent the annotation of some resource with a common tag by two users.

If required, a tripartite graph can be also produced, combining the three bipartite graphs, where all the resource-tag, resource-user, tag-user co-occurrences are represented with simple weighted edges. However, the use of bipartite graphs is more often than the use of tripartite, as most algorithms focus on the correlation between the members of two or one entities. For example, the associations between resources and tags is of most interest for a tag recommendation system, whereas the information about which user tagged a resource is not that interesting in this scenario. However, the associations between resources and users would be useful for an application e.g. that recommends resources that may interest users.

5.2 Social Tagging Systems: Analysis Tasks

Ontology extraction: One of the first expectations of researchers was to take advantage of the emerging folksonomies in order to construct ontologies for the Semantic Web [40]. However, early works indicated that the derivation of ontologies from folksonomies presented some serious difficulties, especially because tagging is not necessarily hierarchical such as the ontology structure, meaning that unless it is otherwise stated, an assignment of tags to a resource signifies that the latter is equally characterized by all tags, but it does not imply a hierarchical relationship between the tags. Moreover, there is the widely-discussed problem of tag *ambiguity* and *polysemy*, i.e. tags that have ambiguous meaning and are used by users to annotate resources that are not relevant to each other. Another issue is the existence of synonyms, that should be identified as tags with a common meaning [26].

Tag meaning disambiguation: Tag ambiguity poses serious challenges to applications that analyze the information included in the STS structure. The annotation of two resources that belong to semantically different categories with common polesymous tags creates a relation between them that is not intended. Therefore, recent research has attempted to address the problem of ambiguous tags. An early effort tried to discover the different dimensions of knowledge in a folksonomy, and after calculating the conditional probabilities of tags in different conceptual dimensions, ambiguous tags were found to have high probabilities on more than one dimensions [52]. In [53], a clustering technique based on the community identification algorithm of Girvan and Newman [42] was employed to find clusters of tags that indicate the different meanings of ambiguous tags in a folksonomy. This work was continued in [55] where the different contexts in which a tag can be used were again on focus, and therefore analysis was conducted for every tag on the associated subset of the folksonomy. Several kinds of network representation were tested and experimental results indicated that tag co-occurrence networks that explicitly incorporate the user-tag associations provide better results in identifying the different contexts a tag can appear in. The results of the proposed automatic tag clustering technique were successfully applied to classify documents retrieved by Web searches.

Study of usage dynamics: Research was also directed towards unveiling the dynamics that characterize the evolution of an STS. Research on the users' behavior in *delicious* showed that users tag collections are growing and evolving over time, due to new interests [26]. However, it was discovered that the set of tags that were used for annotating most of the bookmarks (the resources in delicious) tended to stabilize after a while, exhibiting a stable pattern with fixed frequencies for each tag. This indicates the existence of shared knowledge amongst users, as well as imitation. In addition, the tags that were used to annotate a bookmark by larger numbers of users (the most popular) and also the ones that were used earlier were found to be more representative of the larger category the resource belongs in, therefore have great significance for further analysis. In [29] it was proven, based again on data from delicious, that the distribution of tags is indeed stabilized after some time, following a power law distribution. Moreover, it appeared that after stabilization, analysis of the high-frequency tags of an STS can reveal the collective categorization scheme. Similar results have also been found in [51], where it is stated that tags used to annotate a specific resource are relatively strongly semantically related.

Statistics analysis: It is also interesting to find out the distribution of the user participation in an STS. Earlier research results in social networks in general indicated that user participation follows a power law [50], however subsequent works showed that there were more users contributing content in social networks than those expected from a power law distribution [34]. A recent work [27] showed that the distribution of different users participation follows the stretched exponential distribution, which means that top users are distributed much flatter than those in power law networks.

However, this distribution depends also on the type of content; for example, the distribution of user contribution on content that is more "difficult" to create is more skewed towards a few core users. It should be noted though that the results from this last work have not been based on results from STS.

Clustering: Another direction that has won vivid research interest is clustering, either in users, resources or tags of an STS. The discovery of clusters within a STS has been mainly approached as a *community identification* problem in a graph-structured network. There are different approaches, however, that use either: (a) a bipartite or tripartite graph representation [21], or (b) a simplified tag-tag, user-user, or resource-resource co-occurrence graph. On the first case, the resulting communities are strongly-knit connected components that exist in the graph and are formed by two (or three) kinds of entities, whereas on the second case communities comprise of members of just one type of entity (e.g. tags). Due to their complexity, there have been few methods that have applied clustering for community identification to the induced tripartite hypergraph [9,11].

Tag clustering is a research subject with numerous interesting applications. E.g., the tag clusters resulting from a tag-tag network can be used in a system that recommends to users tags for annotating resources, as they comprise of tags that are semantically "close". Similarly, resource clusters can be used to group objects belonging to the same category, whereas user clusters group people that have exhibited similar behavior patterns in an STS. A tag clustering approach is based on the application of classical community identification methods in the implied graph featuring tag relationships, such as in [16] where a spectral community identification method is employed, in [48] which identifies communities based on graph *modularity* [42] and in [44,45] where a *seed-based community expansion* method has been applied. Moreover, some efforts dealt with the problem of tag clustering, using vector-based agglomerative hierarchical clustering methods rather than the structural properties of the STS graph [12,47]; however they are very slow for large sets of tags. Apart from clustering methods, tags have also been used for the classification of web resources, using optimization techniques that use tag annotations as a feature space for resources and also exploit the link relationships between resources and tags [56].

5.3 Application Setting

All the analysis and mining tasks that are applied to STS and have been discussed in the previous subsection, require a robust graph management infrastructure providing a number of features, dictated by the special characteristics of these systems. The sizes of the graphs formed in the context of STS render them an excellent application setting and motivation for massive graph storage and access frameworks. In the following, the most characteristic STS properties are summarized in order to derive the requirements for a framework developed for their analysis and storage.

- STS users are increasing in numbers and also tend to contribute more in either providing new or annotating existing content. This results in the gradual development of massive folksonomies from STS data that can be available for analysis. Folksonomy data are encoded in graph structures of hundreds of millions of nodes and ten times or even more edges. delicious, for example, was estimated in 2008 to have 462,168,833 bookmarks and 1,632,204 monthly visitors [43]. These numbers combined with the number of tags used for annotation in delicious is indicative of the massive size of tripartite graphs induce from STS (where both resources, users and tags are considered as nodes).

- STS entities follow power law or skewed distributions. This means that the induced graph exhibits scale free characteristics, i.e there are few nodes that have high frequency and many nodes that are infrequent, thus the network is on its larger proportion rather sparse.
- Information in an STS is updated on a daily basis. However, the number of tags that have been used for annotating a resource is not constantly increasing. On the contrary, after some time the tag distribution stabilizes and each tag used to annotate a resource is characterized by a stable frequency [29,26]. This implies that users often follow common tagging patterns [26].
- STS graphs are often used as an application area for various mining tasks, such as community identification. During graph analysis, algorithms need to access random nodes, extract information that is related to them (e.g. the name of a resource), and also find the edges that are attached to them along with their destination nodes. Taking into consideration the size of the graphs and also the frequency of node and edge accesses that are required in mining algorithms, it is evident that these operations should happen as fast as possible.
- When the induced tripartite hypergraph is simplified in a simple e.g. tag-tag co-occurrence network, some information is lost and cannot be recreated. This information, however, may be useful or necessary for some applications. For example, it is possible that community identification in a bipartite graph will result in more semantically "correct" communities than when using its projection in a simplified entity-entity graph [44]. There is also evidence that explicit information about e.g. user contribution [55] helps dealing with the problem of tag ambiguity.
- Depending on the STS analysis application, the graph representation may include directed on undirected edges. For example, maybe an edge with a resource node as source and a tag node as destination is desirable but the reverse edge does not need to be stored, because it is not useful for the application.
- The STS related data include a number of parameters that may differ, e.g. resources in Flickr may have different attributes than resources in delicious (Flickr resources are images that may have attributes like dimensions, file type, and file size, apart form their URL, whereas delicious resources are bookmarks that may have less attributes such as a title).

On the basis of the above characteristics, the requirements of the framework for the analysis and storage of STS graphs are formulated below.

Graph access methods. The basic graph access operations should be supported, namely node and edge lookup, insertion/update and deletion. Since the stored graphs represent an STS, specializations of the above access operations depending on node and edge types (U/R/T and UR/UT/RT respectively) should be exposed. In addition, specializations of neighbourhood access operations should be available (i.e. get all neighbour tags for a given user). Finally, the framework should provide graph nodes and edges iterators (predicated with the type of node/edge). In addition, node and edge properties (e.g. frequency values) should be possible to store and access along with the corresponding nodes/edges.

Memory constraints. The framework should support storage and analysis of graphs that do not fit in the main memory of a typical workstation. Partial graph load, external node and edge indices, as well as caching schemes are desired attributes for the foreseen framework.

Support for graph analysis. Since most graph mining techniques require fast access to the graph's structure, it is necessary to hold in memory the largest possible portion of the graph's structure in order to support fast random node and edge access. Such information takes precedence over additional node/edge property values which can be stored in external memory.

6 STS Data Management Framework Benchmark

In order to test the performance of different infrastructures when used as underlying technologies for the management of STS data, we implemented three STS data management frameworks. The design of the frameworks was based to some extent on the requirements stated in the previous section. The developed frameworks, that can be characterized as transactional graph databases, are based on H2, Lucene, and Neo4j, representing the categories of RDBMS, custom, and native graph stores, respectively. The rest of the section is structured as follows. Subsection 6.1 describes the three implemented frameworks in details, subsection 6.2 presents the benchmark tests that have been designed in order to evaluate the frameworks' performance, and subsection 6.3 presents and discusses and results of the benchmarking procedure.

6.1 Participating Frameworks Description

In general, the interesting information that can be drawn from an STS can be expressed as statements, with a given statement representing the assignment of an online resource with a tag by a given user. We made the assumption that the STS-related information (statements) is provided to the data management frameworks in the format of triplets consisting of labels. The first label of each triplet refers to the username of the user that characterized a resource, the second refers to the hash value of the URI of the resource, whereas the third refers to the tag that was assigned by the specific user to the resource. Triplets of STS data can be provided as input to the frameworks either separately (one at a time) or in batches.

Graph Model Description. All frameworks support the management of a graph consisting of *user* (U), *resource* (R) and *tag* (T) nodes. Each node entity includes: (i) a string value (label) denoting a username, the hash value of a URI or a keyword if the node represents a user, a resource or a tag respectively, and also (ii) an arithmetic value that denotes the node's frequency of appearance in the STS dataset. For example, if a certain user has made 10 tag assignments to resources then the respective node's frequency would be equal to 10.

The nodes of the graph are interconnected via three types of directed edges: (a) User-to-resource edges (UR), (b) User-to-tag edges (UT), and (c) Resource-to-tag edges (RT). Each edge entity also includes an arithmetic value denoting its frequency, e.g. if an edge starting from a resource R and ending to a tag T has frequency 10, this means that tag T has been assigned 10 times to resource R.

Supported Functionality. The developed frameworks support node-, edge-, and graph-based operations. The main operations are lookup, insert/update and delete. More specifically:

- Node lookup, insertion/update and deletion. A node can be of any of the three supported types (U/R/T). A node insertion entails a node lookup (in case of existence, instead of node insertion, a node frequency update is performed). A node deletion entails the deletion of the node's inlinks and outlinks.
- Node neighborhood iteration.
- Edge lookup, insertion/update and deletion. An edge can be of any of the three supported types (UR/UT/RT). An edge insertion entails an edge lookup (in case of existence, instead of edge insertion, an edge frequency update is performed.
- Graph node/edge iteration.
- Graph statistics (number of nodes/edges per type of node/edge.

H2-based Framework. This RDBMS-based framework uses three SQL tables for the storage of the graph's nodes: the USER, RESOURCE and TAG tables. Each table includes three fields: (a) an integer identifier, (b) a string label, and (c) an integer frequency value. All tables storing nodes support the ON DELETE CASCADE SQL feature, so that in case a node is deleted, its outlinks and/or inlinks are also automatically deleted. Moreover, three tables are dedicated to the storage of the graph's edges: the USER-RESOURCE, USER-TAG and RESOURCE-TAG tables. Each table includes three fields: (a) the integer identifier of the source node, (b) the integer identifier of the destination node and (c) an integer frequency value.

Apart from the functionalities mentioned in the previous paragraph, the framework supports also the retrieval of the integer identifier of a node for a given label. Integer identifiers are used in general in order to follow the classical relational database model, and most of all, to reduce the required amount of space for the storage of the graph data. Each communication with the database, whether it is a read, write or delete operation is handled as a separate SQL transaction.

Lucene-based Framework. The Lucene framework uses three separate indexes for indexing and storing the U/R/T nodes and also three indexes for indexing and storing the three types of directed edges. Each entity (either node of edge) is represented in Lucene as a document that contains a number of fields. In our implementation each node document contains a *key* field that stores the node's label and is indexed so that it can be used for retrieving the document when needed. Moreover, each document contains a *frequency* field to store the number of node's occurrences.

The structure of an edge document includes: (i) a key field created by combining the labels of the source and destination nodes, (ii) a field storing the label of the source node (iii) a field storing the label of the destination node, and (iv) a field storing the edge's frequency. The key field is indexed to enable efficient queries for determining whether a specific edge exists. However, the fields storing the labels of the source and destination nodes are also indexed in order for the implementation to support the retrieval of the outlinks and inlinks of a given node. Writes are committed to the indexes in batch, in order to limit the time consuming disk accesses. During subsequent commits the intermediate writes are stored in a cache memory.

Neo4j-based Framework. The Neo4j-based Framework stores all the graph nodes and edges in a common database. However, it allows the definition of a number of relationships types that in our implementation allow distinguishing the category of the graph entities. In total, six relationship types are defined for characterizing UR, UT, and RT edges, and also defining that a node is a user, resource, or tag[11]. Each node includes two properties: the node's label and frequency. Each edge is represented with a relationship that also includes a property storing the edge's frequency. Nodes are indexed using the Lucene-based index implementation provided by Neo4j, so as to allow retrieving a node with its label. Moreover, in order to increase performance the most frequently queried nodes are cached, and also multiple database operations (reads, writes, updates, deletes) are grouped in a database transaction.

6.2 Benchmark Tests Description

The frameworks described in the previous subsection participate in a number of benchmark tests. These tests have been designed to provide an indication of the frameworks' performance with respect to various operations. In particular, the proposed benchmark suite includes the following measurements:

- graph load time (from a triples file)
- disk space usage
- node/edge insertion time (for batches of 1,000 insertions)
- node/edge deletion time (in case of nodes, their in-/out-links are also deleted)
- batch random node query execution times
- batch random edge query execution times (for existing and non-existing edges)
- graph node/edge iteration times
- neighborhood fetch and iteration for a number of randomly selected nodes.

The tests described above are conducted on graphs that contain: (i) real data from a well known STS (Flickr), or (ii) synthetic random data generated by the Erdős-Rényi model [22]. For the synthetic random graph a string generator is used that allows the generation of strings of up to 10 characters. The main difference between the synthetic and real graph data is that the nodes of the synthetic graph are connected with a fixed probability value, whereas the edges of a real STS graph follow

[11] User nodes are connected via the **user** relationship type to a special root node. Similar connections are created for the resource and tag nodes.

the power law distribution. Apart from the type of data there are also some other differences between the loading and insertion tests on real and synthetic graphs. In particular, when the tests are executed on synthetic graphs, a given node or edge is supplied as input only once (along with a frequency value), therefore no updates take place during the testing procedure, and thus there is no need for checking whether the input node or edge exists. Therefore, the nodes and edges are simply added to the graph with the specified frequency parameter as soon as they appear as input. The experiments described above are summarized in Table 1, which also presents the notation that will be used for each type of experiment throughout the rest of the chapter. For example, the notation for a node iteration experiment that runs on a synthetic graph with 1 million edges would be IN-S-1M.

Table 1. Benchmark test notation

Symbol	Position	Meaning	Comments
L	1	Load graph	Load a graph into the graph store.
DN	1	Delete graph nodes	Deletes 10,000 nodes (and their associated edges) from the graph.
DN	1	Delete graph edges	Deletes 10,000 edges from the graph.
QN	1	Query nodes	Executes 10,000 random node queries on the graph.
QEx	1	Query edges 1	Executes 10,000 random edge queries for existing edges on the graph.
QEn	1	Query edges 2	Executes 10,000 random edge queries for non-existing edges on the graph.
DS	1	Disk space	Reports the disk space usage by the graph under test.
IN	1	Node iteration	Iterates over all nodes of the graph.
IE	1	Edge iteration	Iterates over all edges of the graph.
INN	1	Node neighborhood iteration	Iterates over 10,000 random node neighborhoods of the graph.
R	2	Real	A graph created from real-world data is used.
S	2	Synthetic	A graph generated based on the E-R model is used.
K	3	Thousands	Quantifies the size of the graph under test.
M	3	Millions	Quantifies the size of the graph under test.

6.3 Benchmark Results

In the following paragraphs the performance of the developed frameworks based on the results of the benchmark tests will be discussed. Tables 2, 3, 4 and 5 present the experimental results for the disk usage, load, delete, and query experiments, respectively, whereas Figures 3, 4, 5, and 6 illustrate in a a diagrammatic way the results for the node and edge insertion experiments.

The disk usage test results (Table 2) indicate that the H2-based framework has the lowest disk usage for all sizes of real as well as synthetic graphs. This however was somewhat expected as an edge entity stored in the H2-based framework includes the integer identifiers of the source and destination nodes rather than their string labels that naturally occupy more disk space. Between the other two frameworks, the one based on Neo4j seems to be more compact for real graph data. However, when synthetic data are used the disk space usage remains the same for the Neo4j-based framework, whereas it is reduced for the Lucene-based framework. One difference between the synthetic and real graph data is that the label's length for the synthetic graph nodes is limited to 10 characters the most, whereas the labels of the real graph nodes can contain more characters. For example, the URI hash values that are used as labels for the R nodes have a high possibility of containing more than 10 characters. From the above, it can be concluded that the disk space required for the storage of the documents of the Lucene-based framework has a stronger dependency on the labels' length in relation to the space required for the storage of the entities of the Neo4j-based framework.

Table 2. Disk space usage results

			disk space usage (in Mbytes)	
Disk space test	nodes	H2	LUCENE	NEO4J
DS-R-100K	28,388	6,9	14,5	13,2
DS-R-500K	125,942	36,3	72	61,8
DS-R-1M	235,984	72,8	143,1	119,6
DS-R-5M	1,032,947	379,3	712	559,9
DS-R-10M	1,983,803	766,7	1433,6	1126,4
DS-S-100K	28,388	5,1	8,1	12,8
DS-S-500K	125,942	28	38,9	60
DS-S-1M	235,984	56,2	78,4	116,1
DS-S-5M	1,032,947	283	391,5	542,9
DS-S-10M	1,983,803	570,9	782,4	1126,4

Table 3 presents the total time required to build a graph given either a set of triples of U/R/T labels (real graph), or a set of synthetically generated U/R/T nodes with the respective edges (synthetic graph). According to the benchmark results, the Lucene-based framework is the fastest, whereas the Neo4j-based framework is the slowest of the three, with the Lucene-based framework loading: (i) the largest real graph (10 million edges) 6 times quicker than the Neo4j-based one, and (ii) the largest synthetic graph 4.5 times quicker. In general, synthetic graphs are built in less time than real graphs of the same size which is explained by the differences in the experimental procedure (as it has been stated in subsection 6.2), when loading a synthetic graph there is no need to check whether a node or edge has already been added to the graph). Another observation is that the H2-based framework seems to perform relatively well, with the time required to build a graph being almost proportional to the graph size on most occasions.

Table 3. Load test results

Load test	nodes	H2	LUCENE	NEO4J
L-R-100K	28,388	24,277sec	2,166sec	1,44min
L-R-500K	125,942	2,19min	42,326sec	9,19min
L-R-1M	235,984	5,2min	2,1min	19,13min
L-R-5M	1,032,947	28,49min	17,37min	1,55hr
L-R-10M	1,983,803	1,1hr	43,34min	4,25hr
L-S-100K	28,388	7,670sec	874,700ms	21,816sec
L-S-500K	125,942	37,869sec	9,601sec	2,27min
L-S-1M	235,984	1,15min	35,807sec	5,4min
L-S-5M	1,032,947	7,9min	5,5min	30,6min
L-S-10M	1,983,803	15,10min	14,6min	1,5hr

The results of the node and edge insertion experiments for the real graph data are presented as diagrams in Figures 3 and 5, respectively, whereas the respective results for the synthetic graph data are presented in Figures 4, and 6. Each figure includes three diagrams (one for each framework) that plot the average time for the insertion of a new node or edge calculated for every 1,000 insertions. The diagrams also illustrate the dispersion of the values around the average, with use of the standard deviation values that have also been calculated for every 1,000 insertions. The results of the node insertion benchmark for both real and synthetic graph data indicate that the H2-based framework is the most effective for node insertion requiring on average less than 20msec for the insertion of a node, and managing to maintain a rather stable performance as the size of the graph increases. The slowest solution is the Neo4j-based framework, which also exhibits the highest values of standard deviation. The Lucene-based framework, on the other hand, has the lowest values of standard deviation, thus proving to be very stable. As it can be observed from the diagrams, the average node insertion time for the Lucene-based framework seems to be rising until a number of insertions is reached, and then rapidly fall. This rapid fall indicates that at this point the framework cache is full so the Lucene writer commits the changes that have been cached so far to the disk. Afterwards, the cache is emptied to enable the storage of the new insertions (and possible updates), so the performance of the framework improves. The results for the edge insertion tests (Figures 5, and 6) yield similar findings. Both the H2-based and Lucene-based frameworks are much faster than the Neo4j-based framework. It is noticable that the average times per edge insertion for the latter framework reach very large values as the size of the graph increases. This serious delay of the Neo4j-based framework did not allow us to complete the edge insertion experiment. Between the H2-based and Lucene-based frameworks, although their performance is comparable, the H2-based framework seems to maintain a more stable performance regardless of the graph size. The Neo4j-based framework had an even worse performance for the edge insertion test on synthetic graph data. Apart from the average times calculated for the first 2,000 insertions of edges, the successive results for the first next thousands of insertions were approximately 10 times larger than the respective results for the H2-based and Lucene-based frameworks, however they soon began to increase exponentially. In order to present the comparative results for the three frameworks, Figure 6 has been included using a logarithmic scale for the time axis.

Table 4 presents the results for the node and edge deletion experiments. The average times for node deletion indicate that the H2-based framework performs better than the other two frameworks, whereas the slowest framework for every test has proved to be the one based on Lucene. The Neo4j-based framework maintained the same performance for every graph size. In general, the deletion of a node causes the deletion of all edges that are adjacent to it. Therefore, the time measured for a node deletion includes an extra delay required for fetching and deleting the node's neighbors. The deletion of the neighbors of a node seems to be a serious overhead for the Lucene-based framework, whereas it is automatically executed in the H2-based framework due to the SQL ON DELETE CASCADE constraint. However, it can be observed that the times measured for the Lucene-based framework have the

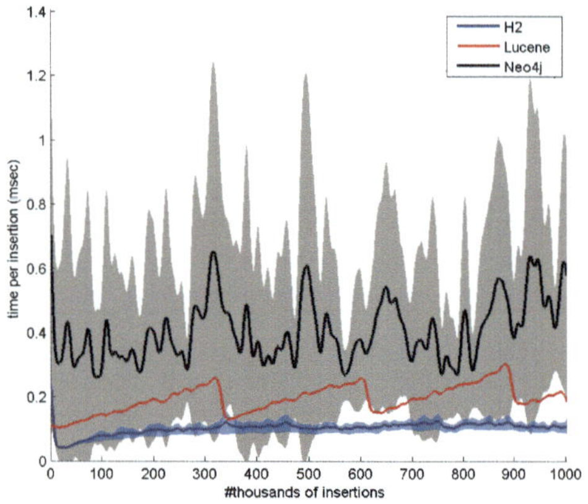

Fig. 3. Diagram of mean time per node insertion in a real graph (per 1,000 insertions)

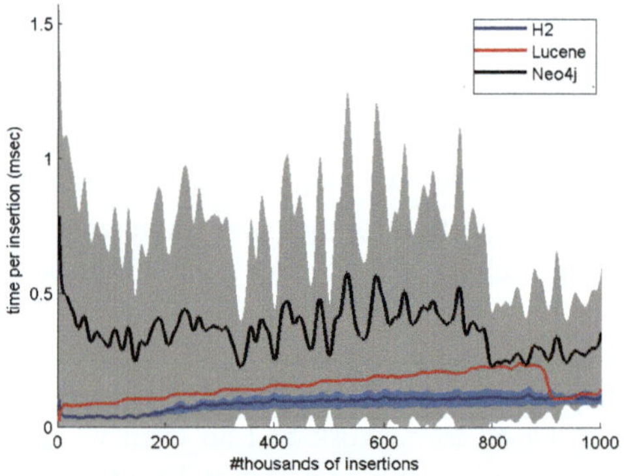

Fig. 4. Diagram of mean time per node insertion in a synthetic graph (per 1,000 insertions)

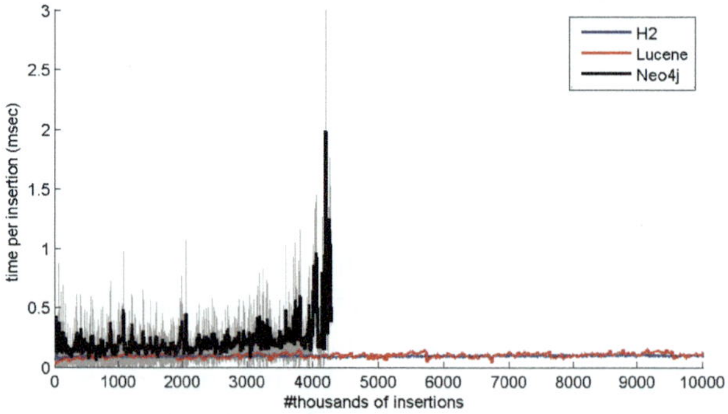

Fig. 5. Diagram of mean time per edge insertion in a real graph (per 1,000 insertions)

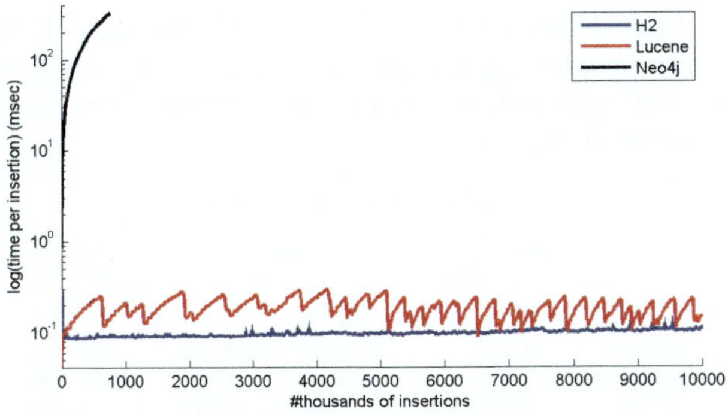

Fig. 6. Diagram of mean time per edge insertion in a synthetic graph (per 1,000 insertions). The time axis is plotted in a logarithmic scale as the time values for the Neo4j-based framework are much higher than the corresponding times for the other two frameworks.

Table 4. Delete test results

	mean time			standard deviation		
Delete test	H2	LUCENE	NEO4J	H2	LUCENE	NEO4J
DN-R-100K	357,947us	567,391ms	3,349ms	5,287ms	0ns	30,187ms
DN-R-500K	1,427ms	1,768sec	3,574ms	13,500ms	0ns	30,162ms
DN-R-1M	1,502ms	3,290sec	3,481ms	7,164ms	0ns	30,171ms
DN-S-100K	182,778us	82,178ms	2,407ms	4,846ms	0ns	30,277ms
DN-S-500K	717,940us	102,943ms	1,766ms	26,585ms	0ns	26,614ms
DN-S-1M	1,102ms	119,799ms	2,602ms	30,351ms	0ns	30,259ms
DE-R-100K	140,674us	1,778us	484,522us	959,511us	9,493us	8,697ms
DE-R-500K	214,985us	2,313us	405,613us	1,735ms	17,285us	8,244ms
DE-R-1M	87,791us	1,808us	348,789us	704,322us	5,711us	7,416ms
DE-S-100K	109,413us	1,624us	704,770us	989,403us	4,371us	8,969ms
DE-S-500K	165,929us	1,638us	465,800us	1,219ms	4,338us	7,418ms
DE-S-1M	62,773us	1,564us	455,313us	17,778us	4,7us	8,47ms

lowest values of standard deviation among all frameworks, something that has been observed in the results of the previous test as well.

The results for the edge deletion experiments clearly show the superiority of the Lucene-based framework. The results of these experiments in combination with the results for the node deletion experiments indicate that this framework is particularly efficient in deleting an edge (and probably a node as they are indexed in a similar way), however it is not very efficient in retrieving the neighbors of a node. Moreover, the average time for an edge deletion does not seem to change linearly with the graph size, but it is affected by the randomness of the edge selection.

The results for the query experiments are presented in Table 5. The experimental results of the tests that involve querying random nodes of real graphs show that the Lucene-based framework has the best performance between all frameworks. Comparing the performance of the other two frameworks, it can be observed that the Neo4j-based framework scales better than the one based on H2 for large graphs. In addition, the results concerning the H2-based framework show a particularly high value of standard deviation for the graph of 1 million edges. However, the same experiments on the synthetic graph generated results that show that the H2-based framework conducts queries faster than the other two frameworks, whereas Neo4j has the worse performance.

Table 5. Query test results

	mean time			standard deviation		
Query test	H2	LUCENE	NEO4J	H2	LUCENE	NEO4J
QN-R-100K	162,16us	94,74us	271,48us	991,29us	562,32us	1,48ms
QN-R-500K	352,649us	132,264us	374,78us	1,504ms	615,906us	1,155ms
QN-R-1M	1,281ms	182,273us	476,680us	3,247ms	1,151ms	1,966ms
QN-S-100K	28,620us	63,881us	203,272us	179,664us	205,149us	691,344us
QN-S-500K	60,307us	93,835us	226,427us	599,284us	370,230us	528,356us
QN-S-1M	99,469us	115,996us	253,828us	1,103ms	642,891us	942,900us
QEx-R-100K	126,499us	170,284us	96,998us	14,933us	686,622us	70,622us
QEx-R-500K	280,310us	259,337us	211,398us	1,361ms	1,161ms	277,977us
QEx-R-1M	4,483ms	397,403us	233,614us	8,536ms	2,739ms	119,956us
QEx-S-100K	181,804us	222,825us	7,200ms	489,827us	3,99ms	5,95ms
QEx-S-500K	302,62us	180,291us	30,204ms	1,375ms	685,576us	3,178ms
QEx-S-1M	2,329ms	288,105us	55,556ms	7,214ms	1,391ms	0ns
QEn-R-100K	72,585us	4,576us	30,95us	445,929us	1,416us	56,208us
QEn-R-500K	70,521us	6,160us	21,57us	171,65us	1,677us	53,164us
QEn-R-1M	73,515us	7,225us	21,774us	149,983us	2,164us	54,961us
QEn-S-100K	61,904us	7,147us	12,530us	39,135us	6,77us	49,602us
QEn-S-500K	52,799us	6,391us	16,113us	30,711us	28,855us	167,289us
QEn-S-1M	53,358us	3,675us	21,633us	64,305us	1,721us	709,977us
IN-R-100K	44,425us	9,131us	7,848us	48,857us	147,153us	41,66us
IN-R-500K	30,842us	1,599us	4,926us	0ns	29,726us	85,871us
IN-R-1M	30,859us	1,387us	4,789us	0ns	24,403us	155,118us
IN-S-100K	5,181us	1,901us	6,668us	19,959us	74,364us	45,639us
IN-S-500K	4,408us	1,544us	5,947us	28,378us	41,969us	86,139us
IN-S-1M	3,824us	1,362us	6,53us	15,509us	34,185us	125,898us
IE-R-100K	6,174us	2,587us	2,740us	17,906us	78,859us	5,81us
IE-R-500K	7,77us	2,466us	3,862us	13,814us	58,637us	964,668us
IE-R-1M	7,611us	2,592us	2,849us	11,598us	107,551us	13,986us
IE-S-100K	6,129us	1,390us	2,638us	16,993us	2,432us	86,702us
IE-S-500K	4,452us	1,468us	2,434us	14,588us	65,639us	128,164us
IE-S-1M	3,110us	1,410us	3,160us	12,598us	45,841us	696,530us
INN-R-100K	124,460us	329,298us	166,693us	528,406us	593,44us	412,642us
INN-R-500K	171,186us	445,346us	270,393us	831,722us	1,49ms	124,973us
INN-R-1M	194,627us	522,203us	281,274us	939,848us	1,884ms	241,676us
INN-S-100K	100,471us	280,704us	171,670us	314,960us	726,270us	351,630us
INN-S-500K	145,330us	384,964s	195,835us	823,617us	2,335ms	386,801us
INN-S-1M	276,626us	422,135us	226,636us	1,192ms	1,649ms	1,521ms

The results of the tests involving the query of existing edges from real graphs indicate that the Neo4j-based framework has the best performance, as it conducts queries faster than its competitors and also has the lowest value of standard deviation. The H2-based framework, on the contrary, was proven to be very slow for larger graphs, having also a high value of standard deviation. However, the same experiment on synthetic graphs generated completely different results for the Neo4j-based framework, as it was the slowest, with the measured average times being much larger than the times measured for the real graph tests. The framework that had the best performance in these experiments for the larger graphs is the Lucene-based one.

The Lucene-based framework was proved to be the fastest when querying non-existing edges for both real and synthetic graphs, as well as the most stable as it had the lowest standard deviation values. The slowest framework for both types of tests was the one based on H2. Another observation is that the standard deviation for the Neo4j-based framework was much larger for the synthetic graph tests, whereas quite the opposite is true for the H2-based framework.

In all iteration tests (including both node as well as edge iteration) the Lucene-based framework had the best performance. For the node iteration tests it is worth noticing that this framework generated the same average times for all real and synthetic graphs, on the exception of the smallest real graph for which it generated a larger average time. Between the other two frameworks the Neo4j-based proved to be faster for real graphs, whereas their performance was comparable for synthetic graph tests. In general, both the Neo4j-based and H2-based frameworks were faster when querying nodes of synthetic rather than real graphs. However, the Neo4j-based framework was observed to have a very high value of standard deviation for the larger graphs. The Neo4j-based framework performed relatively well in edge iteration tests, with the calculated average times for the real graph tests being comparable to the respective average times for the Lucene-based framework. An observation about the edge iteration results for the Lucene-based framework is that it performed two times faster for the synthetic graph tests in relation to real graph tests.

Finally, the tests that involved querying the neighbors of random nodes, indicate that the H2-based framework is the most efficient for such type of queries for most of the tests, whereas the Lucene-based framework has the worst performance. The single test for which the H2-based framework was outperformed by the Neo4j-based framework is the test conducted on the largest synthetic graph (1,000,000 edges). An observation that gives proof of the poor performance of the Lucene-based framework in relation to the other frameworks is that its performance, when the test was conducted on the largest real graph, was 2.5 times worse than the best performance for the same test, whereas it was approximately 2 times worse than than best performance when the test was conducted on the largest synthetic graph. This again indicates that the Lucene-based framework is particularly slow when retrieving the neighbors of a node.

7 Conclusions and Outlook

The abundance of Web data has created the need for more efficient scalable graph data management structures. With this problem in mind, we presented various solutions for the management of massive web graphs. We considered the special case of STS as an application setting and we defined the requirements that STS data impose on the underlying management framework. Moreover, we developed three different STS data management frameworks and presented their structure and functionality. The developed frameworks were benchmarked in terms of the disk space required for data storage, as well as in terms of how fast they perform data insertion, update, and deletion operations. The experimental results showed that both frameworks have their pros and cons, and that the choice of a suitable framework for the management of STS data (or web graph data in general) depends on the type of operations that are expected to be performed more often. In general, the custom Lucene-based framework seems to be an efficient solution for the majority of operations, except for those that involve accessing the neighbors of a node, where the H2-based and Neo4j-based frameworks proved to be better solutions. Moreover, although the proposed frameworks have been designed for the case of the STS, they can be easily adjusted so that they are applicable for other types of Web and Web 2.0 graph data. The possibility of testing the performance of frameworks that use other technologies as underlying infrastructures (such as object databases, presented in subsection 3.2, or data mining-based solutions, presented in Section 4) is also worthwhile to be explored as future work.

From the above, it appears that there are different requirements for the management of Web and Web 2.0 graph data, depending on the reasons why their storage, and management in general, is desirable. A Web graph management framework should consequently either be centered on a specific application, or be adaptable to suit many application requirements. An interesting vision would be to combine the characteristics of data mining-based solutions, such as the compression-based databases, with the update functionalities of transactional databases, in a framework able to support applications both for static graph analysis and for managing time-varying graphs. In general, graph data management frameworks should be developed to maximally exploit the available main memory. Approaches towards this direction would be to e.g. store part of the graph structure in main memory and the rest of the data in external memory, or use a computer cluster to increase the size of available main memory to fit the entire graph structure. An alternative approach would be to differentiate between the way the adjacency lists of high-degree and low-degree nodes are stored, so that e.g. the adjacency lists of high-degree nodes are stored in main memory to decrease the time required for their access.

The possibility of developing a framework for handling temporal graphs that would maintain information about the graph's state in different time steps constitutes another interesting future work area. Such functionality could be included in a graph management framework e.g. information about when a node or edge was added to (or deleted from) the graph is stored as an extra attribute of the node or edge. Managing temporal graph data with such a framework would enable querying about e.g. when an edge was added to the graph, which nodes were adjacent to a node at a given time, etc.

References

1. Abello, J., Buchsbaum, A.L., Westbrook, J.: A Functional Approach to External Graph Algorithms. Algorithmica 32(3), 437–458 (1998)
2. Ahn, K.J., Guha, S.: Graph Sparsification in the Semi-streaming Model. In: ICALP(2), pp. 328–338 (2009)
3. Bader, D., Madduri, K.: Designing multithreaded algorithms for breadth-first search and st-connectivity on the Cray MTA-2. In: Proceedings of the ICPP 2006. IEEE Computer Society, Los Alamitos (2006)

4. Becchetti, L., Boldi, P., Castillo, C., Gionis, A.: Efficient semi-streaming algorithms for local triangle counting in massive graphs. In: Proceeding of the KDD 2008, pp. 16–24. ACM Press, New York (2008)
5. Beynon, M.D., Kurc, T., Catalyurek, U., Chang, C., Sussman, A., Saltz, J.: Distributed processing of very large datasets with DataCutter. Parallel Comput. 27(11), 1457–1478 (2001)
6. Boldi, P., Vigna, S.: The WebGraph Framework I: Compression Techniques. In: Proceedings of the WWW 2004, pp. 595–602. ACM, New York (2004)
7. Boldi, P., Vigna, S.: The WebGraph Framework II: Codes For The World-Wide Web. In: Proceedings of the DCC 2004, vol. 528. IEEE Computer Society, Los Alamitos (2004)
8. Boldi, P., Santini, M., Vigna, S.: Permuting Web Graphs. In: Avrachenkov, K., Donato, D., Litvak, N. (eds.) WAW 2009. LNCS, vol. 5427, pp. 116–126. Springer, Heidelberg (2009)
9. Bothorel, C., Bouklit, M.: An algorithm for detecting communities in folksonomy hypergraphs. Appeared in I2CS 2008, Schoelcher, Martinique, Sponsored by IEEE (2008)
10. Brin, S., Page, L.: The anatomy of a large-scale hypertextual Web search engine. Comput. Netw. ISDN Syst. 30(1-7), 107–117 (1998)
11. Brinkmeier, M., Werner, J., Recknagel, S.: Communities in graphs and hypergraphs. In: Proceedings of CIKM 2007, Lisbon, Portugal, pp. 869–872. ACM, New York (2007)
12. Brooks, C.H., Montanez, N.: Improved annotation of the blogosphere via autotagging and hierarchical clustering. In: Proceedings of the WWW 2006, pp. 625–632. ACM, New York (2006)
13. Buchsbaum, A.L., Giancarlo, R., Racz, B.: New results for finding common neighborhoods in massive graphs in the data stream model. Theor. Comput. Sci. 407(1-3), 302–309 (2008)
14. Buehrer, G., Chellapilla, K.: A scalable pattern mining approach to web graph compression with communities. In: Proceedings of the WSDM 2008. ACM, New York (2008)
15. Buriol, L.S., Frahling, G., Leonardi, S., Marchetti-Spaccamela, A., Sohler, C.: Counting triangles in data streams. Proceedings of the PODS 2006, pp. 253–262. ACM, New York (2006)
16. Cattuto, C., Baldassarri, A., Servedio, D.P.V., Loreto, V.: Emergent Community Structure In Social Tagging Systems. Advances in Complex Systems (ACS) 11(4), 597–608 (2008)
17. Chierichetti, F., Kumar, R., Lattanzi, S., Mitzenmacher, M., Panconesi, A., Raghavan, P.: On compressing social networks. In: Proceedings of the KDD 2009, pp. 219–228. ACM, New York (2009)
18. Claude, F., Navarro, G.: A Fast and Compact Web Graph Representation. In: Ziviani, N., Baeza-Yates, R. (eds.) SPIRE 2007. LNCS, vol. 4726, pp. 105–116. Springer, Heidelberg (2007)
19. Cohen, J.: Graph Twiddling in a MapReduce World. Computing in Science & Engineering 11(4), 29–41 (2009)
20. Dean, J., Ghemawat, S.: MapReduce: simplified data processing on large clusters. Commun. ACM 51(1), 107–113 (2008)
21. Du, N., Wang, B., Wu, B., Wang, Y.: Overlapping Community Detection in Bipartite Networks. In: Proceedings of the WI-IAT 2008, pp. 176–179. IEEE Computer Society, Los Alamitos (2008)
22. Erdős, P., Rényi, A.: On Random Graphs I. Publicationes Mathematicae 6, 290–297 (1959)
23. Feigenbaum, J., Kannan, S., McGregor, A., Suri, S., Zhang, J.: On graph problems in a semi-streaming model. Theor. Comput. Sci. 348(2), 207–216 (2005)

24. Feigenbaum, J., Kannan, S., McGregor, A., Suri, S., Zhang, J.: Graph distances in the streaming model. SIAM J. Comput. 38(5), 1709–1727 (2008)
25. Furtado, P.: Evolving Application Domains of Data Warehousing and Mining: Trends and Solutions. IGI Publishing (2009)
26. Golder, S.A., Huberman, B.A.: Usage patterns of collaborative tagging systems. J. Inf. Sci. 32(2), 198–208 (2006)
27. Guo, L., Tan, E., Chen, S., Zhang, X., Zhao, Y.: Analyzing patterns of user content generation in online social networks. In: Proceedings of the KDD 2009, pp. 369–378. ACM, New York (2009)
28. Guozhu, D., Leonid, L., Jianwen, S., Limsoon, W.: Maintaining Transitive Closure of Graphs in SQL. Int. J. Information Technology 5 (1999)
29. Halpin, H., Robu, V., Shepherd, H.: The complex dynamics of collaborative tagging. In: Proceedings of the WWW 2007. ACM, New York (2007)
30. Hartley, T.D.R., Çatalyürek, Ü.V., Özgüner, F., Yoo, A., Kohn, S., Henderson, K.W.: MSSG: A Framework for Massive-Scale Semantic Graphs. In: Proceedings of the 2006 IEEE International Conference on Cluster Computing, pp. 1–10. IEEE, Los Alamitos (2006)
31. Hotho, A., Robert, J., Christoph, S., Gerd, S.: Emergent Semantics in BibSonomy. GI Jahrestagung P-94, 305–312 (2006)
32. Karande, C., Chellapilla, K., Andersen, R.: Speeding up algorithms on compressed web graphs. In: Proceedings of the WSDM 2009, pp. 272–281. ACM, New York (2009)
33. Keith, H.R., Raymie, S., Janet, L.W., Rajiv, G.W.: The Link Database: Fast Access to Graphs of the Web. In: Data Compression Conference, vol. 0, p. 122. IEEE Computer Society, Los Alamitos (2002)
34. Kittur, A., Chi, E., Pendleton, B.A., Suh, B., Mytkowicz, T.: Power of the Few vs. Wisdom of the Crowd: Wikipedia and the Rise of the Bourgeoisie. World Wide Web 1, 2,19 (2007)
35. Kleinberg, J.M.: Authoritative sources in a hyperlinked environment. J. ACM 46(5), 604–632 (1999)
36. Larsson, N.J., Moffat, A.: Offline Dictionary-Based Compression. In: Data Compression Conference, vol. 0, p. 296. IEEE Computer Society, Los Alamitos (1999)
37. Lawrence, P., Sergey, B., Motwani, R., Winograd, T.: The PageRank Citation Ranking: Bringing Order to the Web. Technical Report. Stanford University (1998)
38. Lempel, R., Moran, S.: The stochastic approach for link-structure analysis (SALSA) and the TKC effect. Comput. Netw. 33(1-6), 387–401 (2000)
39. Madduri, K., Bader, D.A.: Compact graph representations and parallel connectivity algorithms for massive dynamic network analysis. In: Proceedings of the IPDPS 2009, pp. 1–11. IEEE Computer Society, Los Alamitos (2009)
40. Mika, P.: Ontologies Are Us: A Unified Model of Social Networks and Semantics. In: International Semantic Web Conference, pp. 522–536 (2005)
41. Muthukrishnan, S.: Data streams: algorithms and applications. In: Proceedings of the SODA 2003, pp. 413–413 (2003)
42. Newman, M.E.J., Girvan, M.: Finding and evaluating community structure in networks. Physical Review E 69(2), 26113+ (2004)
43. Papadopoulos, S., Menemenis, F., Vakali, A., Kompatsiaris, Y.: Analysis of Content Popularity in Social Bookmarking Systems. In: Evolving Application Domains of Data Warehousing and Mining: Trends and Solutions. IGI Publishing (2009)
44. Papadopoulos, S., Kompatsiaris, Y., Vakali, A.: Leveraging Collective Intelligence through Community Detection in Tag Networks. In: Proceedings of the CKCaR 2009 (2009)

45. Papadopoulos, S., Kompatsiaris, Y., Vakali, A.: A Graph-based Clustering Scheme for Identifying Related Tags in Folksonomies. In: Proceedings of the DaWaK 2010 (2010)
46. Seidel, R., Aragon, C.: Randomized search trees. Algorithmica 16, 464–497 (1996)
47. Shepitsen, A., Gemmell, J., Mobasher, B., Burke, R.: Personalized recommendation in social tagging systems using hierarchical clustering. In: Proceedings of the RecSys 2008, pp. 259–266. ACM, New York (2008)
48. Simpson, E.: Clustering Tags in Enterprise and Web Folksonomies. Technical Report. HP Labs (2008)
49. Stephens, S., Rung, J., Lopez, X.: Graph Data Representation in Oracle Database 10*g*: Case Studies in Life Sciences. IEEE Data Eng. Bull. 27(4), 61–66 (2004)
50. Voss, J.: Measuring Wikipedia. In: The 10th International Conference of the International Society for Scientometrics and Informetrics (2005)
51. Wu, C., Zhou, B.: Analysis of tag within online social networks. In: Proceedings of the GROUP 2009, pp. 21–30. ACM, New York (2009)
52. Wu, X., Zhang, L., Yu, Y.: Exploring social annotations for the semantic web. In: Proceedings of the WWW 2006, pp. 417–426. ACM, New York (2006)
53. Yeung, C.A., Gibbins, N., Shadbolt, N.: Tag Meaning Disambiguation through Analysis of Tripartite Structure of Folksonomies. In: Proceedings of the WI-IATW 2007, pp. 3–6. IEEE Computer Society, Los Alamitos (2007)
54. Yeung, C.A., Gibbins, N., Shadbolt, N.: Collective User Behaviour and Tag Contextualisation in Folksonomies. In: Proceedings of the WI-IAT 2008, pp. 659–662. IEEE Computer Society, Los Alamitos (2008)
55. Yeung, C.A., Gibbins, N., Shadbolt, N.: Contextualising tags in collaborative tagging systems. In: Proceedings of the HT 2009, pp. 251–260. ACM, New York (2009)
56. Yin, Z., Li, R., Mei, Q., Han, J.: Exploring social tagging graph for web object classification. In: Proceedings of the KDD 2009, pp. 957–966. ACM, New York (2009)
57. Alberton, L.: Graphs in the database: SQL meets social networks (2009), `http://techportal.ibuildings.com/2009/09/07/graphs-in-the-database-sql-meets-social-networks`
58. Bergman, M.K.: Scalability of the Semantic Web (2006), `http://www.mkbergman.com/227/scalability-of-the-semantic-web`
59. Bergman, M.K.: Enterprise Semantic Webs Demand New Database Paradigms (2006), `http://www.mkbergman.com/185/enterprise-semantic-webs-esw-demand-new-database-paradigms`
60. Obasanjo, D.: An Exploration of Object Oriented Database Management Systems (2001), `http://www.25hoursaday.com/WhyArentYouUsingAnOODBMS.html`
61. Staken, K.: Introduction to Native XML Databases (2001), `http://www.xml.com/pub/a/2001/10/31/nativexmldb.html`
62. Wang, J.C., Huiling, G., Betsy, G.: Oracle White Paper? A Load-On-Demand Approach to Handling Large Networks in the Oracle Spatial Network Data Model (2009), `http://www.oracle.com/technology/products/spatial/pdf/11gr2_collateral/_ndmlod11gr2_wp_1009.pdf`
63. Apache Xindice, `http://xml.apache.org/xindice`
64. AllegroGraph RDF store, `http://www.franz.com/agraph/allegrograph`
65. Benchmarks: Performance advantages to store complex object structures, `http://www.db4o.com/about/productinformation/benchmarks`
66. db4o, `http://www.db4o.com/about/productinformation/db4o`

67. Facebook Statistics (2010), http://www.facebook.com/press/info.php?statistics
68. Getting Started with Berkeley DB for Java - Release 4.8, http://www.oracle.com/technology/documentation/berkeley-db/db/gsg/_JAVA/BerkeleyDB-Core-JAVA-GSG.pdf
69. H2 database, http://www.h2database.com
70. How ODB Works, http://wiki.neodatis.org/how-odb-works
71. Jena Semantic Web Framework, http://jena.sourceforge.net
72. JUNG Graph Framework, http://jung.sourceforge.net
73. Neo4j graph database, http://neo4j.org
74. Object-relational impedance mismatch, http://en.wikipedia.org/wiki/Object-relational_impedance_mismatch
75. Oracle Berkeley DB, http://www.oracle.com/technology/products/berkeley-db/index.html
76. OWLIM Repository, http://www.ontotext.com/owlim
77. PolePosition Benchmark NeoDatis1.9, http://switch.dl.sourceforge.net/project/neodatis-odb/NeoDatis%20ODB%20Performance/NeoDatis%201.9/PolePosition_NeoDatis-1.9.pdf
78. Sesame Framework, http://www.openrdf.org
79. Tamino XML Server, http://www.softwareag.com/corporate/products/wm/tamino
80. Virtuoso Server platform, http://www.openlinksw.com/virtuoso

Chapter 3
Web Engineering and Metrics

Emilia Mendes

The University of Auckland, Computer Science Department,
Private Bag 92019, Auckland, New Zealand
emilia@cs.auckland.ac.nz
http://www.cs.auckland.ac.nz/~emilia

Abstract. The objective of this chapter is three-fold. First, it provides an introduction to Web Engineering, and discusses the need for empirical investigations in this area. Second, it defines concepts such as metrics and measurement, and details the types of quantitative metrics that can be gathered when carrying out empirical investigations in Web Engineering. Third, it presents the three main types of empirical investigations – surveys, case studies, and formal experiments.

1 Introduction

Despite being originally conceived back in 1989 as an environment to allow for the sharing of information amongst geographically dispersed individuals (e.g. research reports, databases, user manuals), the World Wide Web (Web) has been transformed into an environment where numerous applications of varying types are delivered. These applications, named Web applications, range from small-scale information-dissemination-like applications, typically developed by writers and artists, to large-scale commercial,[1] enterprise-planning and scheduling, collaborative-work applications. The latter are developed by multidisciplinary teams of people with diverse skills and backgrounds using cutting-edge, diverse technologies [5][7] [18]. A large number of the Web applications that are presently developed are fully functional systems that provide business-to-customer and business-to-business e-commerce, and numerous services to numerous users [18].

Web applications are employed by numerous industries (e.g. travel and hospitality, manufacturing, banking, education, and government) to improve and increase their operations [7]. In addition, the client-server nature of the Web facilitates the development of corporate intranet Web applications, for use within the

[1] The increase in the use of the Web to provide commercial applications has been motivated by several factors, such as the possible increase of an organisation's competitive position, and the opportunity for small organisations to project their corporate presence in the same way as that of larger organisations.

A. Vakali & L.C. Jain (Eds.): New Directions in Web Data Management 1, SCI 331, pp. 59–82.
springerlink.com

boundaries of a single organisation [10]. In addition, the use of Web applications in areas such as communication and commerce makes it one of the leading and most important branches of the information and communication technologies industry [18].

Unfortunately, the development of industrial Web applications has been to date reported as generally ad hoc, resulting in poor-quality applications difficult to maintain [16]. The main reasons for such problems are unawareness of suitable design and development processes, and poor project management practices [6]. However, as the reliance on larger and more complex Web applications increases so does the need for using methodologies/standards/best practice guidelines that enable such applications to be delivered on time, within budget, and with quality [13][22][23]. The goal of Web engineering is therefore to provide Web development teams with the means to develop such applications by supplying sound methodologies, systematic techniques, quality assurance, rigorous, disciplined and repeatable processes, better tools, and baselines [7].

Web engineering is defined as [16]:

> "*the use of scientific, **engineering**, and management principles and systematic approaches with the aim of successfully developing, deploying and maintaining high quality Web-based systems and applications*".

Engineering is widely taken as a disciplined application of scientific knowledge for the solution of practical problems:

> "*Engineering is the application of science to the needs of humanity. This is accomplished through knowledge, mathematics, and practical experience applied to the design of useful objects or processes.*" [24]

> "*Engineering is the application of scientific principles to practical ends, as the design, manufacture, and operation of structures and machines.*" [10]

> "*The profession of applying scientific principles to the design, construction, and maintenance of engines, cars, machines, etc. (mechanical engineering), buildings, bridges, roads, etc. (civil engineering), electrical machines and communication systems (electrical engineering), chemical plant and machinery (chemical engineering), or aircraft (aeronautical engineering).*" [9]

In addition, Engineering purports the need for applying scientific principles, which are the result of applying a scientific process [8]. A process in this context means that our current understanding, i.e. our theory (hypothesis), of how best to develop, deploy, and maintain high-quality Web applications, may be modified or replaced as new evidence is found through the accumulation of data and knowledge. This process is illustrated in Fig. 1 and described below [8][15]:

- *Observation*: To observe or read about a phenomenon or set of facts. In most cases the motivation for such observation is to identify cause & effect relationships between observed items, since these entail predictable results. For example, we can observe that an increase in the development of new Web pages seems to also increase the corresponding development effort.
- *Hypothesis*: The formulation of a hypothesis represents an attempt to explain an *Observation*. It is a tentative theory or assumption that is believed to explain the behaviour under investigation [3]. The items that participate in the *Observation* are represented by variables (e.g. number of new Web pages, development effort) and the hypothesis indicates what is expected to happen to these variables (e.g. there is a linear relationship between number of Web pages and development effort, showing that as the number of new Web pages increases so does the effort to develop these pages). However, these variables first need to be measured and to do so we need an underlying Measurement Theory.
- *Prediction*: To predict means to forecast results that are to be found if the rationale used in the hypothesis formulation is correct (e.g. Web applications with a larger number of new Web pages will use a larger development effort).
- *Validation*: To validate requires experimentation to provide evidence either to support or refute the initial hypothesis. If the evidence refutes the hypothesis then the hypothesis should be revised or replaced. If the evidence is in support of the hypothesis, then many more replications of the experiment need to be carried out in order to build a better understanding of how variables relate to each other and their cause and effect relationships. Validating a hypothesis represents the gathering of data from measuring the variables abovementioned. Such data gathering occurs by means of an empirical investigation (empirical study).

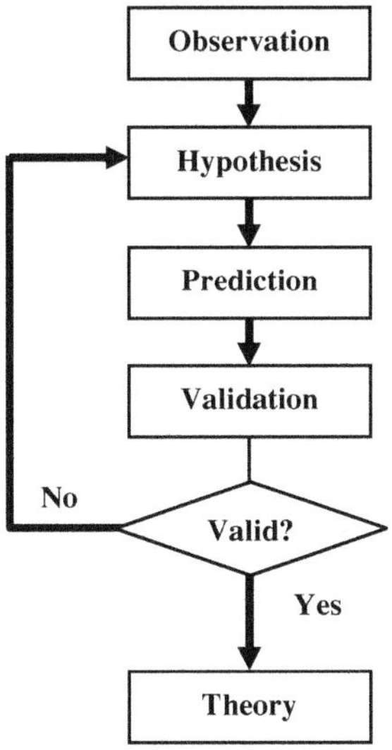

Fig. 1. The Scientific Process (Mendes, 2007)

In summary, the scientific process supports knowledge building, which in turn involves the use of empirical studies to test hypotheses previously proposed, and to assess if the current understanding of the discipline is correct. Therefore experimentation in Web engineering is essential [1][2].

The application by organisations of scientific principles to developing and maintaining Web applications is likely to vary depending on their level of maturity. Maturity reflects an organisation's use of sound development processes and practices [3], where more mature organisations generally tend to apply scientific principles to a larger extent than less mature organisations. In addition, some organisations have clearly defined processes that remain unchanged regardless of the people who work on their projects. For such organisations, success is dictated by following a well-defined process, where feedback is constantly obtained using product, process and resource measures. Other organisations have processes that are not so clearly defined (ad hoc) and therefore the success of a project is often determined by the expertise of the development team. In such case, product, process, and resource measures are rarely used and each project represents a potential risk that may lead an organisation, if it gets it wrong, to bankruptcy [21].

The variables used in the formulation of hypotheses represent the attributes of real-world entities that we observe. An entity represents a process, product, or resource. A process is defined as a software-related activity, where examples include Web development, Web maintenance, Web design, Web testing, and Web project management. A product is defined as an artefact, deliverable, or document that results from a process activity. Examples of products are Web application, design document, testing scripts, and fault reports. Finally, a resource represents an entity required by a process activity. Examples of resources are Web developers, development tools, and programming languages [3].

In addition, for each attribute that is to be measured, it is also useful to identify if it is *internal* or *external*. Internal attributes can be measured by examining the product, process, or resource on its own, separate from its behaviour. Conversely, external attributes can only be measured with respect to how the product, process, or resource relates to its environment [3]. For example, usability is in general an external attribute since its measurement often depends upon the interaction between user and application. An example classifying entities is presented in Table 1 [15].

The measurement of attributes generates quantitative descriptions of key processes, products, and resources, enabling us to understand behaviour and result. This understanding in turn lets us select better techniques and tools to control and improve our processes, products, and resources [19].

Table 1. Classification of process, product, and resources for the Tukutuku[2] dataset

ENTITY	ATTRIBUTE	DESCRIPTION
PROCESS ENTITIES		
PROJECT		
	TYPEPROJ	Type of project (new or enhancement).
	LANGS	Implementation languages used.
	DOCPROC	If project followed defined and documented process.
	PROIMPR	If project team involved in a process improvement programme.
	METRICS	If project team part of a software metrics programme.
	DEVTEAM	Size of project's development team.
WEB DEVELOPMENT		
	TOTEFF	Actual total effort used to develop the Web application.
	ESTEFF	Estimated total effort necessary to develop the Web application.
	ACCURACY	Procedure used to record effort data.
PRODUCT ENTITY		
WEB APPLICATION		
	TYPEAPP	Type of Web application developed.
	TOTWP	Total number of Web pages (new and reused).
	NEWWP	Total number of new Web pages.
	TOTIMG	Total number of images (new and reused).
	NEWIMG	Total number of new images your company created.
	HEFFDEV	Minimum number of hours to develop a single function/feature by one experienced developer that is considered high (above average).
	HEFFADPT	Minimum number of hours to adapt a single function/feature by one experienced developer that is considered high (above average).
	HFOTS	Number of reused high-effort features/functions without adaptation.
	HFOTSA	Number of adapted high-effort features/functions.
	HNEW	Number of new high-effort features/functions.
	FOTS	Number of low-effort features off the shelf.
	FOTSA	Number of low-effort features off the shelf adapted.
	NEW	Number of new low-effort features/functions.
RESOURCE ENTITY		
DEVELOPMENT TEAM		
	TEAMEXP	Average team experience with the development language(s) employed.

The measurement theory that has been adopted in this chapter is the Representational Theory of Measurement [3], which drives the definition of measurement scales presented in the next Section.

[2] The Tukutuku project collects data on industrial Web projects, for the development of effort estimation models and to benchmark productivity across and within Web companies. See http://www.cs.auckland.ac.nz/tukutuku.

2 Measurement Scales

When we gather data associated with the attributes that characterise the Entities we wish to measure, they can be collected using a different measurement scale. The characteristics of each scale type determine the choice of methods and statistics that can be used to analyse the data that was measured using that scale type, and how to interpret their corresponding measures. In this Section we describe the five main scale types [3] [15]:

- Nominal
- Ordinal
- Interval
- Ratio
- Absolute

2.1 Nominal Scale Type

The Nominal scale type represents the most primitive form of measurement. It identifies classes or categories where each category groups a set of Entities based on their attribute's value. Within this context, Entities can only be organised into classes or categories, without any notion of ranking between classes. In addition, classes can be represented as either symbols or numbers; however, if within this context numbers do not have any numerical meaning.

Examples using a Nominal scale are given in Table 2.

Table 2. Examples of Nominal Scale Measures

Entity	Attribute	Categories
Web application	type	e-Commerce, Academic, Corporate, Entertainment
Programming language	type	ASP (VBScript, .Net), Coldfusion, J2EE (JSP, Servlet, EJB), PHP
Web Project	type	New, Enhancement, Re-development
Web company	service type	1, 4, 5, 7, 9, 34, 502, 8

2.2 Ordinal Scale Type

The Ordinal scale supplements the Nominal scale with information about the ranking of classes or categories. As with the Nominal scale, it also identifies classes or categories where each category groups a set of Entities based on their attribute's value. The difference between an Ordinal scale and a Nominal scale is that an Ordinal scale assumes that there is some sort of ranking between classes. The same way as with the Nominal scale, classes can be represented as symbols or numbers, however if we use numbers they do not have any numerical meaning and represent ranking only. This means that addition, subtraction and other arithmetic operations cannot be applied to classes.

Examples of attributes measured using Ordinal scales are given in Table 3.

Table 3. Examples of Ordinal Scale Measures

Entity	Attribute	Categories
Web application	complexity	Very low, Low, Average, High, Very high
Web page	design quality	Very poor, Poor, Average, Good, Very good
Web Project	priority	1,2,3,4,5,6,7

2.3 Interval Scale Type

The Interval scale supplements the Ordinal scale with information about the size of the intervals that separate the classes or categories. As with the Nominal and Ordinal scales, it also identifies classes or categories, where each category groups a set of Entities based on their attribute's value. As with the Ordinal scale, there are ranks between classes or categories. The difference between an Interval scale and an Ordinal scale is that here there is the notion that the size of intervals between classes or categories remains constant. Although the Interval scale is a numerical scale and numbers have a numerical meaning, the class zero does not mean the complete absence of the attribute being measured. To illustrate that, let's look at temperatures measured using the Celsius scale. The difference between 1°C and 2°C is the same as the difference between 6°C and 7°C: exactly 1°. There is a ranking between two classes, thus 1°C has a lower rank than 2°C, and so on. Finally, the temperature 0°C does not represent the complete absence of temperature, where molecular motion stops. In this example, 0°C was arbitrarily chosen to represent the freezing point of water. This means that operations such as addition and subtraction between two categories is permitted (e.g. 50°C - 20°C = 70°C - 40°C; 5°C + 25°C = 20°C + 10°C), however calculating the ratio of two categories (e.g. 40°C/20°C) is not meaningful (40°C is not twice as hot as 20°C) so multiplication and division cannot be calculated directly from categories. If ratios are to be calculated, they need to be based on the differences between categories (e.g. 50°C - 20°C is twice 25°C - 10°C).

Examples using an Interval scale are given in Table 4.

Table 4. Examples of Interval Scale Measures

Entity	Attribute	Categories
Web project	Number of days relative to the starting point of a project	0,1,2,3,4,5,...
Human body	Temperature (Celsius or Fahrenheit)	Decimal numbers

2.4 Ratio Scale Type

The Ratio scale supplements the Interval scale with the existence of a zero element, representing total absence of the attribute being measured. As with the Interval scale, it also provides information about the size of the intervals that separate the classes or categories. As with the Interval and Ordinal scales, there are ranks between classes or categories. As with the Interval, Ordinal and Nominal scales, it also identifies classes or categories, where each category groups a set of Entities based on their attribute's value. The difference between a Ratio scale and an Interval scale is the existence of an absolute zero. The Ratio scale is also a numerical scale and numbers have a numerical meaning. This means that any arithmetic operations between two categories are permitted.

Examples using a Ratio scale are given in Table 5.

Table 5. Examples of Ratio Scale Measures

Entity	Attribute	Categories
Web project	Effort	Decimal numbers
Web application	Size	Integer numbers
Human body	Temperature in Kelvin	Decimal numbers

2.5 Absolute Scale Type

The Absolute scale supplements the Ratio scale with restricting the classes or categories to a specific unit of measurement. As with the Ratio scale, it also has a zero element, representing total absence of the attribute being measured. As with the Ratio and Interval scales, it also provides information about the size of the intervals that separate the classes or categories. As with the Interval and Ordinal scales, there are ranks between classes or categories. As with the Ratio, Interval, Ordinal and Nominal scales, it also identifies classes or categories, where each category groups a set of Entities based on their attribute's value.

The difference between the Absolute and the Ratio scales is the existence in the Absolute scale of a *fixed unit of measurement* associated with the attribute being measured. For example, using a Ratio scale, if we were to measure the attribute *effort* of a *Web project* we could obtain an effort value that could represent effort in number of hours, or effort in number of days, and so on. In case we want all effort measures to be kept using number of hours we can convert effort in number of days to effort in number of hours, or effort in number of weeks to effort in number of hours. Thus, an attribute measured using a given unit of measurement (e.g. number of weeks) can have its class converted into another using a different unit of measurement, but keeping the meaning of the obtained data unchanged. Therefore, assuming a single developer, a Web project's effort of 40 hours is equivalent to a Web project effort's of a week. Thus, the unit of measurement changes however the data that has been gathered remains unaffected. If we were to measure the attribute *effort* of a *Web project* using an *Absolute scale* we would need to determine in advance the unit of measurement to be used. Therefore, once this unit of measurement is determined, it is the one used when effort data is being gathered. Using our example on *Web project's effort*, had the unit of measurement associated with the *attribute effort* chosen to be *number of hours* then all the effort data gathered would represent effort in number of hours only.

Finally, as with the Ratio scale, operations between two categories, such as addition, subtraction, multiplication and division, are also permitted.

Examples using an Absolute scale are given in Table 6.

Table 6. Examples of Absolute Scale Measures

Entity	Attribute	Categories
Web project	Effort, in number of hours	Decimal numbers
Web application	Size, in number of html files	Integer numbers
Web developer	Experience developing Web applications, in number of years	Integer numbers

2.6 Summary of Scale Types

Table 7 presents one of the summaries we are providing regarding Scale types. It has been adapted from [14]. It is also important to note that the Nominal and Ordinal scales do not provide classes or categories that have numerical meaning, and for this reason their attributes are called Categorical or Qualitative. Conversely, given that the Interval, Ratio and Absolute scales provide classes or categories that have numerical meaning, their attributes are called Numerical or Quantitative [14].

Table 7. Summary of Scale Type Definitions

Scale type	Is ranking meaningful?	Are distances between classes the same?	Does the class include an absolute zero?
Nominal	No	No	No
Ordinal	Yes	No	No
Interval	Yes	Yes	No
Ratio	Yes	Yes	Yes
Absolute	Yes	Yes	Yes

In relation to the statistics relevant to each measurement scale type, Table 8 presents a summary adapted from [3].

Table 8. Summary of Scale Type Definitions

Scale type	Examples of suitable statistics	Suitable statistical tests
Nominal	Mode Frequency	Non-parametric
Ordinal	Median Percentile	Non-parametric
Interval	Mean Standard deviation	Non-parametric and parametric
Ratio	Mean Geometric mean Standard deviation	Non-parametric and parametric
Absolute	Mean Geometric mean Standard deviation	Non-parametric and parametric

3 Overview of Empirical Investigations

Validating a hypothesis or research question encompasses experimentation, which is carried out using an empirical investigation (empirical study). This Section details the three different types of empirical investigations that are most often carried out, which are: survey, case study or formal experiment [3][15].

- *Survey*: a retrospective investigation of an activity in order to confirm relationships and outcomes [3]. It is also known as "research-in-the-large" as it often samples over large groups of projects. A survey should always be carried out after the activity under focus has occurred [11]. When performing a survey, a researcher has no control over the situation at hand, i.e. the situation can be documented, compared to other similar situations, but none of the variables being investigated can be manipulated. Within the scope of Web engineering, surveys are often used to validate the response of organisations and developers to a new development method, tool, or technique, or to reveal trends or relationships between relevant variables. For example, a survey can be used to measure the success of changing from Sun's J2EE to Microsoft's ASP.NET throughout an organisation, because it can gather data from numerous projects. The downside of surveys is time. Gathering data can take many months or even years, and the outcome may only be available after several projects have been completed [11].
- *Case study*: an investigation that examines the trends and relationships using as its basis a typical project within an organisation. It is also known as "research-in-the-typical" [11]. Although a case study can be used to investigate a retrospective event, this is not the usual trend. This type of study is the investigation of choice when wishing to examine an event that has not yet occurred and for which there is little or no control over the variables. For example, if an organisation wants to investigate the effect of a programming framework on the quality of the resulting Web application however cannot develop the same project using numerous frameworks simultaneously, the investigative choice is to use a case study. If the quality of the resulting Web application is higher than the organisation's quality baseline, it may be due to many different reasons (e.g. chance, or perhaps bias from enthusiastic developers). Even if the programming framework had a legitimate effect on quality, no conclusions outside the boundaries of the case study can be drawn, i.e. the results of a case study cannot be generalised to every possible situation. Had the same application been developed several times, each time using a different programming framework[3] (as in a formal experiment, described later) then it would be possible to have better understanding of the relationship between framework and quality, given that these variables were controlled. A case study samples *from the variables*, rather than over them. This means that, in

[3] The values for all other attributes should remain the same (e.g. developers, programming experience, development tools, computing power, and type of application).

relation to the variable programming framework, a value that represents the framework usually used on most projects will be the one chosen (e.g. J2EE). A case study is easier to plan than a formal experiment, but its results are harder to explain and in addition, as previously mentioned, cannot be generalised outside the scope of the study [11].

- *Formal experiment*: rigorous and controlled investigation of an event where important variables are identified and manipulated such that their effect on the outcome can be validated [3]. It is also known as "research-in-the-small" since it is very difficult to carry out formal experiments in Web engineering using numerous projects and resources. A formal experiment samples *over the variable that is being manipulated*, such that all possible values that variable may have are validated, i.e. there is a single case representing each possible situation. If we apply the same example used when explaining case studies above, this means that several projects would be developed, each using a different object-oriented programming language. If one aims to obtain results that are largely applicable across various types of projects and processes, the choice of investigation is a formal experiment. However, despite the control that needs to be exerted when planning and running a formal experiment, its results cannot be generalised outside the experimental conditions. For example, if an experiment demonstrates that J2EE improves the quality of e-commerce Web applications, one cannot guarantee that J2EE will also improve the quality of educational Web applications [11]. This type of investigation is most suited to the Web engineering research community as it enables the building of theories to be applied when engineering Web applications.

There are other concrete issues related to using a formal experiment or a case study that may impact the choice of study. It may be feasible to control the variables, but at the expense of a very high cost or high degree of risk. If replicating a study *is* possible however at a prohibitive cost, a case study should then be the type of study used [3]. A summary of the characteristics of each type of empirical investigation is given in Table 9.

Table 9. Summary Characteristics of the Three Types of Empirical Investigations

Characteristic	Survey	Case study	Formal experiment
Scale	Research-in-the-large	Research-in-the-typical	Research-in-the-small
Control	No control	Low level of control	High level of control
Replication	No	Low	High
Generalisation	Results representative of sampled population	Only applicable to other projects of similar type and size	Can be generalised within the experimental conditions

There are a set of steps broadly common to all three types of investigations, which are described below:

Define the Goal(s) of Your Investigation and Its Context

Goals are crucial for the success of all activities part of an empirical investigation. Thus, it is important to allow enough time to fully understand and set the goals so that each one is clear and measurable. Goals represent the research questions, which may also correspond to a number of hypotheses. By setting the research questions or hypotheses it becomes easier to identify the dependent and independent variables that are to be measured as part of the investigation [3]. A dependent variable is a variable whose behaviour we want to predict or explain, and an independent variable is a variable believed to have a causal relationship with, or have influence upon, the dependent variable [25]. For example, if we wish to estimate the effort needed to develop a Web application and we believe that Web application's size and the number of Web developers influence development effort, then we have development effort as our single dependent variable, and Web application size and number of developers as the two independent variables.

Goals also help determine what the investigation will do, and what data is to be collected. Finally, by understanding the goals we can also confirm if the type of investigation chosen is the most suitable type to employ [3].

Each of the hypotheses being validated by means of an empirical investigation will later be either supported or rejected. When stating the hypothesis/es being validated it is customary to present them in two different forms – null and alternative hypotheses [25], as below:

H_0 Using J2EE produces the same quality of Web applications as using ASP.NET.

H_1 Using J2EE produces a different quality of Web applications than using ASP.NET.

H_0 is called the null hypothesis, and assumes the quality of Web applications developed using J2EE is similar to that of Web applications developed using ASP.NET. In other words, it assumes that data samples for both groups of applications come from the same population. In this instance, we have two samples, one representing quality values for Web applications developed using J2EE, and the other, quality values for Web applications developed using ASP.NET. Here, quality is our dependent variable, and the choice of programming framework (e.g. J2EE or ASP.NET), the independent variable.

H_1 is called the alternative or research hypothesis, and represents what is believed to be true if the null hypothesis is false. The alternative hypothesis assumes that samples do not come from the same sample population. Sometimes the direction of the relationship between dependent and independent variables is also presented as part of an alternative hypothesis. If H_1 also suggested a direction for the relationship, it could be described as:

H_1 Using J2EE produces a better quality of Web applications than using ASP.NET.

To support H_1 it is first necessary to reject the null hypothesis and, second, show that quality values for Web applications developed using J2EE are significantly higher than quality values for Web applications developed using ASP.NET.

We have presented both null and alternative hypotheses since they are both equally important when presenting the results of an empirical investigation, and, as such, both should be documented.

To see if the data justify rejecting H_0 we need to perform a statistical analysis. Before carrying out a statistical analysis it is important to decide the level of confidence we have that the data sample we gathered truly represents our population of interest. If we have 95% confidence that the data sample we are using truly represents the general population there still remains a 5% chance that H_0 will be rejected when in fact it truly represents the current situation. Rejecting H_0 incorrectly is called the *Type I error*, and the probability of this occurring is called the *Significance level* (α). Every statistical analysis test uses α when testing if H_0 should be rejected or not.

4 Issues to Consider When Conducting Empirical Studies

In addition to defining the goals of an investigation, it is also important to document the context of the investigation [12]. One suggested way to achieve this is to provide a table, similar to Table 1, describing the entities, attributes, and measures that are the focus of the investigation.

Prepare the Investigation

It is important to prepare an investigation carefully to obtain results from which one can draw valid conclusions, even if these conclusions cannot be scaled up. For case studies and formal experiments it is important to define the variables that can influence the results, and once defined, decide how much control one can have over them [3].

Consider the following case study which would represent a *poorly prepared investigation*.

The case study aims to investigate, within a given organisation, the effect of using the programming framework J2EE on the quality of the resulting Web application. Most Web projects in this organisation are developed using ASP.NET, and consequently all the development team has experience with this language. The type of application representative of the majority of applications this organisation undertakes is in electronic commerce (e-commerce), and a typical development team has two developers. Therefore, as part of the case study, an e-commerce application is to be developed by two developers using J2EE. Because we have stated this is a poorly executed case study, we will assume that no other variables have been considered, or measured (e.g. developers' experience, development environment).

The e-commerce application is developed, and the results of the case study show that the quality of the delivered application, measured as the number of faults per Web page, is worse than that for the other similar Web applications developed using ASP.NET. When questioned as to why these were the results obtained, the investigator seemed puzzled, and without a clear explanation.

What is missing?

The investigator should have anticipated that other variables can also have an effect on the results of an investigation, and should therefore also be taken into account. One such variable is developers' programming experience. Without measuring experience prior to the case study, it is impossible to discern if the lower quality is due to J2EE or to the effects of learning J2EE as the investigation proceeds. It is possible that one or both developers did not have experience with J2EE, and lack of experience has interfered with the benefits of its use.

Variables such as developers' experience should have been anticipated and if possible controlled, or risk obtaining results that will be incorrect.

To control a variable is to determine a subset of values for use within the context of the investigation from the complete set of possible values for that variable. For example, using the same case study presented above, if the investigator had measured developers' experience with J2EE (e.g. low, medium, high), and was able to control this variable, then (s)he could have determined that two developers experienced with J2EE should participate in the case study. If there were no developers with experience in J2EE, two would be selected and trained.

If, when conducting a case study, it is not possible to control certain variables, they should still be measured, and the results documented.

If, however, all variables are controllable, then the type of investigation to use is a formal experiment.

Another important issue is to identify the population being studied and the sampling technique used. For example, if a survey was designed to investigate the extent to which project managers use automatic project management tools, then a data sample of software programmers is not going to be representative of the population that has been initially specified.

With formal experiments, it is important to describe the process by which experimental subjects and objects are selected and assigned to treatments [12], where a treatment represents the new tool, programming language, or methodology you want to evaluate. The experimental object, also known as experimental unit, represents the object to which the treatment is to be applied (e.g. development project, Web application, code). The control object does not use or is not affected by the treatment [3]. In Web engineering it is difficult to have a control in the same way as in, say, formal medical experiments. For example, if you are investigating the effect of a programming framework on quality, and your treatment is J2EE, you cannot have a control that is "no programming framework" [12]. Therefore, many formal experiments use as their control a baseline representing what is typical in an organisation. Using the example given previously, our control would be ASP.NET since it represents the typical programming framework used in the organisation. The experimental subject is the "who" applying the treatment [3].

As part of the preparation of an investigation we also include the preparation and validation of data collection instruments. Examples are questionnaires, automatic measurement tools, timing sheets, etc. Each has to be prepared carefully such that it clearly and unambiguously identifies what is to be measured. For each variable it is also important to identify its measurement scale and measurement unit. So, if you are measuring effort, then you should also document its measurement unit (e.g. person hours, person months) or else obtain incorrect and conflicting data. It is also important to document at which stage during the investigation the data collection takes place. If an investigation gathers data on developers' programming experience (before they develop a Web application), size and effort used to design the application, and size and effort used to implement the application, then a diagram, such as the one in Fig. 2, may be provided to all participants to help clarify what instrument(s) to use and when to use them.

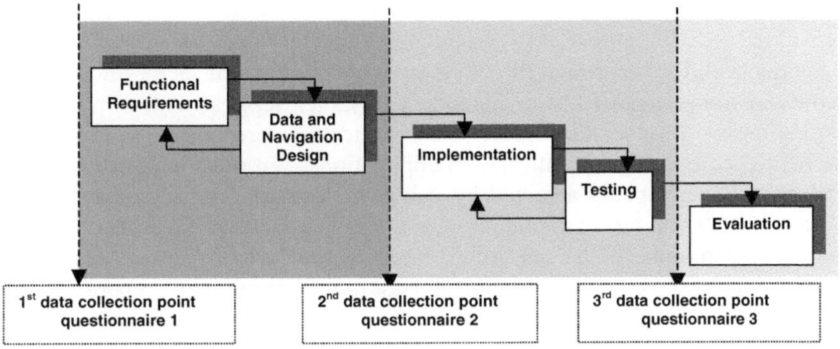

Fig. 2. Plan Detailing When to Apply Each Project [15]

It is usual for instruments to be validated using pilot studies. A pilot study uses similar conditions to those planned for the real investigation, such that any possible problems can be anticipated. It is highly recommended that those conducting any empirical investigations use pilot studies as they can provide very useful feedback and reduce or remove any problems not previously anticipated.

Finally, it is also important to document the methods used to reduce any bias.

Analysing the Data and Reporting the Results

The main aspect of this final step is to understand the data collected and to apply statistical techniques that are suitable for the research questions or hypotheses of the investigation. For example, if the data was measured using a nominal or ordinal scale then statistical techniques that use the mean cannot be applied as this would violate the principles of the representational theory of measurement. If the data is not normally distributed then it is possible to use non-parametric or robust techniques, or transform the data to conform to the normal distribution [3]. Further details on data analysis are presented later in this Chapter.

When interpreting and reporting the results of an empirical study it is also important to consider and discuss the validity of the results obtained. There are three types of threats to the validity of empirical investigations [11][20]: *construct validity*, *internal validity* and *external validity*. Each is described below.

Construct validity: represents the extent to which the measures you are using in your investigation really measure the attributes of Entities being investigated. For example, if you are measuring the size of a Web application using IFPUG function points, can you say that the use of IFPUG function points is really measuring the size of a Web application? How valid will the results of your investigation be if you use IFPUG function points to measure a Web application's size? Another example, if you want to measure the experience of Web developers developing Web applications and you use as a measure the number of years they worked for their current employer, it is unlikely that you are using an appropriate measure since your measure does not take into account as well their previous experience developing Web applications.

Internal validity: represents the extent to which external factors not controlled by the researcher can affect the dependent variable. Suppose that, as part of an investigation, we observe that larger Web applications are related to more productive teams, compared to smaller Web applications. We must make sure that team productivity is not being affected by using, for example, highly experienced developers to develop larger applications and less experienced developers to develop smaller applications. If the researcher is unaware of developers' experience it is impossible to discern whether the results are due to developers' experience or due to legitimate economies of scale. Typical factors that can affect the internal validity of investigations are variations in human performance, learning effects where participants' skills improve as the investigation progresses, and differences in treatments, data collection forms used or other experimental materials.

External validity: represents the extent to which we can generalise the results of our investigation to our population of interest. In most empirical investigations in Web engineering the population of interest often represents industrial practice. Suppose you carried out a formal experiment with postgraduate students to compare J2EE to ASP.NET, using as experimental object a small Web application. If this application is not representative of industrial practice you cannot generalise the results of your investigation beyond the context in which it took place. Another possible problem with this investigation might be the use of students as subject population. If you have not used Web development professionals, it will also be difficult to generalise the results to industrial practice. Within the context of this example, even if you had used Web development professionals in your investigation, if they did not represent a random sample of your population of interest you would also be unable to generalise the results to your entire population of interest.

5 Detailing Formal Experiments

A formal experiment is considered the most difficult type of investigation to carry out since it has to be planned very carefully such that all the important factors are controlled and documented, enabling its further replication. Due to the amount of control that formal experiments use they can be further replicated and, when replicated under identical conditions, if results are repeatable, they provide better basis for building theories that explain our current understanding of a phenomenon of interest. Another important point related to formal experiments is that the effects of uncontrolled variables upon the results must be minimised. The way to minimise such effect is to use randomisation. Randomisation represents the random assignment of treatments and experimental objects to experimental subjects.

The following sub-Sections discuss the typical experimental designs used with formal experiments [26][15]; for each typical design, the types of statistical analysis tests that can be used to examine the data gathered from such experiments are also introduced.

5.1 Typical Design 1

There is one independent variable (factor) with two values and one dependent variable. Suppose you are comparing the productivity between Web applications developed using J2EE (treatment) and Web applications developed using ASP.NET (control). 50 subjects are participating in the experiment and the experimental object is the same for both groups. Assuming other variables are constant, subjects are randomly assigned to J2EE or ASP .NET (see Fig. 3).

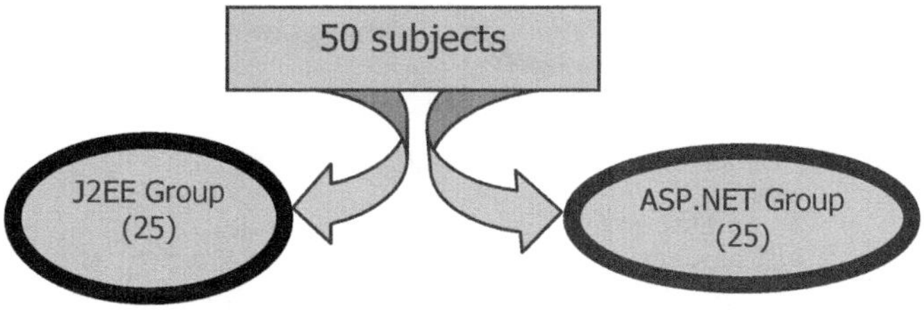

Fig. 3. Example of One-Factor Design

Once productivity data is gathered for both groups the next step is to compare the productivity data to check if productivity values for both development frameworks come from the same population (H_0) or from different populations (H_1). If the subjects in this experiment represent a large random sample or the productivity data for each group is normally distributed you can use the independent samples t-test statistical technique to compare the productivity between both groups. This is a parametric test and as such it assumed that the data is normally distributed or the sample is large and random. Otherwise, the statistical technique to use would be the independent samples Mann-Whitney test, a non-parametric equivalent to the t-test. Non-parametric tests make no assumptions related to the distribution of the data and that is why they are used if you cannot guarantee that your data is normally distributed or represent a large random sample.

5.2 Typical Design 1: One Factor and One Confounding Factor

There is one independent variable (factor) with two values and one dependent variable. Suppose you are comparing the productivity between Web applications developed using J2EE (treatment) and Web applications developed using ASP.NET (control). 50 subjects are participating in the experiment and the experimental object is the same for both groups. A second factor (confounding factor) – gender, is believed to have an effect on productivity however you are only interested in comparing different development frameworks and their effect on productivity, not the interaction between gender and framework type on productivity. The solution is to create two blocks (see Fig. 4), one with all the female subjects, and another with all the male subjects, and then, within each block, randomly assign a similar number subjects to J2EE or ASP .NET (balancing).

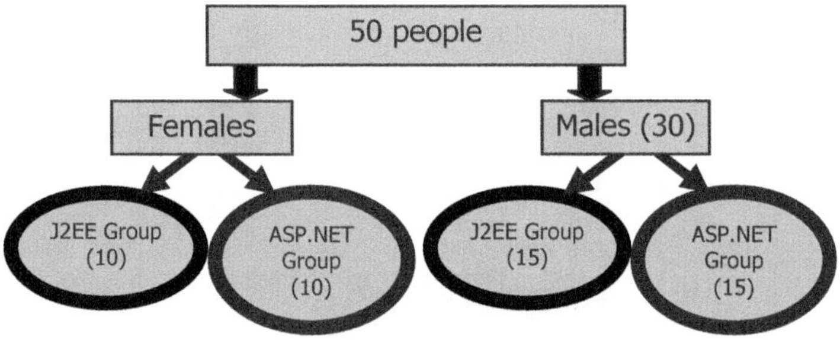

Fig. 4. Example of Blocking and Balancing with One-Factor Design

Once productivity data is gathered for both groups the next step is to compare the productivity data to check if productivity values for both groups come from the same population (H_0) or come from different populations (H_1). The mechanism used to analyse the data would be the same one presented previously. Two sets of productivity values are compared, one containing productivity values for the 10 females and the 15 males who used J2EE, and the other containing productivity values for the 10 females and the 15 males who used ASP.NET. If the subjects in this experiment represent a large random sample or the productivity data for each group is normally distributed you can use the independent samples t-test statistical technique to compare the productivity between both groups. Otherwise, the statistical technique to use would be the independent samples Mann-Whitney test, a non-parametric equivalent to the t-test.

5.3 Typical Design 2

There is one independent variable (factor) with two values and one dependent variable. Suppose you are comparing the productivity between Web applications developed using J2EE (treatment) and Web applications developed using ASP.NET (control). 50 subjects are participating in the experiment using the experimental object. You also want every subject to be assigned to both the control and the treatment. Assuming other variables are constant, subjects are randomly assigned to the control or the treatment, and then swapped around (see Fig. 5).

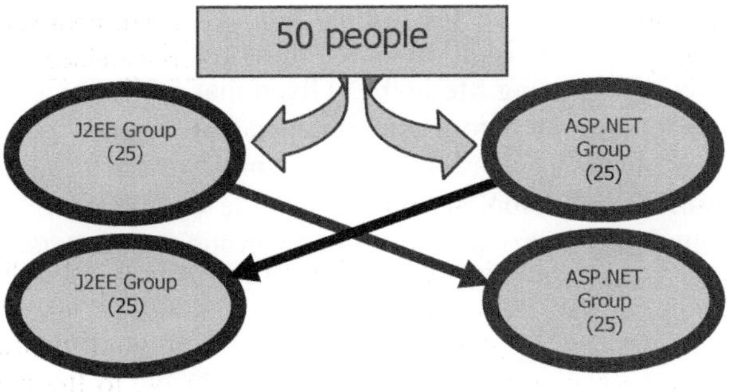

Fig. 5. Example of Typical Design 2

Once productivity data is gathered for both groups the next step is to compare the productivity data to check if productivity values for both groups come from the same population (H_0) or come from different populations (H_1). Two sets of productivity values are compared: the first contains productivity values for 50 subjects when using J2EE; the second contains productivity values for the same 50 subjects, when using ASP.NET. Given that each subject was exposed to both control and treatment you need to use a paired test. If the subjects in this experiment represent a large random sample or the productivity data for each group is normally distributed you can use the paired samples t-test statistical technique to compare the productivity between both groups. Otherwise, the statistical technique to use would be the two related samples Wilcoxon test, a non-parametric equivalent to the paired samples t-test.

5.4 Typical Design 3

There is one independent variable (factor) with more than two values and one dependent variable. Suppose you are comparing the productivity amongst Web applications designed using Methods A, B and C. 60 subjects are participating in the experiment and the experimental object is the same for all groups. Assuming other variables are constant, subjects are randomly assigned to one of the three groups (see Fig. 6).

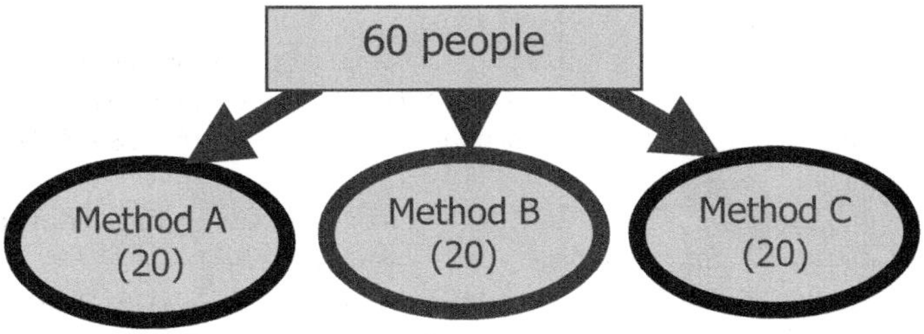

Fig. 6. Example of Typical Design 3

Once productivity data is gathered for all the three groups the next step is to compare the productivity data to check if productivity values for all groups come from the same population (H_0) or come from different populations (H_1). Three sets of productivity values are compared: the first contains productivity values for 20 subjects when using Method A; the second contains productivity values for another 20 subjects when using Method B; the third contains productivity values for another 20 subjects when using Method C. Given that each subject was exposed to only a single method you need to use an independent samples test. If the subjects in this experiment represent a large random sample or the productivity data for each group is normally distributed you can use the One-Way ANOVA statistical technique to compare the productivity among groups. Otherwise, the statistical technique to use would be the Kruskal-Wallis H test, a non-parametric equivalent to the One-Way ANOVA.

5.5 Typical Design 4

There are at least two independent variables (factors) and one dependent variable. Suppose you are comparing the productivity between Web applications developed using J2EE (treatment) and Web applications developed using ASP.NET (control). 60 subjects are participating in the experiment and the experimental object is the same for both groups. A second factor – gender, is believed to have an effect on productivity and you are interested in assessing the interaction between gender and framework type on productivity. The solution is to create four blocks (see Table 10) representing the total number of possible combinations. In this example each factor has two values therefore the total number of combinations would be given by multiplying the number of values in the first factor by the number of values in the second factor (2 multiplied by 2), which is equal to 4. Then, assuming that all subjects have similar experience using both frameworks, within each gender block, subjects are randomly assigned to J2EE or ASP .NET (balancing). In this scenario each block will provide 15 productivity values.

Once productivity data is gathered for all the four blocks the next step is to compare the productivity data to check if productivity values for males come from the same population (H_0) or come from different populations (H_1), and the same has to be done for females. Here productivity values for blocks 2 and 4 are

Table 10. Example of Typical Design 4

		Gender	
		Female	Male
Framework	J2EE	Female, J2EE (15) *Block 1*	Male, J2EE (15) *Block 2*
Framework	ASP.NET	Female, ASP.NET (15) *Block 3*	Male, ASP.NET (15) *Block 4*

compared; and productivity values for blocks 1 and 3 are compared. If the subjects in this experiment represent a large random sample or the productivity data for each group is normally distributed you can use the independent samples t-test statistical technique to compare the productivity between groups. Otherwise, the statistical technique to use would be the Mann-Whitney test, a non-parametric equivalent to the independent samples t-test.

5.6 Summary of Typical Designs

Table 11 summarises the statistical tests to be used with each of the typical designs previously introduced. Each of these tests is explained in detail in statistical books, such as [25].

Table 11. Examples of Statistical Tests for Typical Designs

Typical Design	Parametric test	Non-parametric test
Design 1 no explicit confounding factor	Independent samples t-test	Independent samples Mann-Whitney test
Design 1 explicit confounding factor	Independent samples t-test	Independent samples Mann-Whitney test
Design 2	paired samples t-test	Two-related samples Wilcoxon test
Design 3	One-Way ANOVA	Kruskal-Wallis H test
Design 4	independent samples t-test	Mann-Whitney test

6 Detailing Case Studies

It is often the case that case studies are used in industrial settings to compare two different technologies, tools or development methodologies. One of the technologies, tools or development methodologies represents what is currently used by the company, and the other technology, tool or development methodology represents what is being compared to the company's current situation. Three mechanisms are suggested to organise such comparisons to reduce bias and enforce internal validity [26]:

- To compare the results of using the new technology, tool or development methodology to a company's baseline. A baseline generally represents an average over a set of finished projects. For example, a company may have established a productivity baseline against which to compare projects. This means that productivity data has been gathered from past finished projects and used to obtain an average productivity (productivity baseline). If this is the case then the productivity related to the project that used the new technology, tool or development methodology is compared against the existing productivity baseline, to assess if there was productivity improvement or decline. In addition to productivity other baselines may also be used by a company, e.g. usability baseline, defect rate baseline.

- To compare the results of using the new technology, tool or development methodology to a company's sister project, which is used as a baseline. This means that two similar and comparable projects will be carried out, one using the company's current technology, tool or development methodology, and another using the new technology, tool or development methodology. Once both projects are finished measures such as productivity, usability and actual effort can be used to compare the results.
- Whenever the technology, tool or development methodology applies to individual application components, it is possible to apply at random the new technology, tool or development methodology to some components and not to others. Later measures such as productivity and actual effort can be used to compare the results.

7 Detailing Surveys

There are three important points to stress here. The first is that, similarly to formal experiments and case studies, it is very important to define beforehand what is it that we wish to investigate (hypotheses) and what is the population of interest. For example, if you plan to conduct a survey to understand how Web applications are currently developed the best population to use would be the one of Web project managers as they have the complete understanding of the development process used. Interviewing Web developers may lead to misleading results as it is often the case that they do not see the forest for the trees.

The second point is related to piloting the survey. It is important to ask different users, preferably representative of the population of interest, to read the instrument(s) to be used for data collection to make sure questions are clear and no important questions are missing. It is also important to ask these users to actually answer the questions in order to have a feel for how long it will take them to provide the data being asked for. This should be a similar procedure if you are using interviews.

Finally, the third point relates to the preparation of survey instruments. It is generally the case that instruments will be either questionnaires or interviews. In both cases instruments should be prepared with care and avoid misleading questions that can bias the results. If you use ordinary mail to post questionnaires to users make sure you also include a pre-paid envelope addressed to yourself, to be used to return the questionnaires. You can also alternatively have the same questionnaire available on the Web. Unfortunately the use of electronic mails as means to broadcast a request to participate in a survey has been impaired by the advent of spam emails. Many of us nowadays use filters to stop the receipt of unsolicited junk emails and therefore many survey invitation requests may end up being filtered and deleted.

8 Conclusions

This chapter discussed the need for empirical investigations in Web engineering, and introduced the three main types of empirical investigation – surveys, case studies, and formal experiments. Surveys are typically used when we wish to gather data from events that have already occurred, and over a large sample that if random can be used to generalise the results, to some extent, to the wider population. This type of investigation is also known as research in the large. Case studies represents studies conducted a typical context, and within the context of Web engineering, this context would normally represent a Web company. The data gathered from a case study is obtained as the case study is carried out, and the results can only be generalised to other companies similar to the one where the case study was carried out, or to similar projects to the one that was investigated. This type of investigation is often referred to as research in the typical. Finally, formal experiments aim to gather data that was gathered under a controlled setting, where the variables being investigated can be manipulated. This control helps and is paramount to understand the phenomenon being investigated and to be used as input to building a theory that details the phenomenon. Formal experiments are very difficult to carry out in industrial environments due to the amount of control they need, so it is often the case that they are carried out using students as surrogates of more junior Web developers. This type of investigation is known as research in the small.

References

[1] Basili, V.R.: The role of experimentation in software engineering: past, current, and future. In: Proceedings of the 18th International Conference on Software Engineering, March 25-30, pp. 442–449 (1996)

[2] Basili, V.R., Shull, F., Lanubile, F.: Building knowledge through families of experiments. IEEE Transactions on Software Engineering 25(4), 456–473 (1999)

[3] Fenton, N.E., Pfleeger, S.L.: Software metrics: a rigorous and practical approach, 2nd edn. PWS Publishing Company (1997)

[4] Gellersen, H., Wicke, R., Gaedke, M.: WebComposition: an object-oriented support system for the Web engineering lifecycle. Journal of Computer Networks and ISDN Systems 29(8-13), 865–1553 (1997); Proceedings of the Sixth International World Wide Web Conference, pp. 429–1437 (1996)

[5] Gellersen, H.-W., Gaedke, M.: Object-oriented Web application development. IEEE Internet Computing, 3(1), 60–68 (1999)

[6] Ginige, A.: Workshop on web engineering: Web engineering: managing the complexity of Web systems development. In: Proceedings of the 14th International Conference on Software Engineering and Knowledge Engineering, July 2002, pp. 72–729 (2002)

[7] Ginige, A., Murugesan, S.: Web engineering: an introduction. IEEE Multimedia 8(1), 14–18 (2001)

[8] Goldstein, M., Goldstein, I.F.: How we know: an exploration of the scientific process. Plenum Press, New York (1978)

[9] Collins English Dictionary. Harper Collins Publishers (2000)

[10] The American Heritage Concise Dictionary, 3rd edn. Houghton Mifflin Company, Boston (1994)

[11] Kitchenham, B., Pickard, L., Pfleeger, S.L.: Case studies for method and tool evaluation. IEEE Software 12(4), 52–62 (1995)

[12] Kitchenham, B.A., Pfleeger, S.L., Pickard, L.M., Jones, P.W., Hoaglin, D.C., El Emam, K., Rosenberg, J.: Preliminary guidelines for empirical research in software engineering. IEEE Transactions on Software Engineering 28(8), 721–734 (2002)
[13] Lee, S.C., Shirani, A.I.: A component based methodology for Web application development. J. of Systems and Software 71(1-2), 177–187 (2004)
[14] Maxwell, K.: What you need to know about statistics. In: Mendes, E., Mosley, N. (eds.) Web Engineering, pp. 365–407. Springer, Heidelberg (2005)
[15] Mendes, E.: Cost Estimation Techniques for Web Projects, 424 pages. IGI Global Publishers (2007); ISBN: 978-1-59904-135-3
[16] Murugesan, S., Desphande, Y.: Web Engineering, Managing Diversity and Complexity of Web Application Development LNCS, vol. 2016. Springer, Heidelberg (2001)
[17] Murugesan, S., Deshpande, Y.: Meeting the challenges of web application development: the web engineering approach. In: Proceedings of the 24th International Conference on Software Engineering, May 2002, pp. 687–688 (2002)
[18] Offutt, J.: Quality attributes of Web software applications. IEEE Software 19(2), 25–32 (2002)
[19] Pfleeger, S.L., Jeffery, R., Curtis, B., Kitchenham, B.: Status report on software measurement. IEEE Software 14(2), 33–43 (1997)
[20] Porter, A.A., Siy, H.P., Toman, C.A., Votta, L.G.: An experiment to assess the cost-benefits of code inspections in large scale software development. TSE 23(6), 329–346 (1997)
[21] Pressman, R.S.: Can Internet-based applications be engineered? IEEE Software 15(5), 104–110 (1998)
[22] Ricca, F., Tonella, P.: Analysis and testing of Web applications. In: Proceedings of the 23rd International Conference on Software Engineering, pp. 25–34 (2001)
[23] Taylor, M.J., McWilliam, J., Forsyth, H., Wade, S.: Methodologies and website development: a survey of practice. Information and Software Technology 44(6), 381–391 (2002)
[24] Wikipedia, http://en.wikipedia.org/wiki/Main_Page (accessed on October 25, 2004)
[25] Wild, C., Seber, G.: Chance Encounters: a First Course in Data Analysis and Inference. John Wiley & Sons, New York (2000)
[26] Wohlin, C., Host, M., Henningsson, K.: Empirical Research Methods in Web and Software Engineering. In: Mendes, E., Mosley, N. (eds.) Web engineering, pp. 409–430. Springer, Heidelberg (2005)

Chapter 4
Modern Web Technologies

Leonidas Akritidis, Dimitrios Katsaros, and Panayiotis Bozanis

Department of Computer & Communication Engineering,
University of Thessaly, Volos, Greece

Abstract. Nowadays, World Wide Web is one of the most significant tools that people employ to seek information, locate new sources of knowledge, communicate, share ideas and experiences or even purchase products and make online bookings. The technologies adopted by the modern Web applications are being discussed in this book chapter. We summarize the most fundamental principles employed by the Web such as the client-server model and the http protocol and then we continue by presenting the current trends such as asynchronous communications, distributed applications, cloud computing and mobile Web applications. Finally, we conduct a short discussion regarding the future of the Web and the technologies that are going to play key roles in the deployment of novel applications.

1 Introduction

During the past few years we have witnessed a massive evolution in the applications hosted on the World Wide Web. The obsolete, static Web sites have been replaced by innumerable, novel services that changed dramatically the manner that users navigate, purchase, communicate, think and make decisions. New types of dynamic applications have been developed by using the modern technologies and their participatory features have made them extremely popular, since hundreds of millions of people use them on a daily basis. Blogs, social networks, forums, search engines, wikis, media sharing services and office suites are only a small subset of these applications, which are collectively known as *Web2.0*.

Therefore, understanding the technologies that support the continuous expansion of the Web is of significant importance. Since the field of Web technology constantly changes and evolves, researchers and developers are facing the challenge of being early informed in order to investigate and propose novel solutions to newly posed problems. Note that some of the techniques characterizing the Web since its birth still exist and development must always obey to the traditional rules set by them. One representative example of these technologies is the HTTP protocol which, essentially, has been left unchanged since 1995.

A. Vakali & L.C. Jain (Eds.): New Directions in Web Data Management 1, SCI 331, pp. 83–107.
springerlink.com

However, the majority of the technologies have either been evolved, or completely replaced by novel ones. In this chapter we present the state-of-the-art technologies that are now vital in modern Web computing. Robust design and efficient development of systems and applications deployed on the Web is a topic of critical importance, since it determines the acceptance of a system by the users and affects the commercial success of the product. Here we discuss such topics along with some of the tools and issues involved in this development.

At first, in Sections 2 and 3 we describe the basic networking models adopted by the developers when building Web applications. The Client-Server model is a basic computer networking architecture, which was established before the Web explosion and it sets two types of devices; clients and servers. On the other hand, the Peer-To-Peer model was developed later in order to interconnect users and every device in such a network is treated equally.

Next, we present some of the basic characteristics of *hypertext*, the most popular manner of publishing and distributing information across the entire Internet. At first we provide some definitions and then we discuss how hypertext is transferred and formatted by using HTTP (Section 5) and HTML (Section 6), respectively.

The description of the basic principles of the XML language follows in Section 7. XML is a tool gaining rapid acceptance by both users and developers and it is mainly employed in order to transfer information in a fast and effective manner across different platforms. It is definitely becoming the method of choice for the most modern Web applications such as news portals, blogs, forums and even electronic stores publishing their product lists.

Javascript is one of the most popular scripting languages encountered on the Web and millions of Web sites and applications use it for various purposes. In Section 8 we present the basic characteristics of the scripting languages and especially Javascript. A family of modern technologies, AJAX, is later described in Section 9. Nowadays, AJAX is a quickly expanding tool, since it offers important solutions to the stateless nature of the HTTP and the traditional Client-Server model. By submitting requests to a Web server asynchronously, the AJAX-enabled Web pages can modify their content without requiring to refresh their display. This family of technologies which is one of the most important Web2.0 features, employs Javascript for scripting and XML for transferring data between the client and the server. AJAX is currently being used by numerous Web services, such as email managers, global and planetary mapping services, instant messengers, Web search engines and others.

Finally, we discuss how Web applications are constructed and deployed. In Section 10 we present some of the most popular programming and storage tools and in the sequel, we describe some characteristic applications of the Web, such as social networks, Web communities, office suites, and mobile software.

Since the number of people using the Web constantly increases, one of the main trends encountered today is the utilization of large clusters of computers in order to handle the tremendous workloads. The first category of these applications, discussed in Section 12, includes software which is distributed across the computers of the users themselves and exploit their free resources in order to solve complex scientific problems. In the second category, we mainly encounter the popular *cloud computing* solutions (Section 13), which employ thousands of interconnected servers usually hosted in one or more data centers with the aim of addressing the huge traffic that millions of users produce.

Concluding in Section 16, we provide a brief discussion regarding the presented technologies and the future of the Web.

2 The Client-Server Model

The Client-Server model is one of the most popular architectures for computer networking. It utilizes two types of devices to address the communication requirements of the terminals of a network; Clients and Servers. The model can be used on the Internet as well as local area networks. Examples of client-server systems on the Internet include email services, Web browsers, Web servers and FTP clients and servers. It was originally developed to allow multiple users to share access to database applications and offers improved scalability because connections can be made as needed, rather than being fixed.

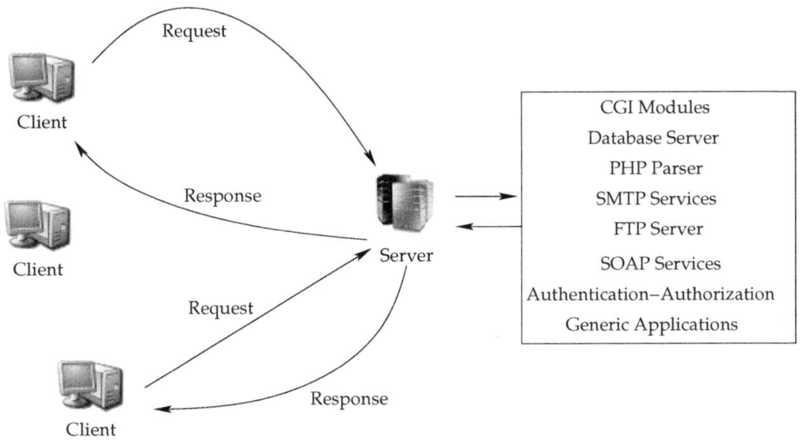

Fig. 1. Typical Architecture of the Client-Server Model

A Client is a device, typically a personal computer or a mobile device with network software applications installed that request and receive information over the network. On the other hand, a Server typically hosts applications (including other servers), or stores files and databases. Server devices often feature higher-powered equipment, greater processing performance and larger storage capabilities than clients.

Figure 1 illustrates a typical communication of several remote machines according to the Client-Server model. Network clients submit requests to a server machine by transmitting messages. On the other side, servers respond to their clients by processing each request and by sending the appropriate response messages. In a typical Client-Server environment one server is utilized to support the traffic originated by multiple clients, whereas numerous servers can be linked together in order to handle and address increased processing load.

The Client-Server model is employed by some of the most widespread applications on the Internet. Such applications include FTP, email and Web services and in each of these, the clients employ software of special type (an Internet Browser, an email management program or an FTP client), allowing them to connect and receive data from the corresponding servers.

3 The Peer-To-Peer (P2P) Model

Peer-To-Peer networks, often abbreviated to P2P, is a network architecture alternative to the Client-Server model. However, instead of requiring central coordination by a machine or software such as a server, all the devices (peers) are treated equally. In such an architecture, every peer makes a portion of its resources directly available to other peers, hence each network participant is both a supplier and consumer. The resources that the peers may share include network bandwidth, processing power and disk storage.

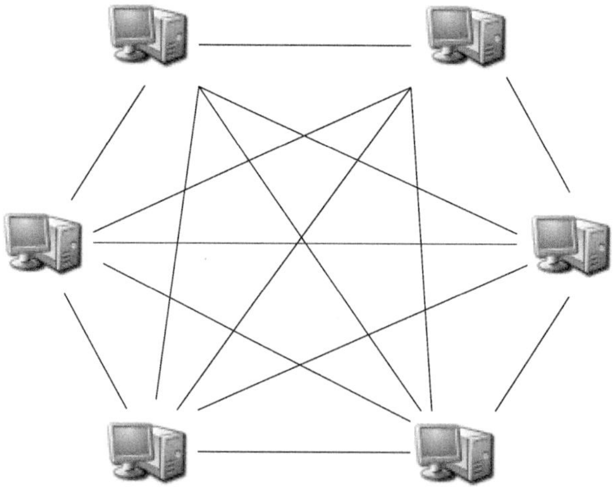

Fig. 2. Typical Architecture of a Peer-To-Peer network

This network architecture became really popular when file sharing services (such as Napster, kazaa eMule and others) appeared on the Internet. Their distributed architecture and the fact that no server is required, provides improved scalability, since computers may dynamically enter or leave the network without significant impact.

Depending on how the connections among the peers are established, Peer-To-Peer networks are divided into two categories: The structured and the unstructured networks. The first category includes topologies where the connections are fixed, whereas in the second category we encounter architectures that do not provide any algorithm for organizing or optimizing these connections.

The most popular application employing Peer-To-Peer technologies is the file sharing services, where millions of users join a community in order to request or provide media files, documents and software. There are also other important types of applications such as instant messaging, online chat and voice over IP where this network architecture finds remarkable utilization.

4 Hypertext

Currently, the information on the Web is mainly provided through documents of special type, which provide inter-linking capabilities. Consequently, every Web document (also known as Web page) may contain references to other documents and resources and the user is able to immediately access these resources by simply following the corresponding links.

All this functionality is provided through *hypertext*, a term coined by Ted Nelson around 1965. Hypertext is a specific form of text containing dynamic references to other resources. Although it is an old invention, it still remains the main way of information propagation within the Web. Apart from running simple text and links, hypertext may also contain headings, lists, tables, images and other presentational devices.

Other means of interaction could also be present, such as a form to complete and submit.

5 Hypertext Transfer

The requirement for retrieving inter-linked resources containing hypertext, led to the establishment of *HTTP (HyperText Transfer Protocol)*. The utilization of HTTP imposed a set of rules and specifications allowing hypertext documents to be transmitted and received. It is a generic, stateless protocol which can be applied on many tasks beyond its use for hypertext, such as name servers and distributed object management systems.

The specifications of the protocol obey the request-response standard, which is typical in client-server computing. The client is usually an application (i.e. the user's browsing software) running on the machine of an end user, whereas the server is a computer program executed by the machine hosting the Web site or service. The communication is conducted through HTTP *requests* submitted by the client and HTTP *responses* returned by the server. The responses are accompanied by specific code numbers which indicate the status of the corresponding response. In other words, response codes reveal whether the request has been served successfully, or whether an error has occurred (including the type of the error).

HTTP is usually implemented on top of the TCP/IP protocol which controls the reliable data transfers. However, this is not a constraint, since HTTP can operate by using any other protocol which guarantees reliable transports.

Nowadays, there are two major versions of the protocol, HTTP/1.0 and HTTP/1.1. According to [13] the latter improves the former by providing a more efficient caching mechanism, and more effective bandwidth usage (the introduced range requests allow a client to request portions of a resource). Furthermore, while HTTP/1.0 included some support for compression, it did not provide adequate mechanisms for negotiating its use, an issue addressed by the newer version.

6 Hypertext Markup

The requirement for building Web documents of unique and personalized style, layout and formatting, led to the introduction of a special markup language which can be used by Web developers to construct their pages according to their personal preferences. Moreover, the requirement for publishing information that can be globally distributed, made the need for a universally understood language imperative.

HTML, which stands for HyperText Markup Language, is now the predominant markup language for Web pages. It provides a set of command (*HTML tags*) to create structured documents by denoting structural semantics for text. Such semantics include tables, paragraphs, headings, lists, links and numerous others. The need for interaction between a Web application and the user is mainly satisfied by the introduction and usage of forms. These forms are now used for conducting transactions with remote services, making reservations, ordering products, etc

Finally, HTML allows the integration of scripts which are usually executed by a Web client (browser) and Cascading Style Sheets (CSS) which is a standard of parallel commands for explicit presentational markup.

There are several releases of HTML, but the most popular among them is HTML 4.01 [1] also recommended by the W3C. In addition to the text, multimedia, and hyperlink features of the previous versions, this version supports more multimedia options (i.e. embedded videos), scripting languages, style sheets, more printing facilities, improved support for right to left and mixed direction text, frames and enhancements to forms, offering improved accessibility for people with disabilities. HTML 4.01 is also more oriented towards the internationalization of the Web documents since they can be written in every language and be transported easily around the world.

7 XML

As we aforementioned in Subsection 6, HTML was designed to display Web pages with focus on the manner that the data is presented to the user. The need for a language that would focus on transportation and storage of data led to the introduction of *Extensible Markup Language* or *XML*. Note that XML is not a replacement for HTML, since the latter is mainly designed for displaying information, whereas XML was created to structure, store, and transport information. The precise design goals of the XML as set by the World Wide Web Consortium [2] emphasize on the wide, fast and easy usability that the standard must provide.

XML is a simple, very flexible text format derived from SGML (Standard Generalized Markup Language). It was originally designed in order to provide convenient, uniform and platform-independent publishing of information. Nowadays it is also playing an increasingly important role in the exchange of a wide variety of data on the Web and elsewhere.

The documents of this type are composed of markup and content and have both a logical and a physical structure. There are six kinds of markup that may occur in an XML document: elements, entity references, comments, processing instructions, marked sections, and document type declarations. An entity may refer to other entities to cause their inclusion in the document and furthermore, the logical and physical structures must nest properly.

To obtain the data contained in an XML document, we usually employ especially designed applications known as *XML parsers* which are capable of translating the structure of the document and generating data nodes. There are many types of XML parsers with various capabilities. One of their most remarkable feature is their ability to examine whether the document is well formed. In general, XML parsers are more strict than HTML renderers and the most sophisticated among them perform *validation* of the XML structure and syntax. Hence, a document which does not conform to the XML grammar or does not contain a proper document type declaration, is not considered valid and cannot be parsed.

7.1 RSS Feeds

Really Simple Syndication (or *RSS*) is one of the most common applications of the XML language. It is a family of *feed* formats mainly employed by authors to publish frequently updated information. Such information includes news headlines, multimedia content, blog posts, product catalogs (accompanied by the corresponding availability data and prices), airline tickets and numerous others. The RSS documents are built by using standard XML syntax and their popularity increased rapidly after 2005.

An RSS document (which is called a *feed*) includes text and meta-data such as publishing dates, authorship and others. The generation and delivery of information by using such feeds is beneficial for both publishers and readers. More specifically, publishers are allowed to syndicate content automatically, whereas readers are able to subscribe to temporal updates from preferred sites, or to aggregate different RSS feeds originating from multiple sources.

A reader can access and extract the information stored in RSS feeds by using specific software called an *RSS reader* or *aggregator*. Most modern Web browsers include built-in RSS readers, whereas some new mobile devices have their own reading software available. A standardized XML format allows the information to be published once and viewed by many different programs. The user subscribes to a feed by informing the corresponding reading software about the desired feed. In turn, the reader checks the subscribed feeds regularly for new publications and updates and informs the user accordingly.

8 Scripting

HTML is a powerful markup language offering the Web developers various tools in order to format the Web pages that they build. However, it lacks programming capabilities such as setting variables, computing values, handling files etc. To cover this drawback, HTML provides the option of composing external *scripts* that is, fragments of code written by programmers in order to solve specific problems.

Scripts can be separated into two main categories: *server-side* and *client-side*. The first category includes programs executed by the server and only the result of the programm processing is returned to the end user. Furthermore, this type of scripts may, or may not require compilation before they can be executed. For example, PHP scripts do not require compilation, but there is a dedicated parser running as a server module that handles runtime or syntax errors. Therefore, in this case, no executable file is created, but the opposite holds for server-side programs written in other programming languages (i.e. ASP.NET).

In the second category we encounter scripts executed by the client (that is, the user's browser) and the code is integrated within the Web page itself. All client-side scripts do not require compilation or parsing. This has the cost of harder and slower debugging, since when an error is triggered, no informative messages are generated and the execution is being terminated silently. Therefore, the programmer has to examine and debug the entire code in order to detect the specific portion causing the problem.

Nowadays, the most popular language on the Web offering client-side scripting is *Javascript*. Javascript is used in millions of Web pages to add functionality, validate form data, build visual effects, open pop-up windows and there are many other applications. Another significant field, where it finds application, is the development of enhanced user interfaces and dynamic Web sites.

Furthermore, it offers event-driven programming that is, it allows the programmer to write code which is executed before or after a specific event is triggered. Hence, a Web page may be modified without having to send any data to the server and receive its response. For this reason, Javascript is ideal for building robust Web applications with modern user interfaces, and it is one of the main characteristics that the services of the next generation of Web include.

9 Asynchronous Transfers and AJAX

The usage of the traditional request/response methods that the classic client-server model imposed, prevented Web sites and browsers from providing a fast and responsive user experience. For example, filling and submitting an online form was inconvenient and time consuming, since all the requested information had to be entered and then submitted to the machine hosting the service (Web server). Then the server performed a validation of the form data and if problems were detected, the user was obliged to refill and resubmit the same form. The flow of information and the resulting experience was inconstant and disconnected, reflecting the stateless nature of HTTP.

The introduction of Java applets offered a different route: this route included asynchronous loading of content and allowed client-side code to load data asynchronously from the Web server after a web page was fetched. Moreover, the IFrame component of the HTML language and some introduced ActiveX controls also enabled this to be achieved. More specifically, these ActiveX controls included a special object, namely XMLHttpRequest[1], which was designed in order to submit requests to a server asynchronously.

AJAX (shorthand for *Asynchronous JavaScript and XML*) is a modern Web technology introduced by the World Wide Web Consortium (W3C) in 2006[2]. It is a group of interrelated web development techniques used on the client-side to create interactive web applications. The applications employing AJAX technologies are capable of retrieving data from the server asynchronously in the background, without having to interfere with the display and behavior of the currently loaded page.

In Figure 3 we illustrate how the usage of AJAX technologies in modern Web applications offers uninterrupted user experience. The left diagram of this Figure depicts how the user and the remote server communicate according to the traditional client-server model. Note the idle times in the side of the client; these are the times required to transfer the desired data between the client and the server and also concern the time the server consumes to process this data and return a response to the end user. In the AJAX environment, the user experiences no idle times, since the asynchronous communications between the browser and the server offer continuous work flow.

---

[1] http://www.w3.org/TR/XMLHttpRequest/

[2] http://www.w3.org/standards/webdesign/script

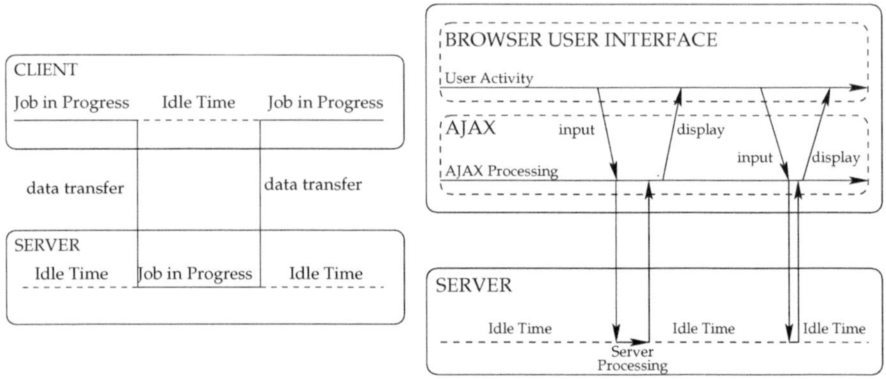

Fig. 3. AJAX asynchronous data transfer model (right) vs traditional client server workflow model (left)

In the sequel, let us provide some of the main benefits deriving from the usage of AJAX techniques. At first, in many cases, some pages on a Web site include content that is common among them. The usage of traditional methods required that this content would have to be reloaded on every request. However, by employing AJAX an application can request only the content that needs to be updated. Therefore, we manage to drastically reduce both the usage of valuable bandwidth usage and load time.

In addition, the utilization of asynchronous requests allows the interface of the client's browser to be more interactive and to respond quickly to inputs. Several portions of pages can be reloaded individually and the users may perceive the application to be faster or more responsive, even if the application has not changed on the server side.

Finally, with AJAX we can minimize connections to the Web server, since external files such as scripts and style sheets only have to be requested once. Programmatically, this means that the local variables will retain their values, because the main container page need not be reloaded.

For all these reasons, the usage of these techniques has led to a significant increase to the applications providing interactive and dynamic user interfaces[2][3]. Some of the most common services employing AJAX techniques are:

- Mailbox management applications, where the entire user interface is designed to allow composition, reading and deletion of messages without refreshing the display. Moreover when a new message arrives it is added to the in-box automatically without requiring the user to refresh the page.

- The new technology allowed the introduction of modern Web instant messengers. These services are constructed in such a way that allow their users to exchange their messages instantly. Their main characteristic is that each time a message is sent or received, only its content is loaded by the client and the entire interface remains unchanged.
- Global Maps Services employ the asynchronous features of AJAX to allow their users navigate through the surface of the planet. They also provide magnification potentials by directly accepting data from satellites.
- The novel translation services now operating on the Web offer their users new functionality. They are capable of accepting words or even sentences and paragraphs written in a specific language and, as the user types, they translate the content into another language.
- A huge amount of other smaller services is now built by using asynchronous technologies. Such services include result retrieval in the major commercial search engines, spelling correction, auto-complete features (as the user types his/her query, current search engines fetch similar entries from their query logs and present them on the fly below the text box),

Nevertheless, the remarkable new features and functionality introduced by the AJAX technologies do not come without costs. The main drawback is that the interfaces constructed by using AJAX are substantially more difficult to develop than static pages. Pages dynamically created using successive AJAX requests do not automatically register themselves with the browser's history engine and this may raise problems regarding the user's navigation on the Web. For example, it is possible that when a user clicks the "Back" button of the Web browser, he/she will not return to an earlier state of the AJAX-enabled page, but may instead return them to the last full page visited before it. Dynamic web page updates also make it difficult for a user to bookmark a particular state of the application.

Web crawlers are computer programs developed by the search engines in order to browse the Web in a methodical, automated manner. However, since the majority of the Web crawlers do not execute Javascript code, applications indexed by search engines should provide means for accessing the content actually retrieved with AJAX. Note that any user whose browser does not support Javascript or XMLHttpRequest, or simply has this functionality disabled, will not be able to properly display and use pages which depend on AJAX. Similarly, devices such as mobile devices and screen readers may not have support for the required technologies.

Finally, like other web technologies, AJAX has its own set of vulnerabilities that developers must address. Developers familiar with other web technologies may have to learn new testing and coding methods to write secure AJAX applications.

10 Application Deployment

In this subsection we present some of the main tools and technologies employed by the modern Web applications.

10.1 Database Servers

Currently, XML is the dominant technology to publish and distribute semi-structured information on the Web. However, there are types of applications that require their data to be organized in a more robust and structured manner. Such applications include electronic stores, social networks, forums, search engines and others which usually have to deal with massive amounts of data.

Database Management Systems (or DBMS) is a tool developed to offer efficient organization, storage, management and retrieval of an application's data. These systems usually reside in dedicated machines and offer database services to other computers and applications. Instead of having to write computer programs to store and extract information, user can ask simple questions in a supported query language. Thus, many DBMS packages provide a structured query language (SQL) and other application development features.

Within a typical DBMS, data is organized into *records* which is a collection of data regarding a physical entity (i.e. an employee, a book, or a product). Each record consists of numerous user-defined *fields*, that are able to store information of different types (text, binary data, integers and floats, time stamps, dates and several others which vary across different DBMSs). Records of the same type are again grouped within *tables*. Databases provide an efficient manner of separating the application logic from the data logic, therefore, different applications can cooperate with the same database.

One of the most important characteristics offered by Database Management Systems is the indexing feature. An index is an auxiliary data structure usually implemented in a form of a tree such as B-Tree, to allow fast and efficient data access and retrieval. The indexes also allow effective sorting of the returned records and offer fast organization (i.e. grouping of records).

Other features commonly offered by database management systems include:

- Restricted access to resources and attributes. Each user of the system is assigned privileges which determine whether he or she has read or write access to several resources and attributes of the database. These privileges are assigned by individuals, or groups of individuals maintaining elevated authority across the entire system.
- Data safety and integrity are of critical importance for every informational system, hence copies of attributes are required to made regularly in case of equipment failure. DBMS usually provide utilities to facilitate the process of extracting and disseminating attribute sets.

– Data retrieval by submitting queries. Instead of composing special software to obtain and format the data stored within a database, most modern systems accept structured queries which usually follow the simple syntax of a structured query language. By submitting queries we request attribute information from various perspectives and combinations of factors (i.e. who are the male clients that purchased a specific product?). Queries can also be submitted to the database in order to insert, update or delete data, according to the privileges each user is granted.

The introduction of World Wide Web in 1995 imposed new challenges for database systems. Researchers realized that the traditional database management techniques were becoming too complex and there was a need for automated configuration and management. For example, online transactions have become extremely popular with the evolution of the e-business world. Consumers and businesses are able to purchase products and make payments securely on corporal Web sites.

In addition, Web search engines have even been remarkably influenced by database management. Using technologies similar to the ones employed by current database systems, these services are able to accept user queries and locate data across the entire the Web.

10.2 Hypertext Preprocessor - PHP

PHP is one of the most widespread scripting languages used to deploy Web applications. The rich features it offers combined with the natural easiness and the open source characteristics, have made it the second most popular scripting language encountered on the Web [4]. Although there is a general intuition that PHP is mainly preferred for constructing small or medium sized applications, several large-scale Web sites serving hundreds of millions of users worldwide, such as Facebook, Wikipedia and Wordpress have been developed with it. Currently, PHP is installed on over 20 million sites and 1 million Web servers [3].

It was originally designed for the development of Web applications, in order to produce dynamic pages. PHP scripts can be embedded into HTML and they generally run on a Web server (server-side scripting), which needs to be configured properly to execute PHP code. It can be deployed on most Web servers and on almost every operating system and platform.

PHP scripts are phrased by following a C-style coding syntax and all the allocated resources are released after the script execution by an automatic garbage collection mechanism. Since its fifth version, it also supports the object-oriented programming style by adopting principles such as abstract data types and information hiding, inheritance and polymorphism. Moreover, it includes features such as variables, arrays and associative arrays setting and manipulation, conditional statements, loops, function setting and file handling. Apart from these classic characteristics, PHP allows programming of the HTTP protocol by providing access to HTTP sessions and cookies and, furthermore, by implementing secure file uploads.

One of the most robust features of PHP is its native support to MySQL, SQLite and PostgresSQL database systems. Through built-in functions and classes, PHP scripts can easily connect to database servers, submit queries and retrieve data. The combination of PHP and MySQL is one of the most common techniques currently employed by the developers when building Web applications.

10.3 Active Server Pages - ASP/ASP.NET

Another popular server-side scripting technology that is competent to PHP is *Active Server Pages*. It has been introduced by Microsoft and provides to the developers robust tools in order to create dynamic and interactive Web applications. Similarly to the PHP documents, an ASP page is a standard HTML document which contains server-side scripts. The scripts are processed by a properly configured Web server which sends the processing output to the user's browser.

In contrast to PHP, ASP is not a scripting language, but rather a technology used to produce dynamic pages when a browser requests ASP files from a Web server. The default scripting language employed for scripts composition is VBScript, although alternative languages like JScript (Microsoft's version of Javascript) can also be used. When an ASP script is called, the server processes the requested file from top to bottom and executes the commands it contains. It the sequel, it generates and formats a standard Web page and sends it to the browser.

During 2002, Microsoft released a large set of coded solutions to common programming problems. This library, known as the *.NET Framework*, includes solutions regarding user interface design, data access and processing, database connectivity and development of dynamic Web applications, whereas the programmers are able combine its classes with their own work.

In addition, the library includes a virtual engine able to execute the software written specifically for the framework. The applications developed with the .NET Framework are deployed in a special environment which manages their runtime requirements. This runtime environment, which is known as the *Common Language Runtime (CLR)*, allows the programmers to work without considering the capabilities or the specifications of the specific machine that will execute their program. The CLR also provides other important services such as security, memory management, and exception handling.

Along with the release of the .NET Framework, Microsoft also introduced ASP.NET, an enhanced version of ASP used to produce dynamic Web sites, applications and services. The new development framework is built upon the

CLR and allows programmers to compose software by employing any of the supported languages such as the VB.NET (Visual Basic .NET), C#, J# and others. Moreover it offers the ability to construct applications by using an event-driven user interface model, in contrast to the conventional scripting environments such as ASP and PHP.

10.4 Java Server Pages - JSP

Java Server Pages or simply *JSP*, is another technology used to deploy dynamic Web applications, by allowing Java code to be embedded into the content of a regular static page. The code is not pre-compiled, but it is actually being compiled on the server at each page request similarly to PHP. The Web pages that are created by using JSP are loaded in the server and handled by a special Java server packet, called the *J2EE* Web Application.

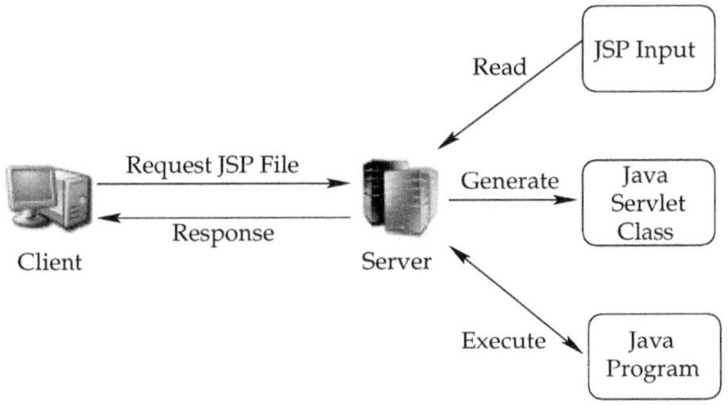

Fig. 4. JSP page translation and processing phases

The processing of the a JSP page is performed in two phases. At first, we employ a typical JSP compiler which converts the input file into a *servlet*, that is, a particular Java class that responds to HTTP requests. In the sequel, the servlet can be either compiled by a Java compiler and generate a standard Java program, or be converted to a directly executable byte code. Figure 4 illustrates the procedure of translating and processing JSP pages.

JSP is currently an alternative method to PHP and ASP, allowing developers to construct dynamic Web sites and services by writing their code in Java. Although it provides rich features and offers almost equivalent possibilities, it is not as popular as the other two aforementioned technologies.

11 SOAP

A protocol which gained attention during the past few years is the *Simple Object Access Protocol*, or simply SOAP. It is a simple XML-based protocol which allows the applications to exchange structured or semi-structured information over HTTP. Its messages follow the standard XML syntax, whereas the trasmission/receive procedure is handled by other application protocols, such as the HTTP. SOAP specifies exactly how to encode an HTTP header and an XML file, so that a program in one terminal can call a program in another terminal and transmit information to it. It also specifies how the called software can return a response.

In details, SOAP messages consist of three parts:

- An envelope which describes the content of the message and instructions about how this content should be processed,
- a set of rules containing the data types defined within the application and
- a convention which represents procedure calls and responses

Some of the applications operating on the Web require the transmission and processing of attached binary files (i.e. images or documents). But since all the parts of a SOAP message must conform to the strict XML standards, binary data cannot be included directly into the message (they contain characters and sequences not allowed by the official XML rules). Furthermore, the inclusion of binary data within the message itself, would render the majority of the parsers inefficient; some of them initially read and process the entire SOAP message before deciding what to do with the contents. This operation requires large amounts of memory and processing power. For all these reasons it was decided that SOAP requires some mechanism for carrying large payloads and binary data as an attachment rather than inside the SOAP message envelope.

To cover such issues, third parties released a set of specifications which determine how binary pfiles should be attached to a SOAP message. The most common set of specifications is the the JSR-67 (Java Specification Request) which included the SAAJ (SOAP with Attachments API for Java) standard.

The main advantage of SOAP is the integrated simplicity and extensibility. The protocol allows easier and more robust communication between proxy servers and firewall applications along with the language and platform independence. In addition, using HTTP is not obligatory, since SOAP also supports the usage of different transfer protocols, such as SMTP. The transmitted packets not only include the content of the message, but also sufficient information describing how this content should be processed by the receiver. However, the verbose XML format that SOAP employs can render it relatively slower than other solutions.

Since one of the most common purposes of Web services is to exchange XML data, SOAP is rapidly becoming the generally accepted protocol for XML-based systems communication. For example, Web search engines APIs make wide use of SOAP. In addition, numerous stock quote services, weather services or news portals, employ it in order to transmit and receive data formatted in the XML language.

12 Distributed Applications

Distributed computing is one of the most discussed and hot topics in computer science. It refers to partitioning a large or complex problem into several smaller parts and assigning each of these parts to a machine that belongs to a wider cluster of processing nodes (also called workers). When each of the processing nodes finishes its computations, the distributed solutions are merged to form the final solution of the problem. In such a distributed environment, a central coordinator is usually employed in order to synchronize and send messages to the processing nodes.

Of course, distributed computing is not a pure Web technology. However, there are some projects which utilize the machines of the Web users in order to solve large scientific problems. These projects exploit the free (or idle) resources (mainly the processing power) of thousands or even millions machines of Web users in order to compute the solution of a small fraction of a huge problem.

Folding@home is one of the world's largest distributed computing projects developed with the official goal of "understanding protein folding and related diseases". It does not rely on powerful supercomputers for processing the available data; instead, the primary contributors to the project are many hundreds of thousands of personal computer users who have installed a client program. The client runs in the background, utilizing the unused resources, whereas it periodically connects to a server in order to retrieve new data and continue the calculations, or send back the produced results.

Seti@Home is a similar project which exploits the computers of Web users with the aim of performing Search for Extraterrestrial Intelligence (SETI) by analyzing radio signals. Similarly to the Folding@Home project, the users download and install a client software which is capable of processing data generated by radio telescopes. The client exploits the unused resources of the machine it is installed on and proves the viability and practicality of the distributed grid computing concept.

13 Cloud Computing

Cloud computing is a recent trend in Computer Science that moves computing and data away from desktop and portable PCs into large data centers. It refers to applications delivered as services over the Internet, as well as to the actual cloud infrastructure. Currently, the main technical characteristics of cloud computing services include virtualization, grid computing technologies, service-oriented software management of large facilities and power efficiency.

Within a cloud computing environment, applications and services are provided in the following three forms:

- *Platform-as-a-Service (PaaS)* The term PaaS denotes the allotment of a large computing platform over the Web. The service enables the developers to create Web applications rapidly, without concerning the cost and complexity of buying and managing the underlying software and hardware, since servers, databases, security software and several frameworks are provided by the service itself. The applications developed and delivered under a PaaS environment are referred as On-Demand or as Software as a Service (SaaS) Applications.

- *Software-as-a-Service (SaaS)* SaaS is a another form of cloud computing services and through it, companies, organizations and users can access software and large amounts of computing power without having to purchase it. Instead, SaaS adopts a business model according to which a client of a cloud computing service pays only the processing power it consumes. The applications are hosted by another hosting company in a large computing cluster, hence the maintenance and setup operations along with the security issues are of no concern for the users. *GMail* is a representative example of an application designed to operate according to the SaaS model.
- *Infrastructure-as-a-Service (IaaS)* This model has its origins in thin computing techniques that have been evolving since the previous decade. In general, the term refers to the principle that instead of having a network infrastructure on its own installation, a company can rent space from a service provider and use it across the internet. With IaaS, customers are provided with the ability to choose the hardware and some basic software servers for their part of the cloud and then transfer their applications and data on these machines. Virtualisation enables IaaS providers to offer almost unlimited instances of servers to customers and make cost-effective use of the hosting hardware.

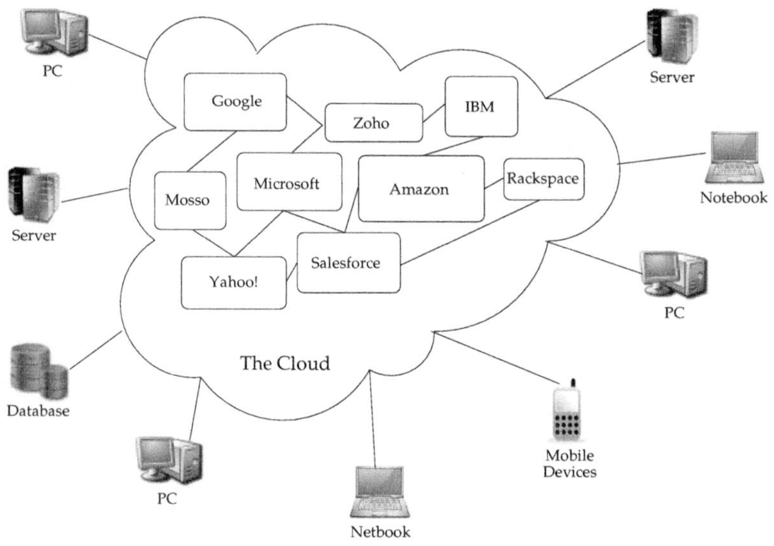

Fig. 5. The Cloud

We can distinguish two different architectural models for the clouds: the first one is designed to scale out by providing additional computing instances on demand. Clouds can use these instances to supply services in the form of SaaS and PaaS. The second architectural model is designed to provide data and compute-intensive applications via scaling capacity. In most cases clouds provide on-demand computing instances by adopting a "pay-as-you-go" economic model.

Regarding service provisioning, the providers supply cloud services by signing service-level agreements (SLAs) with consumers and end-users. These agreements concern the amount of the processing power and bandwidth the user's applications consume at any given time through a specific period (day or month). The estimation of the resource provisioning is a task of critical importance since a possible underestimation would lead to broken SLAs, service interruption and other penalties. On the other hand, overestimating the provision of resources would lead to resource underutilization and, consequently, a decrease in the revenue for the provider.

Deploying an autonomous system to efficiently provision services in a cloud infrastructure is a challenging problem due to the unpredictability of consumer demand, software and hardware failures, heterogenity services, power mangement and conflicting signed SLAs between consumers and service providers.

In terms of cloud economics, the provider should offer resource-economic services. Novel, power efficient schemes for caching, query processing and thermal management are mandatory due to the increasing amount of waste heat that data centers dissipate for Internet-based application services. Moreover, new pricing models based on the pay-as-you-go policy are necessary to address the highly variable demand for cloud resources.

Cloud computing is a disruptive technology with profound implications not only for Internet services but also for the entire IT field. Its emergence promises to streamline the on-demand provisioning of software, hardware and data as a service, achieving economies of scale in IT solutions' deployment and operation.

Still, several outstanding issues exist, particularly related to service-level agreements (SLAs), security and privacy, and power efficiency. Other open issues include ownership, data transfer bottlenecks, performance unpredictability, reliability and software licensing issues. Several companies have already built Internet consumer services such as search, social networking, Web email, and online commerce that use cloud computing infrastructure.

14 The Mobile Web

During the past few years, the mobile devices have played a key role in the market of telecommunications. By offering significant features such as small sizes, light weights, efficient power consumption, ability for direct user-to-user communication, affordable prices and remarkable processing power, they have attracted hundreds of millions of consumers. The market of mobile devices is expected to experience an impressive growth over the following few years [5] and recently, numerous manufacturers have included Web browsing capabilities to their products.

According to the inventor of the Web Tim Berners-Lee, "the Mobile Web initiative goal is to make browsing the Web from mobile devices a reality". By employing browsers particularly designed for mobile devices (mobile browsers), mobile Web access is becoming increasingly popular. Since a large desktop system is not required any more to access the Web, users are provided with the ability to work online at all times in all situations. Curren mobile Web browsers retain the main functionality offered by the respective desktop applications, such as basic browsing, form completion and submission and generic transactions.

However, there are still some problems that need to be confronted and interoperability is the most significant among them. It derives from the the existence of many different platforms with various operating systems and browsers. In addition, the limited size of these devices and the small display sizes raise important usability issues.

Within the Mobile Web, the information is published and delivered via lightweight pages written in XHTML or WML (Wireless Markup Language). The new versions of the mobile browsers raise these limitation by supporting a wider range of Web formats, including variants of HTML commonly found on the desktop Web. In addition, W3C have published a set of recommendations [6] to Web site creators and developers who desire their applications to be fully accessible from mobile devices.

15 Web 2.0 Applications

Web 2.0 is a widespread term which reveals the evolution we have witnessed in the World Wide Web and the applications hosted on it. The definition of Web 2.0 [10] does not refer to an update to the technical specifications characterizing the Web, but rather to fundamental changes in the manner that application developers and users exploit it. Therefore, the Web is currently treated as a platform, where new applications are built upon it, similarly to how applications are developed and deployed upon the desktop platform.

The new version of the Web is usually connected with web applications that offer participatory and sharing features. More generally, the main characteristic that a Web 2.0 application has, is its user-centered design. That is, the information is not simply provided to the users, but the users contribute to it by expressing and publishing their own knowledge, experiences and opinion. In such an environment, the users are not limited in a traditional passive role, but they dynamically determine the content of a Web site.

This design has led to the introduction of numerous novel services such as Web communities, social networks, media sharing services, wikis, blogs, forums, online auctions and numerous others. In this Section we provide brief descriptions of the most popular Web 2.0 sites, applications and services. Of course, we do not intend to provide a complete directory of the most significant Web 2.0 applications, but we rather exhibit some of their major features which made them extremely popular during the past few years.

15.1 Web Communities

One of the major benefits which derived from the introduction of Web 2.0, was the participatory features that were integrated within the traditional Web sites. These features lead to the deployment of novel services with interactive characteristics and generated new architecture models by providing additional possibilities to the Web users. One of these introduced models allowed the users to communicate in environments that are currently known as social networks. Social networks function like online communities, where their members share common interests in hobbies, religion, or politics. Examples of Web communities include social networking sites, forums and community blogs.

Blogs are locations on the Web where individuals (the bloggers) express opinions or experiences about a subject. Such entries are called blog posts and may contain text, images, embedded videos or sounds and hyperlinks to other blog posts and Web pages. On the other hand, the readers are provided with the ability to submit their own comments in order to express their agreement or disagreement to the ideas or opinions contained in the blog post. The comments are usually placed below the post, displayed in reverse chronological order. The virtual universe that contains all blogs is known as the *Blogosphere* and accommodates two types of blogs: a) *individual blogs*, maintained and updated by one blogger (the blog owner), and b) *community blogs*, or multi-authored blogs, where several bloggers may start discussions about a product or event.

In a physical community, people use to consult others about a variety of issues such as which restaurant to choose, which medication to buy, which place to visit or which movie to watch. Similarly, the Blogosphere is a virtual world where bloggers buy, travel and make decisions after they listen to the opinions, knowledge, suggestions and experience of other bloggers.

There is a significant reasearch towards identifying the influential members of the major online communities such as blogs [11,12] and Twitter [14,15]. The problem of identifying the influentials is of remarkable importance, because such members act as direct or indirect advertisers of products, events, companies, brands or even travel locations.

15.2 Social Networks

A social network is a group of individuals sharing common experiences, knowledge and ideas. These individuals are usually grouped within social structures such as communities or neighborhoods. The introduction of Web 2.0 along with the participatory features of the applications it established have led to the creation of numerous social networking Web sites which function as online communities for their members.

Many of these online community members share common interests, beliefs, knowledge, hobbies, religion, or politics. The users who are granted access to a social network are free to construct their profile by filling information regarding their name, email, education and geographic location or describe their habits and personal interests. File uploading is also one of the provided features and these files usually include online games, documents and personal photographes. Furthermore, the members of such a community are free to read the profile pages of other members and contact them.

According to sources published in Wikipedia [7] there are typically several hundreds of such social Web sites. The most popular among them is MySpace[3] and Facebook[4] which accommodate about 471 and 350 millions of registered users respectively. A large fraction of these users connect to their favorite social networking service on a daily basis.

15.3 Office Suites

The vast majority of computer users are somehow familiar to an office suite. Almost everybody have used at least once, a word processor to create a textual document or a spreadsheet software to create documents comprised of enriched and dynamic data.

One vision of the 21st century computing is that a large portion of the applications that now operate offline, will be transferred on the Internet and their users will be able to create, store and distribute information online. By using cloud computing techniques, the dominating Web companies are redesigning their applications in order to provide such functionality. The most ambitious of the existing projects, is the creation of an operating system capable of operating entirely on the Web, within a browsing software.

Another project which has already been released, is the online office suites offered by Web sites in the form of software-as-a-service. Such suites basically include a word processor and a spreadsheet, whereas some of them also offer drawing utilities, graphics editors, presentation applications and even media players.

Nowadays, there are numerous services offering office productivity. The most popular among them is Ajax13[5], Google Docs and Spreadsheets[6], Thinkfree Office Online[7] and Zoho Office Suite[8]. Each of them has its own strong points but generally online office suites offer satisfying capabilities

[3] http://www.myspace.com
[4] http://www.facebook.com
[5] http://us.ajax13.com/en/
[6] http://docs.google.com/
[7] http://www.thinkfree.com
[8] http://www.zoho.com

at low (or no) cost, whereas they do not require to download and install any software. Moreover, the users can access their documents from almost any computer with a connection to the Internet, regardless of which operating system they use. Finally, in 2009 Google introduced Google Wave[9], a web-based communication and collaboration tool using richly formatted text, photos, videos, maps, and the like—currently (Jan, 2010), this application is available only after one gets invited.

Nevertheless, there are still some significant disadvantages which indicate that such tools are only at their infant stages. For instance, there are accessibility issues arising from the fact that in case the remote server or network is unavailable, the content will also be unavailable. Moreover, such applications usually require high bandwidth Internet connections, otherwise speed is limited dramatically. Even in that case, an online office application cannot compete an offline opponent in terms of response speed. Finally, although basic functionality is provided, current online office suites do not provide the more advanced features available on their offline counterparts.

15.4 File and Media Sharing Services

The large communities that have been created on the Web have led to the generation of specific services allowing their register users to share files of any type. The most popular forms of file sharing include applications, electronic forms of books, documents, audio files and videos.

YouTube[10] is currently one of the most popular locations on the Web where users can publish, watch, share and comment videos. The users of the service are divided into two categories: The unregistered users who can just watch videos and the registered ones, who are permitted to upload an unlimited number of files. The latter are also provided with the ability to publish comments about the presented material and judge the quality of the content by voting.

Although each registered user can upload an unlimited number of videos, he/she is not free to publish those which contain defamation, pornography, copyright violations, and material encouraging criminal conduct. These restrictions are all described in the terms of use of the service and videos violating these terms are immediately erased from the database of the site.

15.5 Real-Time Web

The advent of Twitter[11] in 2007, introducing the micro-blogging concept, i.e., the posting and delivering of short messages up to 140 characters long to author's "followers", emphasized the need for the *real-time web.* That is, new technologies for rapid dissemination of information as soon as it gets

---

[9] http://wave.google.com
[10] http://www.youtube.com
[11] http://twitter.com

published by its author on the web. Examples of real-time web are friendfeed[12] and notifixious[13], while the *Extensible Messaging and Presence Protocol (XMPP)* and the *Simple Update Protocol (SUP)* are two protocols for developing instant-messaging-like applications [8,9]. Many believe that this instant-messaging perspective is the next big thing of web.

16 Discussion

In this chapter we have briefly described the core technologies employed by the modern applications of the Web. Knowledge of the modern Web technologies is a key issue for both developers and researchers. The former need to be informed in order to apply the most robust tools when they build their applications, whereas the latter are expected to deeply examine the key issues regarding these technologies in order to provide efficient solutions to newly posed problems.

From one perspective, Web is a society which constantly evolves. At the time these lines are written, novel services are being released and fresh software is composed. Cloud computing services promise to solve current issues and elevate computing. According to the most optimistic judges, the moment at which computers will not require an operating system to work and every transaction will be accomplished through a Web browser is very close.

References

1. `http://www.w3.org/TR/html4/`
2. `http://www.w3.org/TR/REC-xml/`
3. `http://www.php.net/usage.php`
4. `http://www.cio.com/article/446829/`
5. `http://www.abiresearch.com/products/service/`
6. `http://www.w3.org/TR/mobile-bp/`
7. `http://en.wikipedia.org/wiki/List_of_social_networking_websites`
8. `http://xmpp.org/about/`
9. `http://code.google.com/p/simpleupdateprotocol/`
10. O'Reilly, T.: What Is Web 2.0. O'Reilly Network (September- 30 - 2009) (Augut- 06 -2006)
11. Agarwal, N., Liu, H.: Blogosphere: Research issues, tools and applications. ACM SIGKDD Explorations 10(1), 18–31 (2008)
12. Akritidis, L., Katsaros, D., Bozanis, P.: Identifying Influential Bloggers: Time Does Matter. In: Proceedings of the IEEE / WIC / ACM International Conference on Web Intelligence (WI), pp. 76–83 (2009)
13. Krishnamurthy, B., Mogul, J.C., Kristol, D.M.: Key Differences between HTTP/1.0 and HTTP/1.1. Computer Networks-the International Journal of Computer and Telecommunications Networkin 31, 1737 (1999)

---

[12] http://friendfeed.com

[13] http://notifixio.us

14. Weng, J., Lim, E.P., Jiang, J., He, Q.: TwitterRank: Finding Topic-Sensitive Influential Twitterers. In: Proceedings of the third ACM International Conference on Web Search and Data Mining, pp. 261–270 (2010)
15. Mathioudakis, M., Koudas, N.: Efficient Identification of Starters and followers in Social Media. In: Proceedings of the 12th International Conference on Extending Database Technology: Advances in Database Technology, pp. 708–719 (2009)

Chapter 5
Federated Data Management and Query Optimization for Linked Open Data

Olaf Görlitz and Steffen Staab

Institute for Web Science and Technologies,
University of Koblenz-Landau, Germany
{goerlitz,staab}@uni-koblenz.de

Abstract. Linked Open Data provides data on the web in a machine readable way with typed links between related entities. Means of accessing Linked Open Data include crawling, searching, and querying. Search in Linked Open Data allows for more than just keyword-based, document-oriented data retrieval. Only complex queries across different data source can leverage the full potential of Linked Open Data. In this sense Linked Open Data is more similar to distributed/federated databases, but with less cooperation between the data sources, which are maintained independently and may update their data without notice. Since Linked Open Data is based on standards like the RDF format and the SPARQL query language, it is possible to implement a federation infrastructure without the need for specific data wrappers. However, some design issues of the current SPARQL standard limit the efficiency and applicability of query execution strategies. In this chapter we consider some details and implications of these limitations and presents an improved query optimization approach based on dynamic programming.

1 Introduction

The automatic processing of information from the World Wide Web requires that data is available in a structured and machine readable format. The *Linking Open Data* initiative[1] actively promotes and supports the publication and interlinking of so called *Linked Open Data* from various sources and domains. Its main objective is to open up data silos and to publish the contents in a semi-structured format with typed links between related data entities. As a result a growing number of Linked Open Data sources are made available which can be freely browsed and searched to find and extract useful information.

The network of Linked Open Data (or simply Linked Data) is very similar to the World Wide Web's structure of web pages connected by hyperlinks. Linked Data entities are identified by URIs. A relationship between two data entities is commonly expressed with the *Resource Description Framework* (RDF) [37] as a triple consisting of *subject*, *predicate*, and *object*, where the predicate denotes the type of the relation between subject and object. Linking a data entity from one data

[1] http://esw.w3.org/SweoIG/TaskForces/CommunityProjects/LinkingOpenData

A. Vakali & L.C. Jain (Eds.): New Directions in Web Data Management 1, SCI 331, pp. 109–137.
springerlink.com

source to a data entity in a different data source is very simple. It only requires a new RDF triple to be placed in one data source with the URI of the referenced data entity in the triple's object position. Additionally, the *linked data principles* [5] require that the referenced URI is resolvable, i.e. a common HTTP GET request on the URI returns *useful information* about the referenced data entity. A good overview about the foundations of Linked Open Data and current research directions is given in [8].

A Linked Open Data infrastructure can have different characteristics, namely central or distributed data storage, central or distributed data indexing, and independent or cooperative data sources, which influence how query processing can be implemented on top of it. Not all of the eight potential combinations are reasonable, in fact, three main paradigms can be identified (c.f. Fig. 1).

Central Repository. RDF data can be obtained from data sources, either by crawling them or by using data dumps, and put into a single RDF store. A centralized solution has the advantage that query evaluation can be implemented more efficiently due to optimized index structures. The original data sources are not involved in the query evaluation. Hence, it is irrelevant if they are cooperative or not. Note also that the combination of a central data storage with a distributed index does not make sense at all.

Federation. The integration of distributed data sources via a central mediator implies a federation infrastructure. The mediator maintains a global index with statistical information which are used for mapping queries to data sources and also for query optimization. Cooperation between data sources allows for reducing the query processing overhead at the mediator, as certain query evaluation steps can be moved to the data sources. Otherwise, the whole query execution, including join computations, has to be done at the mediator.

Peer-to-peer Data Management. RDF data and index data can both be maintained in a distributed fashion if all data sources are cooperative. The result is a peer-to-peer network where all data sources share the responsibility for managing parts of the RDF data and the index data. Hence, no central mediator is necessary and the data storage and processing load can be better balanced across all involved data sources. Without cooperation among the data sources it is not possible to realized a distributed index. Peer-to-peer RDF Management is not further discussed in this chapter.

<table>
<tr><td></td><td colspan="2">Central Data Storage</td><td colspan="2">Distributed Data Storage</td></tr>
<tr><td>Independent Data Sources</td><td>n/a</td><td rowspan="2">Central Repository</td><td rowspan="2">Federation</td><td>n/a</td></tr>
<tr><td>Cooperative Data Sources</td><td>n/a</td><td>P2P Data Management</td></tr>
<tr><td></td><td>Distr. Index</td><td colspan="2">Central Index</td><td>Distr. Index</td></tr>
</table>

Fig. 1. Combinations of characteristics which yield different infrastructure paradigms

This chapter focuses on a federation infrastructure for Linked Open Data and the optimization of federated SPARQL queries. Section 2 illustrates our running example. Section 3 motivates why a federation infrastructure is necessary for Linked Open Data and presents some basic requirements. The related work is presented in Sect. 4 before the detailed architecture of our federation infrastructure is discussed in Sect. 5. The elaboration of optimization strategies follows in Sect. 6. Further improvements for federating Linked Open Data are shortly mentioned in Sect. 7. Finally, Sect. 8 discusses the evaluation of federation infrastructures for Linked Open Data and Sect. 9 concludes the chapter.

2 Example

A common albeit tedious task is to get a general overview of an arbitrary research area exploring the most important research questions and the current state-of-the-art. It requires (i) finding important publications and good scientific overviews of the research area, (ii) identifying the most relevant conferences and journals, (iii) finding influential researchers based on number of publications or activity in conference committees etc., (iv) taking social aspects into account like personal relations and joint work on research projects, and (v) filtering and ranking all the information based on individual criteria, like recent activities and hot topics. Especially, students often struggle with this task due to the vast amount of information which can be found on the World Wide Web that has to be collected and connected to form the big picture. Ideally, an easy to use tool takes the description of a research topic and appropriate constraints as input and automatically gathers and merges all relevant information from the different data sources. The results are also valuable when looking for a new job or just to rank the impact or activity of researchers from different institutions/countries.

For the area of computer science, much of this information is already publicly available on the web, in semi-structured and machine readable formats. General information about research areas can be obtained from Wikipedia and, in a semi-structured format, from *DBpedia* [4]. Over 1.3 million publications can be found in the *DBLP Computer Science Bibliography*[2], including information about authors, conferences, and journals. Additionally, some conference web sites also provide semi-structured data about their conference program, like accepted papers, speakers, and members of the program committee. Other information like the acquisition of projects by research institutions and the amount of funding can be retrieved from funding agencies, e.g. from the EU and the *Community Research and Development Service* (CORDIS), or national funding programs. Last but not least, individual researchers offer information about their affiliation, research interests, and social relations in so called *friend-of-a-friend* (FOAF) profiles [12]. The *CS AKTive Space*[3] project, for example, already integrates some of this information for the UK Computer Science research domain.

[2] http://dblp.uni-trier.de/

[3] http://www.aktors.org/technologies/csaktivespace/

The example in Fig. 2 illustrates three Linked Open Data sources, i.e. DBLP, DBPedia, and Freebase, which contain RDF data about the mathematician Paul Erdős. The DBLP data describes a publication written by Paul Erdős and a coauthor. DBpedia and Freebase contain information about his nationality. The similarity of data entities is expressed via `owl:sameAs` relations.

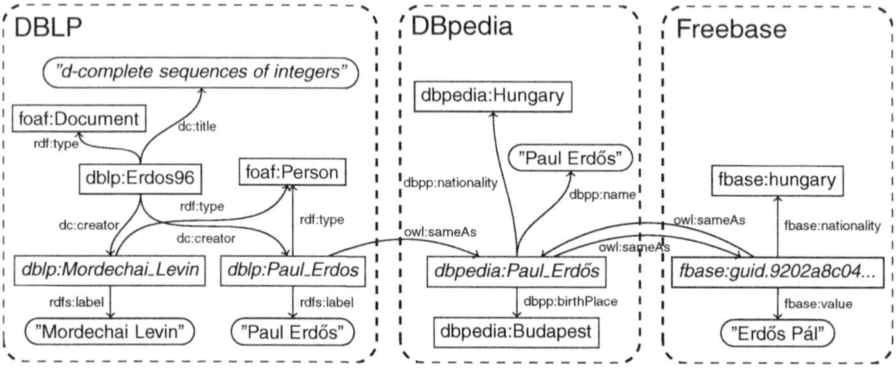

Fig. 2. Example RDF data located at three different Linked Open Data sources, namely DBLP, DBpedia, and Freebase (namespaces are omitted for readability)

3 Linked Open Data Search

Browsing is a natural way of exploring Linked Open Data. Starting with an initial URI that identifies a data entity outgoing links can be followed in any direction, to any data source, to discover new information and further links. However, there is no guarantee that the desired information can be found within a certain number of steps and that all relevant data is reached along the browsing path. Moreover, since links are directed the usefulness of results found by browsing depends heavily on the choice of a good starting point and the followed path.

In contrast, search based on complex queries provides a high flexibility in terms of expressing the desired properties and relations of RDF data entities to be retrieved. However, query evaluation is based on finding exact matches. Hence the user has to have good knowledge about the structure of the data sources and the used vocabularies. Obviously, this is not feasible for a large number of diverse linked data sets. Instead, queries can be formulated based on some standard vocabulary and the query processor applies query relaxation, approximate search, inferencing, and other techniques to improve the quality and quantity of search results. In the ideal case one benefits from higher expressiveness of queries and an abstraction from the actual data sources, i.e. a query does not need to specify which linked data sources to ask for collecting the results.

3.1 Requirements

A federation infrastructure for Linked Open Data needs basic components and functionalities:

A declarative query language is required for concise formulation of complex queries. Constraints on data entities and relations between them need to be expressible in a flexible way. For RDF there is SPARQL [48], a query language similar to SQL, based on graph pattern matching.

A data catalog is required in order to map query expressions to Linked Open Data sources which contain data that satisfies the query. Moreover, mappings are also needed between vocabularies in order to extend queries with similar terms.

A query optimizer is required, like in (distributed) databases, to optimize the query execution in order to minimize processing cost and the communication cost involved when retrieving data from the different Linked Data sources.

A data protocol is required to define how queries and results are exchanged between all involved Linked Data sources. SPARQL already defines such a protocol [15] including result formats.

Result ranking should be used for Linked Open Data search but is not directly applicable on RDF data since query evaluation is based on exact match. Additional information has to be taken in to account to rank search results.

Provenance information should be integrated, especially for ranking results. With a growing number of data sources it becomes more important to trace the origin of result items and establish trust in different data sources.

3.2 Architecture Variations

The query-based search on Linked Open Data can be implemented in different ways, all of which have certain advantages and disadvantages.

Centralized repositories are the common approach for querying RDF triples. All available datasets are retrieved (e.g. crawled) and stored in a central repository. This approach has the advantage that optimized index structures can be created locally for efficient query answering. However, the local copies and the index data need to be updated whenever something changes in the original data source. Otherwise, old and inconsistent results may be returned.

Explorative query processing is based on the browsing principle. A query is first evaluated on an initial data set to find matching data entities and also interesting links pointing to other data sets which may contain more data entities satisfying the query. In an iterative manner, the links are resolved and newly discovered data is fed as a stream into the query evaluation chain. Results are also returned in a streamed fashion as soon as they are available. The search terminates when there are no more links with potential results to follow. This approach evaluates the queries directly on the linked data and does not require any data copies or additional index structures. However, this also implies an increased communication effort to process all interesting links pointing to other data sources. Moreover, the choice of the starting point can significantly influence the completeness of the result.

Data source federation combines the advantages of both approaches mentioned above, namely the evaluation of queries directly on the original data source and using data indices and statistics for efficient query execution and returning complete results. A federated Linked Open Data infrastructure only maintains the meta information about available data sources, delegates queries to data sources which can answer at least parts of them, and aggregates the results. Storing the data statistics requires less space than keeping data copies. Moreover, changes in the original data have less influence on the metadata since the structure of the data source, i.e. the used vocabulary and the interlinks between them, does not change much. Thus, data changes are more likely to affect the statistics about data entities which influences mostly the quality of the query optimization than the correctness or completeness of the result.

Table 1. Comparison of the three architecture variations for querying Linked Open Data with respect to the most relevant characteristics

	central repository	link exploration	data source federation
data location	local copies	at data sources	at data sources
meta data	data statistics	none	data statistics
query processing	local	local+remote	local+remote
requires updates	yes (data)	no	yes (index)
complete results	yes	no	yes
up-to-date results	maybe	yes	yes

3.3 Federation Challenges

A federation infrastructure offers a great flexibility and scalability for querying Linked Data. However, there are some major differences to federated and distributed databases. Linked Data sources are loosely coupled and typically controlled by independent third party providers which means that data schemata usually differ and the raw data may not be directly accessible. Hence, the base requirement for the federation infrastructure is that all data sources offer a standard (SPARQL) query interface to retrieve the desired data in RDF format. Additionally, we assume that each Linked Data source also provides some data statistics, like the number of occurrences of a term within the dataset, which are used to (i) identify suitable data sources for a given query and (ii) to optimize the query execution. Hence, the responsibility of the federation infrastructure is to maintain the data statistics in a *federation index* and to coordinate the interaction with the Linked Data sources.

Scalability is without doubt the most important aspect of the infrastructure due to the large and growing number of Linked Data sources. That implies two main challenges – an efficient statistics management and an effective query optimization and execution.

3.3.1 Statistics Management

Data statistics are collected from all known Linked Data sources and stored in a combined index structure.

Accuracy vs. index size. The best query optimization results can be achieved with detailed data statistics. However, the more statistics are collected the larger the required size for storing the index structure. Hence, the challenge is to find the right tradeoff between rich statistical data and low resource consumption.

Updating statistics. Linked Data sources will change over time. Hence, the stored statistical information needs to be updated. However, such changes may not be detected easily if not announced. Sophisticated solutions may perform updates on the fly based on statistical data extracted from query results.

3.3.2 Query Optimization and Execution

The execution order of query operators significantly influences the overall query evaluation cost. Besides the important query execution time there are also other aspects in the federated scenario which are relevant for the query optimization:

Minimizing communication cost. The number of contacted data sources directly influences the performance of the query execution due to the communication overhead. However, reducing the number of involved data source trades off against completeness of results.

Optimizing execution localization. The standard query interfaces of linked data sources are generally only capable of answering queries on their provided data. The join of results obtained from different sources needs to be done at the query issuer. If possible at all, a better strategy will move parts of the result merging operations to the data sources, especially if they can be executed in parallel.

Streaming results. Retrieving a complete result when evaluating a query on a large dataset may take a while even with a well optimized execution strategy. Thus one can return results as soon as they become available, which can be optimized by trying to return relevant results first.

4 Related Work

The federation of heterogeneous data sources has been a popular topic in database research for a long time. A large variety of optimizations strategies has been proposed for federated and distributed databases [33,55,30,32]. In fact, the challenges for federated databases are very similar to the ones in the federated Linked Data scenario. But there are also significant differences. Distributed databases typically use wrappers to abstract from diverse schema used in different database instances.

Such wrappers are not required for Linked Data as Linked Data sources provide a SPARQL endpoint which returns the results in one data format, i.e. RDF. However, database wrappers are typically involved in the estimation of result cardinalities and processing cost. Without them, the estimation has to be handled differently. Optimization strategies used in federated and distributed databases rely on the cooperation of the individual database instances, e.g. parts of the query execution can be delegated to specific instances in order to make use of data locality or to improve load balancing. Query evaluation in Linked Data sources can not be optimized in the same way since the SPARQL protocol only defines how a query and the results are exchanged with the endpoints. It does not allow for cooperation between them.

Semantic web search engines like Sindice [45], Watson [16], SWSE [26], and Falcons [14] allow for document-oriented, keyword-based search. Like typical web search engines, RDF data is crawled from the web and indexed in a central index structure. Frequent re-crawling is necessary to keep the index up-to-date. Support for complex queries is limited or not available at all.

Some general investigations on the complexity of SPARQL query optimization, i.e. identifying the most complex elements and proposing specific rewriting rules, have been done by [53,46]. A recent trend is the development of highly scalable RDF repositories. Implementations like RDF3X [41], Hexastore [60], and BitMatrix [3] focus on optimal index structures and efficient join ordering which allows answering queries directly from the indexed data. Other systems, e.g. YARS2 [28], 4Store [24], and Virtuoso [17] use clustering techniques or federation in order to achieve high scalability, but in an environment with full control over the data storage. Federation of distributed RDF data source has been investigated in systems like DARQ [49] and SemWIQ [35]. Details of these and likewise approaches are presented in Sect. 5.

5 Federation Infrastructure for Linked Open Data

The architecture of a federation infrastructure for Linked Data differs not much from the architecture of a federation system for relational data sources, like Garlic or Disco [23,57]. In fact, it is simplified due to the use of SPARQL as common query protocol. First, customized data source wrappers that provide a common data format are not needed. Second, the increasing reuse of ontologies, such as FOAF [12], SIOC [11], and SKOS [38], lessens the need for conceptual mappings (c.f. [51] for ways of integrating conceptual mappings ex post). Figure 3 depicts the main components of a generic federation infrastructure for Linked Open Data.

All data sources are accessible via a SPARQL endpoint, i.e. a web interface supporting the SPARQL protocol. The actual data does not necessarily need to be stored in a native RDF repository. The data source may also use a relational database with a RDF wrapper like the D2R-Server [7].

The Resource Description Framework (RDF) is a widely accepted standard in the Semantic Web for semi-structured data representation. RDF defines a graph structure where data entities are represented as nodes and relations between them as edges.

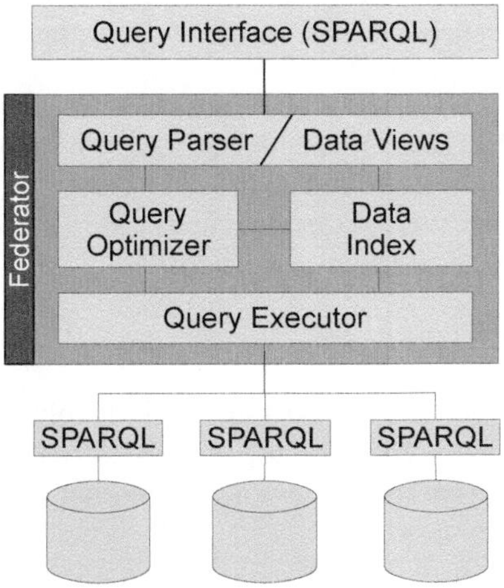

Fig. 3. Architecture of the federation infrastructure

Definition 1 (RDF Graph[4]). Let U, L, B be the pairwise disjoint sets of URIs, Literals, and Blank Nodes. Let $T = U \cup L \cup B$ be the set of RDF terms. A triple $S = (s, p, o) \in T \times U \times T$ is called a statement, where s is the subject, p is the property, and o is the object of the statement.

The example in Fig. 4 depicts a set of RDF triples describing a publication with the title "d-complete sequences of integers" written by Paul Erdős and Mordechai Levin (c.f. Fig. 2 for the graph representation). The prefix definitions in line 1-4 simplify the URI notation of the RDF triples in lines 6-13 and improve the readability.

```
1  @prefix dc:   <http://purl.org/dc/elements/1.1/>.
2  @prefix rdf:  <http://www.w3.org/1999/02/22-rdf-syntax-ns#>
3  @prefix dblp: <http://dblp.l3s.de/d2r/resource/>
4  @prefix foaf: <http://xmlns.com/foaf/0.1/>
5
6  dblp:ErdosL96 rdf:type foaf:Document.
7  dblp:ErdosL96 dc:title "d-complete sequences of integers".
8  dblp:ErdosL96 dc:creator dblp:Paul_Erdos.
9  dblp:ErdosL96 dc:creator dblp:Mordechai_Levin.
10 dblp:Paul_Erdos rdf:type foaf:Person.
11 dblp:Paul_Erdos foaf:name "Paul Erdos".
12 dblp:Mordechai_Levin foaf:name "Mordechai Levin".
13 dblp:Mordechai_Levin rdf:type foaf:Person.
```

Fig. 4. Example RDF data from DBLP in Turtle notation

[4] This definition slightly differs from the original W3C Recommendation [37] but is in line with the SPARQL query protocol [48] in allowing literals in subject position.

The SPARQL query language [48] defines graph patterns which are matched against an RDF graph such that the variables from the graph patterns are bound to concrete RDF terms.

Definition 2 (SPARQL Query). A query $Q : \{A\}$ is a set of query expressions. Let V be the set of query variables which is disjoint from T and let $\mathscr{P} \in (T \cup V) \times (U \cup V) \times (T \cup V)$ be a triple pattern. A triple pattern $\mathscr{P}$ is a query expression. If A and B are query expressions, then the compositions

$$(i)\ A\ .\ B\quad (ii)\ A\ \texttt{UNION}\ B\quad (iii)\ A\ \texttt{OPTIONAL}\ B\quad (iv)\ A\ \texttt{FILTER}\,(exp)$$

are query expressions, too. The dot operator in (i) denotes a join in SPARQL.

A function, such as $(a = b)$, $(a < b)$, or $(a > b)$, with a and b being variables or constants, is a filter expression. If e_1 and e_2 are filter expressions, then following expressions are filter expressions, too.

$$(i)\ e_1 \,||\, e_2\quad (ii)\ e_1\ \&\&\ e_2\quad (iii)\ !e$$

The SPARQL example in Fig. 5 selects German co-authors of Paul Erdős. The graph pattern, which needs to be matched, is defined in the WHERE clause in lines two to twelve and also depicted as a graph on the right side. Co-authorship is defined via the creator relation between people and articles. The German nationality is a property of a person. Line 13 defines an order on the results and line 14 restricts the number of results to 10. The first line specifies the final projection of variables. Namespaces are omitted for readability.

```
1  SELECT ?name ?workplace
2  WHERE {
3     ?author foaf:name "Paul␣Erdos".
4     ?article dc:creator ?author.
5     ?article dc:creator ?coauthor.
6     ?article rdf:type foaf:Document.
7     ?coauthor foaf:name ?name.
8     ?coauthor dbpprop:nationality dbpedia:German.
9     OPTIONAL {
10       ?coauthor dbpprop:workplaces ?workplace.
11    }
12 }
13 ORDER BY ?name
14 LIMIT 10
```

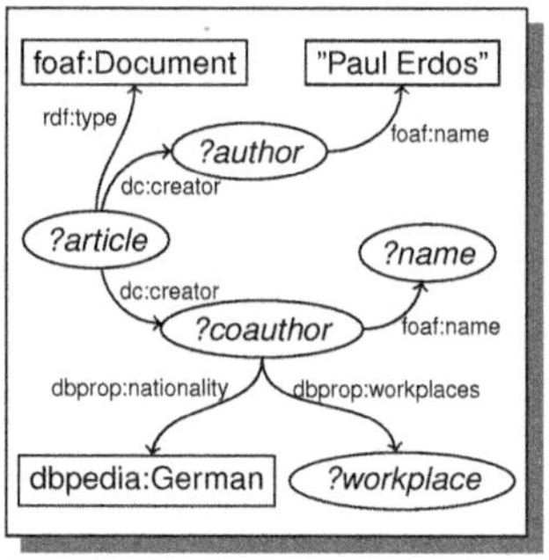

Fig. 5. SPARQL example: Find 10 German co-authors of Paul Erdős

Some common elements of SQL, like insert, aggregates, sub-queries, and grouping, are not part of the original SPARQL standard but only supported in version 1.1 [25].

5.1 Federator

The main component of the infrastructure is the *federator*. It is responsible for maintaining meta data about known data sources and for managing the whole query evaluation process. The federator offers an interface to the user which allows for keyword-based search and for submitting SPARQL queries. The keyword-based search is ideal for users without a good knowledge about the federated data sets. Although keyword-based search is usually entity centric, it can also be used to derive complex SPARQL queries, as shown in Hermes [59].

The actual query processing includes query parsing, query adaptation, query mapping, query optimization, and query execution. Query adaptation, which will not be further discussed here, is about modifying and extending a query, e.g. by including similar or broader terms and relations from other vocabularies, in order to broaden the search space and to obtain more results. The query mapping is about selecting data sources which can return results for the expressions contained in a query. During query optimization different join strategies for combining results are evaluated. Finally, the query execution implements the communication with the data sources and the processing of the optimized query execution plans.

5.1.1 Data Source Selection

Queries may describe complex data relations across different domains. Hence, it is very likely that a single data source may only be able to return results for parts of the query, as it is the case in the example query in Fig. 5. The first part about the co-authorship can be answered by DBLP, the second part about the nationality only by DBpedia. Thus, the respective query fragments have to be send to different data sources and the individual results have to be merged. The SemaPlorer [50], an application for location-based search on top of Linked Open Data, employs explicit data source mappings. This works well for the application's specific scenario but is not flexible enough for a general federation of Linked Open Data.

There have been discussions on how to implement federation in SPARQL, but it has not been included (yet) in the SPARQL standard. The proposed solution [47] is an extension of the SPARQL syntax with the `SERVICE` keyword, which should be used for specifying the designated data source. But an explicit definition of query endpoints requires that the user knows where the data is located. Hence, that solution is impractical for a flexible Linked Data infrastructure. Instead, the query federation should remain transparent, i.e. the federator should determine which data sources to contact based on a data source catalog. Such a catalog is an index which maps RDF terms or even sub-graph structures of a query to matching data sources. The catalog information is tightly coupled with the maintained data statistics (c.f. Sect. 5.3).

```
SELECT ?name ?workplace
WHERE {
  SERVICE <http://dblp.uni-trier.de> {
    ?author foaf:name "Paul␣Erdos".
    ?article dc:creator ?author.
    ?article dc:creator ?coauthor.
    article rdf:type foaf:Document.
    ?coauthor foaf:name ?name.
  }
  SERVICE <http://dbpedia.org> {
    ?coauthor dbpprop:nationality dbpedia:German.
    OPTIONAL {
      ?coauthor dbpprop:workplaces ?workplace.
    }
  }
}
ORDER BY ?name
LIMIT 10
```

Fig. 6. Example SPARQL query with explicit service endpoints

5.1.2 Join Strategies

In a federated system, the main cost factor is the communication overhead for contacting data sources and transferring data over the network. Not only the execution order of join operators, as discussed is Sect. 6, influences the processing cost, but also the execution strategy for transferring the query and the result data. In the following the different approaches an their limitations are discussed in more detail.

Remote Join. Executing joins directly at the data sources is the most efficient way as the processing overhead for the federator is minimized and the communication costs are reduced because of smaller intermediate result sets. However, remote joins are only applicable if all parts of the join can be satisfied by the data source. A problem arises when one data source can satisfy more parts of a join expression than another data source. In that case, the data source which satisfies less join parts may be able to generate results when joining with partial results from the first data source. Hence, even if the first data source evaluates a join completely, it also has to return the results for its intermediate join parts.

Mediator Join. Executing the join in the federator (or mediator) after receiving the intermediate results from the data sources is a common approach in existing RDF federation systems [49,56,35]. The join is typically realized as nested-loop-join or hash join. Since the SPARQL protocol allows for streaming result sets, i.e. by reading the HTTP response stream, it is possible to start joining a result set just after the smaller result set has been received completely. Other join variants like *merge-join* can only be used when the intermediate results are ordered, which is usually not the case. However, the mediator join can significantly increase the communication cost and also the processing cost at the federator if a data source returns a large intermediate result set which is part of a highly selective join.

Semi Join. The communication cost and the processing cost of the federator can be significantly reduced with semi-joins [6]. First, the federator retrieves the smaller intermediate result set of the two joins parts. Then a projection of the join variables is done and the extracted variable bindings are attached to the query fragment which is sent to the second data source. The intermediate results of the second data source are filtered with the transmitted bindings and only the reduced result set is returned to the federator. Thus, the federator has to join smaller intermediate results and also less data has to be transmitted over the network. Essentially, a semi-join realizes a pipelined execution, in contrast to the centralized join where each join part can be executed in parallel.

Unfortunately, the current SPARQL standard does not support the inclusion of variable bindings in a query. The SPARQL federation draft [47] proposes a solution by extending the syntax with the `BINDINGS` keywords. Since this extension is not yet accepted or supported by current SPARQL endpoints the only alternative is to include variable bindings as filter expressions in a query. Both variants are shown in Fig. 7. Although the second approach can be realized with standard SPARQL it is not optimal as the query can be blown up for a large number of bindings, thus increasing the communication cost. DARQ [49] realizes bind joins by sending a query multiple times with all the different bindings. However, this increases the communication overhead as for each binding a separate query and result message has to be sent.

Finally, the proposed `SERVICE` keyword for referencing other datasets may be used within a query fragment to define a bind join. When resolved by a SPARQL endpoint the intermediate result could be directly transferred between the data source without involving the federator. But if multiple data sources include the same sub query it will be sent multiple times to the remote data source and the intermediate results will also be generated more than once if no caching is applied.

```
SELECT ?article ?author ?coauthor
WHERE {
  ?article dc:creator ?author.
  ?article dc:creator ?coauthor.
}
BINDINGS ?author ?coauthor {
  ("Paul Erdos" "Andreas Blass")
  ("Paul Erdos" "Walter Deuber")

}
```

```
SELECT ?article ?author ?coauthor
WHERE {
  ?article dc:creator ?author.
  ?article dc:creator ?coauthor.
  FILTER (
  (?author = "Paul Erdos" &&
   ?coauthor = "Andreas Blass") ||
  (?author = "Paul Erdos" &&
   ?coauthor = "Walter Deuber"))
}
```

Fig. 7. Variable bindings in SPARQL: via `BINDINGS` syntax extension or as `FILTER` expression

Bind Join. The bind-join [23] is an improvement of the semi-join. It is executed as a nested loop join that passes bindings from the intermediate results of the outer part to the inner part, i.e. to the other data source, which uses them to filter its results. This is comparable to a prepared query where variable bindings are provided after the query template has already been optimized. However, SPARQL is also missing a suitable mechanism for defining prepared queries and does not support the streaming of bindings. A common SPARQL endpoint will only start evaluating a query after all query data has been transmitted as the query can only be optimized when all query information is available.

Filter Chain. A completely different approach is implemented in [29]. The presented query engine operates in a pipelined fashion by resolving linked data references on-the-fly. Thus, it follows the exploration scheme of linked data. Specific optimizations for speeding up the retrieval are implemented, namely cascaded iterators which operate on the data stream and return results as soon as they become available.

5.2 Data Catalog

The data catalog stores two different kinds of data mappings. The first mapping captures relations between RDF terms, like similarity defined with the `owl:sameAs` and `rdfs:seeAlso` predicate. Such information can be used to adapt a query to different data schemata, or simply to broaden the search space. The second mapping associates RDF terms or even complex graph structures with data sources. During the data source selection phase this information is used to identify all relevant data sources which can provide results for query fragments.

The data catalog may be combined with the data statistics described below. Additional statistical information is indeed quite useful for ranking data sources and mappings between RDF terms. Popular predicates, like `rdf:type` and `rdfs:label`, occur in almost every data source. Thus, a ranking helps to avoid querying too many less relevant data sources.

The most common constants in query patterns are predicates. The number of different predicates in a data set is usually limited since predicates are part of the data schema. In contrast, the number of data entities, i.e. RDF terms occurring in a RDF triple's subject or object position, can be quite large. However, they may have just one occurrence. Hence, there is a trade-off between storing many item counts for detailed mappings and minimizing the catalog size. It should also be noted that literals are best stored in a full text index. They will usually occur only a few times in a data set. Moreover, this also allows for searching the data based on string matching.

5.3 Data Statistics

Statistical information is used by the query optimizer to estimate the size of intermediate result sets. The cost of join operations and the amount of data which needs to be transmitted over the network is estimated based on these statistics. Very accurate and fine grained data statistics allow for better query optimization results but also require much more space for storing them. Therefore, the main objective is to find the optimal trade-off between accuracy of the statistics and the required space for storing them. However, precise requirements will depend on the application scenario.

Item counts. The finest granularity level with the most exact statistical information is implemented by counting data items. In RDF, such counts typically comprise the overall number of triples as well as the number of individual instances of subject, predicate, and object. Counts for combinations of subject, predicate, and object are useful statistics, too. State-of-the-art triple store implementations like RDF3X[41,42], Hexastore[60], and BitMatrix [3] employ full triple indexing, i.e. all triple variations (S, P, O, SP, PO, SO) are indexed, which allows for generating query answers directly from the index.

Full text indexing. The RDF graph of a data source can also be seen as a document which can be indexed with techniques known from information retrieval, like stop word removal and stemming. As a result, data entities may be searched efficiently via keyword-based search. The difference is that typically only literals, as objects in RDF triples, are really suitable for indexing. If URIs should be indexed as well building a prefix-tree is usually a good choice.

Schema level indexing. Instead of maintaining statistics for individual data instances one may also restrict the index to the data schema, i.e. the type of instances and relations. This can reduce the overall index size but it also has disadvantages. Certain types and properties, like `foaf:Person` and `rdfs:label`, are widely used in RDF data sets and can be a bad choice for discrimination. Moreover, queries must contain the indexed types. Otherwise, all data sources have to be contacted.

Flesca et al.[18] built equivalence classes for RDF terms using a similarity measure based on types and properties. Counts and references to data sources with equivalent entities are attached to the equivalence classes. This approach can resolve identical data instances across data sources. However, there is no information about the scalability of this approach.

Structural indexing. Join statistics contain information about the combination of certain triple patterns. They are much better for estimation the join cardinality since the existence of results for two individual triple pattern in a data source does not automatically imply that there are also results for the joined pattern. However, since there is an exponentially large number of join combinations, not all of them can be stored in an index.

Therefore, as RDF data represents a graph and SPARQL queries describe graph patterns, it makes sense to identify and index only the most common graph structures found in the Linked Data sets. An early proposal for federating SPARQL [56] was based on indexing path structures extracted from data graphs. However, since star-shaped queries are very common for SPARQL queries the path-based approach is not optimal. Instead, generic or frequent sub graph indexing [36,58] can be used but requires sophisticated algorithms for efficiently identifying sub graphs patterns.

A major limitation of structural indexing is its restriction to a single data source. Query federation would benefit significantly from the identification of graph structures across data sources, as certain combinations of data sources could be excluded from the query execution. Moreover, structural indexing is costly and a typical offline pre-processing step. Hence it is not easily applicable for life Linked Data.

5.3.1 Index Size Reduction

Ideally, all required index data should fit into main memory to avoid frequent disk access. Hence, for a large data sets it is necessary to reduce the size of the index data.

Histograms are commonly used for restricting index data to a fixed size while still retaining a high accuracy. Similar data items are put into so called buckets which count the number of items they contain. Space reduction is achieved through a fixed number of buckets. The typically used histogram type (in the database world) is the equi-depth histogram [40], since it provides a good balance of items per bucket – even for skewed data distributions. QTrees, which are a combination of histograms and R-Trees, are used in [27] with three dimensions representing the three components of an RDF triple. Buckets are used as leave nodes to store item counts and the list of respective data sources which fall into the region covered by the bucket.

Alternatively, Bloom Filters [10] are also suitable for reducing index data to a fixed size. Items are hashed to a bit vector which represents an item set. Membership of items in the set can be efficiently checked with low error rate. However, Bloom filters are not keeping track of the number of items in the represented set. If such information is needed extensions as presented in [44] are necessary.

If a high accuracy of statistics is necessary, large index structures and disk access are not avoidable. An optimal index layout and common data compression techniques, e.g. as applied in RDF3X [41], can be employed to reduce the required disk space and the frequency of disk access.

5.3.2 Obtaining and Maintaining Data Source Statistics

In order to build a federation index with detailed statistics about all involved data sources it is necessary to first pull the statistical details from the data sources. But not all data sources are capable or willing to disclose the desired information. Hence, additional effort may be necessary to acquire and maintain the data source statistics.

Data Dump Analysis. Several popular Linked Open Data Sources provide RDF dumps of their data. Analyzing the dumps is the easiest way to extract data statistics and if a new version of the dataset becomes available the analysis on the new data dump is simply redone. However, new data versions are not always advertised and only a few large datasets provide dumps. In the future it will probably be more common to have many small frequently changing datasets, e.g. on product information.

Source Descriptions. A Linked Data Source may publish additionally some information about the data it contains. For the federator it is necessary to obtain at least some basic statistics like the overall number of triples, the used vocabulary, and individual counts for RDF properties and entities. A suitable format for publishing a data source's statistical information is the *Vocabulary of interlinked Datasets* (voiD) [2]. Additionally, voiD also allows to express statistics about data subsets, like the interlinks to other datasets, which is quite useful when retrieving data from different data source. However, complex information, like frequent graph patterns, can not be expressed with voiD.

DARQ [49] employs so called *service descriptions* describing data source capabilities. A service description contains information about the RDF predicates found in the data source and it can also include additional statistical information like counts and the average selectivity for predicates in combination with bound subjects or objects. However, the explicit definition of service description for each involved data source is not feasible for a large number of Linked Open Data sources. Moreover, DARQ restricts the query expressiveness as predicates always need to be bound.

Source Inspection. If no dump nor data description is offered by a data source but only a SPARQL endpoint it is still possible to retrieve some statistics by sending specifically crafted queries. But this can be tedious work as initially no knowledge about data set size and data structure is available. So the first step is to explore the data structure (schema) by querying for data type and properties. Additionally, SPARQL aggregates are required to retrieve counts for individual data instances, which are not supported before SPARQL 1.1. But even with counts, crawling data is expensive and SPARQL endpoints typically restrict the number of requests and the size of the returned results.

Result-based Refinement. As an alternative for data inspection, it is possible to extract data statistics from the results returned for a query. This approach has little extra processing overhead and it is non-intrusive. However, the lack of exact statistics in the beginning results in increased communication cost and longer execution time as more or even all data sources have to be queried. But with every returned result the statistics will be refined. Moreover, changes in the remote data sets are adapted automatically. Ideally, this approach is combined with some initial basic statistics setup which minimizes the number of inefficient queries in the beginning.

5.3.3 Index Localization

The index management can also be viewed from the perspective of where an index is located, i.e. at the data source or at the federator, and which statistical information it maintains, i.e. local data source statistics or global federation statistics.

Data Source Index. A data source usually maintains its own statistics for local query optimization. These statistics are not accessible from the outside due to their integration with the data source's query engine. Publicly exposed data statistics, e.g. as voiD descriptions, are explicitly generated and need to be updated when the actual data changes.

Virtual Data Source Index. If a data source does not offer any data statistics publicly they have to be collected by other means, as mentioned above. The result is essentially a virtual index which is created and managed at the federator as part of the federation index.

Federation index. The federator maintains a centralized index where the statistical information of all known data sources is collected and updated. The stored statistical information ranges from basic item counts to complex information like frequent graph patterns. All statistical data is connected to the data source where the data can be found.

Distributed Federation Index. In a cooperative environment, a federation index may be partitioned and distributed among all data sources to improve scalability. Like in peer-to-peer systems, a data source would additionally keep a part of the global index and references to data sources with similar information. In such an environment, the query execution can take advantage of the localized index information and less central coordination is needed. The interlinks in Linked Data Sources can already be seen as a step in that direction. To support cooperation, a new protocol would be required to exchange information among linked data sources about mutual links and data statistics. The usage of a distributed index for data federation is outlined in [18] but it involves a large number of update messages when data changes. An efficient solution for updating a distributed index is presented in [20].

6 Query Optimization

The objective of the query optimization is to find a query execution plan which minimizes the processing cost of the query and the communication cost for transmitting query and results between mediator and Linked Data sources. In the following, basic constraints for the federation of RDF data sources and the structure of query execution plans will be presented before optimization strategies are discussed.

Existing federation approaches for RDF mainly use centralized joins for merging intermediate results at the mediator. The application of semi-joins has not yet been considered for the optimization of distributed RDF queries. We will show that optimization based on dynamic programming, which is presented in more detail, can easily be applied for semi-join optimizations of federated SPARQL queries.

6.1 Data Source Mappings

The mapping of multiple data sources to different query fragments implies specific constraints for the query optimization. Consider the SPARQL query in Fig. 8 with two triple patterns which can be answered by three data sources, i.e. `foaf:name` is matched by {`http://dblp.uni-trier.de/`, `http://dbpedia.org/`} and the combination of `dbprop:nationality` and `dbpedia:German` by {`http://dbpedia.org/`,`http://rdf.freebase.com/`}.

Fig. 8. Query fragment with data source mappings

Although DBpedia could answer the whole query it is not possible to just merge the result from DBpedia with the joined results of the other two sources since a partial result set of DBpedia, i.e. results for a single triple pattern, may also be joined with an intermediate result from `http://dblp.uni-trier.de/` or `http://rdf.freebase.com/`. Hence, each triple pattern has to be evaluated individually at the respective data sources. The results for every pattern are combined via UNION and finally joined. For a SPARQL endpoint, which could resolve remote graphs, the rewritten SPARQL query would look like in Fig. 9.

```
SELECT ?coauthor ?name
WHERE {
  {
    SERVICE <http://dblp.uni-trier.de/> {
      ?coauthor foaf:name ?name
    }
    UNION
    SERVICE <http://dbpedia.org/> {
      ?coauthor foaf:name ?name
    }
  }.
  {
    SERVICE <http://dbpedia.org/> {
      ?coauthor dbprop:nationality dbpedia:German
    }
    UNION
    SERVICE <http://rdf.freebase.com> {
      ?coauthor dbprop:nationality dbpedia:German
    }
  }
}
```

Fig. 9. Rewritten SPARQL query which maps triple patterns to different data graphs and merges the results via UNION

6.2 Query Execution Plans

The output of the query parser is an *abstract syntax tree* containing all logical query operators, e.g. join, union, triple pattern, that make up the query. The tree structure defines the order in which the query operators have to be executed, i.e. child nodes have to be evaluated first. A *query execution plan* is an executable query plan where logical operators are replaced by physical operators, e.g. a join may be implemented as *nested loop join*, *sort-merge join*, *hash-join*, or, as explained in Sect. 5.1.2, as *semi-join* or *bind-join*.

The structure of the query execution plan is also important. The two main types are *left-deep trees* and *bushy trees*. Figure 10 depicts both variations for the running example. Left-deep trees imply basically a pipelined execution. Starting with the leftmost leaf node operators are evaluated one after another and results are passed as input to the parent node until the root node is reached. In contrast, bushy trees allow for parallel execution as sub trees can be evaluated concurrently.

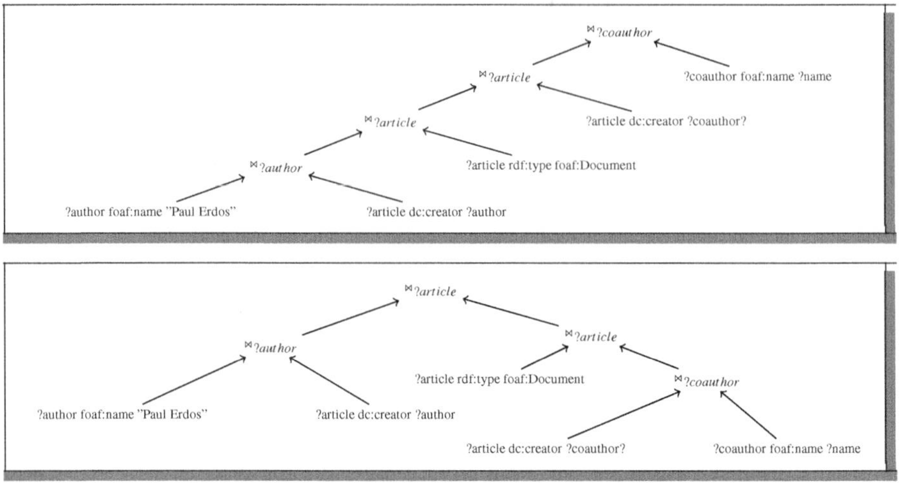

Fig. 10. Left deep and bushy query execution plan

The choice of physical operators also affects the execution characteristics. Although the processing and communication cost can be reduced significantly with semi-join it also implies longer query execution times as operators have to wait for the input of preceding operators. Streaming the data between operators can speed up the execution but with the SPARQL standard streaming is only possible for query results. Queries including variable bindings have to be propagated completely before the query execution can start.

The leaf node operators represent *access plans*, i.e. they are wrappers which send the query fragments to the data sources and retrieve and merge the results. If semi-joins are used, the access plan will include the propagated variable bindings in the query fragments. Additionally, a semi join operator needs to project the join variables from the results returned by the child operator, which is executed first, before the variable bindings can be passed on.

Filters are not explicitly mentioned in the execution plans but will be considered during the execution. Generally, filters are pushed down as far in the tree as possible, i.e. a filter will be attached to a leaf node or join node if the node satisfies all variables in the filter and there is no other child node that does. Filters attached to leaf nodes are included in the query fragment sent to remote data sources. In case of a semi-join, the filter will also be propagated down to child nodes along with the variable bindings. Otherwise, the filter will be applied by the federator on the (joined) result set.

6.3 Optimization Fundamentals

The objective of the query optimization is to find a query execution plan with minimal cost in terms of processing cost and communication cost. Typically, the *query execution time* is the main cost measure. It combines both processing cost and communication cost.

The join order has a significant influence on the query execution time. Small intermediate result sets reduce the communication cost as well as the join processing cost. Thus, the join order optimization is often the main focus of the query optimization. Join order optimization in SPARQL is mainly about optimizing *basic graph patterns*, i.e. a set of conjunctively connected triple patterns. Other operators, like `OPTIONAL` and `UNION`, usually have additional constraints which complicate the optimization. They are not further considered here.

There are different optimization strategies which will be discussed shortly. All of them rely on the same two basic measures for estimating the cost of a query execution plan, namely *cardinality* and *selectivity*. Cardinality is the estimated number of elements in a result set which are returned for a query expression. Selectivity defines the estimated fraction of elements which match a query expression. Selectivity values are in the range $[0..1]$ where a selectivity of 0 means most selective and 1 means least selective.

The cardinality and selectivity for RDF is based on triples and formally defined as follows:

Definition 3 (RDF graph cardinality). Let $|G| = |\{S_i \in G\}|$ be the cardinality of a graph, i.e. the graph size in terms of the overall number of triple statements contained in the graph.

Definition 4 (RDF term selectivity). Let $sel_G(t) = \frac{|\{S_t \in G\}|}{|G|}$ be the selectivity of term $t \in T$ in graph G with $t = subj(S_i) \lor t = pred(S_i) \lor t = obj(S_i)$.

Definition 5 (Triple Pattern Selectivity). The selectivity of a triple pattern $\mathscr{P}$ is the product of the selectivities of the contained RDF terms:

$$sel_G(\mathscr{P}) = \prod_{i}^{n} sel_G(t_i);\; t_i \in const(\mathscr{P})$$

Consequently, the pattern cardinality $|\mathscr{P}_G| = |G| \times sel_G(\mathscr{P})$ is the estimated number of matching statements in graph G. This assumes that the RDF terms in triple patterns are independent.

6.4 Optimization Strategies

There are different approaches for query optimization. Usually there is a trade-off between finding the optimal query plan and finding a query plan quickly. Optimization strategies can be classified by static and dynamic optimization. Static optimizers generate one query plan and sticks to it during the whole query execution. Dynamic optimizers may change a query plan during execution due to updated statistics.

Applying heuristics is a common approach to find a good solution fast. Popular heuristics are pushing down filters and sorting query expression by their estimated selectivity. The optimization approach presented in [56] uses iterative improvement and simulated annealing.

Although heuristics can provide good results they will often produce sub optimal execution plans. A guaranteed optimal solution can be found with the *dynamic programming* approach, which is commonly used for query optimization in databases. The SPARQL federation implementation in [49] also uses dynamic programming but details are not given. Dynamic programming will be discussed in more detail in the following section.

Query optimization is a complex topic and there are a lot more approaches, some of which have already been applied for SPARQL. For example, there is RCQ-GA [31], a genetic algorithm for optimizing chain queries. An evolutionary algorithm for approximate querying with anytime behavior is presented in [21].

6.5 Dynamic Programming

Dynamic programming [54] is an optimization strategy in traditional relational databases which ensures to find the optimal query execution plan for any given query. All possible query execution plans are iterated and inferior plans are pruned based on the calculated cost estimates. A cost function is used to estimate the execution cost for each operator based on the cardinality and selectivity of intermediate results.

6.5.1 Query Plan Generation

In Dynamic programming query execution plans are generated in a bottom up fashion. The initialization is done by creating an access plan for each query pattern. Then in each iteration step *n-ary joins* are created by combining partial plans from previous iterations. Joins which yield cross products are deferred until the end. An optimized algorithm for the generation of bushy trees is presented in [39]. If many joins and different alternatives for physical operator are involved in the query plan generation the number of plan variations can rapidly grow too large for the available memory. *Iterative Dynamic Programming* [34] can be used in such a situation to iterate the plans in a divide and conquer fashion.

6.5.2 Query Plan Evaluation

The result of each iteration step is a set of execution plans which includes equivalent plans with different operator order. The plan evaluation step computes for each plan the execution cost, in order to prune inferior plans. The execution cost is computed recursively based on a cost model which uses the cardinality of each query operator to estimate the individual processing and communication cost. The cardinality of query expressions is defined as follows.

Definition 6 (Query expression cardinality). Let A_G and B_G be query expressions applied on graph G. Then, $|A_G|$ is the expression cardinality and $sel_G(A) = \frac{|A_G|}{|G|}$ is the selectivity of expression A_G. Under the assumption that terms in expressions are independent, we define the cardinality of complex expressions as

$$
\begin{aligned}
|A_{G_i} \,.\, B_{G_j}| &= |A_{G_i}| \times |B_{G_j}| \times min(sel_{G_i}(A),\, sel_{G_j}(B)) \\
|A_{G_i}\ \texttt{UNION}\ B_{G_j}| &= |A_{G_i}| + |B_{G_j}| \\
|A_G\ \texttt{FILTER}\ (exp)| &= |A_G| \times sel_G(exp)
\end{aligned}
$$

Definition 7 (RDF filter selectivity). Similar to [54] the selectivity of a filter is defined as:

$$
\begin{aligned}
sel_G(a = x) &= sel_G(x) \\
sel_G(a > x) &= \frac{max_a - x}{max_a - min_a} \text{ or } \frac{1}{3} \text{ if not comparable.} \\
sel_G(a \mid\mid b) &= sel_G(a) + sel_G(b) - sel_G(a) \times sel_G(b). \\
sel_G(a \&\& b) &= sel_G(a) \times sel_G(b) \text{ assuming that } a \text{ and } b \text{ are independent.} \\
sel_G(!a) &= 1 - sel_G(a).
\end{aligned}
$$

6.5.3 Cost Model

Each query operator is evaluated based on the cost estimates for the individual operations. The cost of a query execution plan $c(\mathcal{Q})$ is the sum of the cost of all its operators. The cost c_{op} of an operator applied on a query fragment $\mathcal{Q}^*$ and a set of variable bindings B is defined based on a cost model. The constants $c_{connect}$, $c_{compare}$, c_{hash}, and $c_{transmit}$ define the cost for establishing a connection to a data source and the cost for comparing, hashing, and transmitting a binding. The cost c_{eval} is the cost for evaluating a query fragment at a data source. It depends on the actual implementation which is usually not known and may employ index lookups or full table scans. Hence, c_{eval} is based on rough estimates.

$$
\begin{aligned}
c_{remote\_eval}(\mathcal{Q}^*, B) &= c_{connect} + c_{send}(\mathcal{Q}^*, B) + c_{eval}(\mathcal{Q}^*, B) + c_{send}(B') \\
c_{send}(B) &= |B| \cdot c_{transmit} \\
c_{filter}(B, f) &= |B| \cdot c_{compare} \\
c_{union}(B, \hat{B}) &= |B| + |\hat{B}| \\
c_{nested-loop-join}(B, \hat{B}) &= |B| \cdot |\hat{B}| \cdot c_{compare} \\
c_{hash-join}(B, \hat{B}) &= min(|B|, |\hat{B}|) \cdot c_{compare} + max(|B|, |\hat{B}|) \cdot c_{hash} \\
c_{sort-merge-join}(B, \hat{B}) &= (|B| \cdot log|B| + |\hat{B}| \cdot log|\hat{B}| + |B| + |\hat{B}|) \cdot c_{compare}
\end{aligned}
$$

Parallel Execution Cost. For parallel query execution plans the overall cost is the maximum cost of all individual plans.

$$
c(\mathcal{Q}) = max(\mathcal{Q}_1', \ldots, \mathcal{Q}_n')
$$

7 Improvements for Federation

The presented federation infrastructure and query optimization covers the basic requirements for the federation of Linked Open Data sources. However, there is still room for improvements. Not every optimization technique, which works for distributed and federated databases, may be applied to federated linked data. Some constraints are due to limitations of the SPARQL standard, as pointed out earlier.

7.1 Streaming Results

The execution chain of operators can be a critical bottleneck if large intermediate results are produced or if some data sources have bad response times. The standard SPARQL protocol and its implementation in typical SPARQL endpoints requires that a query, including all filters expressions, must be completely available before the query optimization can be performed. That implies that each query stage in the chain has to be completed before the next one can be executed. In order to speed up the query execution, partial query results may be propagated as soon as they become available. However, such data streaming is not (yet) supported by the SPARQL standard.

7.2 Result Ranking

A SPARQL query does not define an order for a result set, unless it is explicitly defined with the keyword `ORDER_BY`. Hence, the result items have to be considered unordered. Nevertheless, some result items may be more relevant than others (from a user's perspective) and should be returned first. However, the criteria for relevance in a federated infrastructure may also include trust and other factors, like response time and data quality. Existing ranking algorithms for RDF data, like RSS [43] or TripleRank [19], are not directly applicable because they are working on the link structure and do not take other aspects into account.

Ranking is also important for the query optimization. The dynamic programming approach [54] considers so called interesting orders, i.e. orders which are required for the final result and can minimize the join processing cost. Such information is not yet considered for federated queries.

7.3 Views

Views are a common concept in the relational database world. They allow for data abstraction and simplify the querying of complex data relations. For RDF there is no standardized definition of views. With so called named graphs[13] it is possible to define a context for RDF graphs. But this is rather limited and not flexible enough for managing a large number of RDF graphs, as all RDF triples in a graph context have to be explicitly listed.

Networked RDF Graphs [51] extend named graphs with a SPARQL based view mechanism. They allow users to define RDF graphs both, by extensionally listing statements describing the graph or by using views which are defined as SPARQL queries on other graphs. These views can be used to include parts of other graphs, to transform data before including it and to denote rules. Networked Graphs can be evaluated in a distributed setting using existing protocols. The benefits of networked graphs is the easy reuse and exchange of graphs, recursive view definitions and the application for data integration from distributed data sources. Especially the last point is interesting for Linked Open Data.

Views are basically an adequate way to establish an abstraction for underlying data schemata. They also provide transparency concerning data distribution. If data is moved or merged only the respective view definition needs to be adapted while everything else remains unchanged.

8 Performance Evaluation

In order to compare different federation infrastructures, an evaluation scenario is required which can measures the performance based on different criteria. Different benchmarks like LUBM [22], the MIT Barton dataset benchmark [1], or the SP2 benchmark [52] have been developed in recent years, but primarily for evaluating query processing performance of local repositories on a single large data set. Hence, they are not applicable for a distributed infrastructure. Unfortunately, there is no suitable benchmark for evaluating an infrastructure for (federated) Linked Open Data sources. So the problem is to find an evaluation scenario with several linked data sets and a number of complex queries spanning these data sets. Essentially, there is only the option to choose between real world and artificial data sets which both have advantages and disadvantages.

8.1 Real World Datasets

The number of available linked data sets has grown significantly in recent months with DBPedia [9] being one of the most popular ones. Thus, there should be lots of interesting information to be queried. However, formulating meaningful queries involving multiple data sources requires a good understanding of the information provided by the data sources in the first place. A good set of queries should cover different query types and should also produce results of different sizes. Due to the large number and diversity of linked data source, plus the constantly changing data, it requires a lot of effort to create such a consistent set of benchmark queries. But more importantly, the possibility to reproduce results is questionable.

8.2 Artificial Datasets

Most of the above mentioned benchmarks use artificial data sets. The design objective of such artificial datasets is to cover all typical characteristics of data relations and queries that can be evaluated on top of them. Hence, they allow for comparable evaluations of different systems. The only problem with existing artificial benchmarks is that they are not directly applicable for evaluating data federation which requires the existence of multiple data sources. The obvious solution is to split one large data set into several smaller partitions.

8.3 Data Partitioning

The SP2 benchmark [52] is a good basis for creating a data set for benchmarking the federated scenario. It covers a wide range of SPARQL query types and reproduces the characteristics of the DBLP bibliography dataset. Its data generator can be used to create data sets of arbitrary size.

In order to resemble the characteristics of linked data sources the partitioning should be applied vertically and horizontally and also retain a certain overlap between the partitions. Vertical partitioning means splitting the data schema, i.e. different partitions should only share a few common RDF types and predicates to mimic different domains. Horizontal partitioning implies a separation at the instance level, e.g. RDF triples with the same subject are placed in the same partition. Overlap can be realized by placing data instances in multiple different partitions. This usually happens automatically when data instances occurs in subject and object position of RDF triples.

9 Summary

A federated infrastructure was presented in this chapter which allows for transparent querying of distributed Linked Open Data sources. The main components of the architecture, namely the federator, the data catalog, and the data statistics were discusses in details. The SPARQL standard does not support all requirements for an efficient processing of federated queries. Specifically, semi-joins, which can significantly reduce the processing an communication cost, are not well supported.

The optimization of SPARQL queries is mainly focusing on join order optimization. A new optimization strategy using semi-joins and dynamic programming was explained in more detail. There is still room for improving the federation of Linked Open Data, e.g. with data streaming, ranking, and the support for data views. Especially, the efficiency of the query processing is not optimal yet.

References

1. Abadi, D., Marcus, A., Madden, S., Hollenbach, K.: Using the Barton libraries dataset as an RDF benchmark. Tech. rep., Massachusetts Institute of Technology Computer Science and Artificial Intelligence Laboratory (2007)
2. Alexander, K., Cyganiak, R., Hausenblas, M., Zhao, J.: Describing Linked Datasets – On the Design and Usage of voiD, the "Vocabulary Of Interlinked Datasets". In: Proceedings of the Linked Data on the Web Workshop. CEUR Workshop Proceedings, Madrid, Spain (2009); ISSN 1613-0073
3. Atre, M., Chaoji, V., Zaki, M., Hendler, J.: Matrix "Bit" loaded: A Scalable Lightweight Join Query Processor for RDF Data. In: Proceedings of the 19th International World Wide Web Conference, Raleigh, NC, USA, pp. 41–50 (2010)
4. Auer, S., Bizer, C., Kobilarov, G., Lehmann, J., Cyganiak, R., Ives, Z.: DBpedia: A Nucleus for a Web of Open Data. In: Proceedings of the 6th International Semantic Web Conference, Busan, Korea, pp. 722–735 (2007)

5. Berners-Lee, T.: Linked Data Design Issues, http://www.w3.org/DesignIssues/LinkedData.html
6. Bernstein, P., Chiu, D.: Using Semi-Joins to Solve Relational Queries. Journal of the ACM 28(1), 25–40 (1981)
7. Bizer, C., Cyganiak, R.: D2R Server – Publishing Relational Databases on the Semantic Web, http://www4.wiwiss.fu-berlin.de/bizer/d2r-server/
8. Bizer, C., Heath, T., Berners-Lee, T.: Linked Data – The Story So Far. International Journal on Semantic Web and Information Systems 5(3), 1–22 (2009)
9. Bizer, C., Lehmann, J., Kobilarov, G., Auer, S., Becker, C., Cyganiak, R., Hellmann, S.: DBpedia – A Crystallization Point for the Web of Data. Web Semantics: Science, Services and Agents on the World Wide Web 7(3), 154–165 (2009)
10. Bloom, B.: Space/time trade-offs in hash coding with allowable errors. Communications of the ACM 13(7), 422–426 (1970)
11. Breslin, J., Decker, S., Harth, A., Bojars, U.: SIOC: an approach to connect web-based communities. International Journal of Web Based Communities 2(2), 133–142 (2006)
12. Brickley, D., Miller, L.: FOAF Vocabulary Specification 0.97, Namespace Document (January 1, 2010), http://xmlns.com/foaf/spec/
13. Carroll, J., Bizer, C., Hayes, P., Stickler, P.: Named graphs. Web Semantics: Science, Services and Agents on the World Wide Web 3(4), 247–267 (2005)
14. Cheng, G., Qu, Y.: Searching Linked Objects with Falcons: Approach, Implementation and Evaluation. International Journal on Semantic Web and Information Systems 5(3), 49–70 (2009)
15. Clark, K.G., Feigenbaum, L., Torres, E.: SPARQL Protocol for RDF, W3C Recommendation (January 15, 2008), http://www.w3.org/TR/rdf-sparql-protocol/
16. D' Aquin, M., Baldassarre, C., Gridinoc, L., Angeletou, S., Sabou, M., Motta, E.: Characterizing Knowledge on the Semantic Web with Watson. In: Proceedings of the 5th International Workshop on Evaluation of Ontologies and Ontology-based Tools (EON), Busan, Korea, pp. 1–10 (2007)
17. Erling, O., Mikhailov, I.: RDF Support in the Virtuoso DBMS. In: Pellegrini, T., Auer, S., Tochtermann, K., Schaffert, S. (eds.) Networked Knowledge - Networked Media, pp. 7–24. Springer, Heidelberg (2009)
18. Flesca, S., Furfaro, F., Pugliese, A.: A Framework for the Partial Evaluation of SPARQL Queries. In: Proceedings of the 2nd International Conference on Scalable Uncertainty Management, Naples, Italy, pp. 201–214 (2008)
19. Franz, T., Schultz, A., Sizov, S., Staab, S.: TripleRank: Ranking SemanticWeb Data By Tensor Decomposition. In: Proceedings of the 8th International Semantic Web Conference, Chantilly, VA, USA, pp. 213–228 (2009)
20. Görlitz, O., Sizov, S., Staab, S.: PINTS: Peer-to-Peer Infrastructure for Tagging Systems. In: Proceedings of the 7th International Workshop on Peer-to-Peer Systems (IPTPS), Tampa Bay, Florida, USA (2008)
21. Gueret, C., Oren, E., Schlobach, S., Schut, M.: An Evolutionary Perspective on Approximate RDF Query Answering. In: Proceedings of the 2nd International Conference on Scalable Uncertainty Management, Naples, Italy, pp. 215–228 (2008)
22. Guo, Y., Pan, Z., Heflin, J.: LUBM: A benchmark for OWL knowledge base systems. Web Semantics: Science, Services and Agents on the World Wide Web 3(2-3), 158–182 (2005)
23. Haas, L., Kossmann, D., Wimmers, E.L., Yang, J.: Optimizing Queries across Diverse Data Sources. In: Proceedings of the 23rd International Conference on Very Large Data Bases, Athens, Greece, pp. 276–285 (1997)

24. Harris, S., Lamb, N., Shadbolt, N.: 4store: The Design and Implementation of a Clustered RDF Store. In: Proceedings of the 5th International Workshop on Scalable Semantic Web Knowledge Base Systems (SSWS 2009), Chantilly, VA, USA, pp. 94–109 (2009)
25. Harris, S., Seaborne, A.: SPARQL Query Language 1.1, W3C Working Draft (January 26, 2010), `http://www.w3.org/TR/sparql11-query/`
26. Harth, A., Hogan, A., Delbru, R., Umbrich, J., O'Riain, S., Decker, S.: SWSE: Answers Before Links! In: Proceedings of Semantic Web Challenge (2007)
27. Harth, A., Hose, K., Karnstedt, M., Polleres, A., Sattler, K.U., Umbrich, J.: Data Summaries for On-Demand Queries over Linked Data. In: Proceedings of the 19th International World Wide Web Conference, Raleigh, NC, USA, pp. 411–420 (2010)
28. Harth, A., Umbrich, J., Hogan, A., Decker, S.: YARS2: A Federated Repository for Querying Graph Structured Data From The Web. In: Proceedings of the 6th International Semantic Web Conference, Busan, Korea, pp. 211–224 (2007)
29. Hartig, O., Bizer, C., Freytag, J.C.: Executing SPARQL Queries over the Web of Linked Data. In: Proceedings of the 8th International Semantic Web Conference, Chantilly, VA, USA, pp. 293–309 (2009)
30. Heimbigner, D., McLeod, D.: A Federated Architecture for Information Management. ACM Transactions on Information Systems 3(3), 253–278 (1985)
31. Hogenboom, A., Milea, V., Frasincar, F., Kaymak, U.: RCQ-GA: RDF Chain Query Optimization Using Genetic Algorithms. In: Proceedings of the 10th International Conference on E-Commerce and Web Technologies, Linz, Austria, pp. 181–192 (2009)
32. Josifovski, V., Schwarz, P., Haas, L., Lin, E.: Garlic: A New Flavor of Federated Query Processing for DB2. In: Proceedings of the 2002 ACM SIGMOD International Conference on Management of Data, Madison, Wisconsin, pp. 524–532 (2002)
33. Kossmann, D.: The State of the Art in Distributed Query Processing. ACM Computing Surveys 32(4), 422–469 (2000)
34. Kossmann, D., Stocker, K.: Iterative dynamic programming: a new class of query optimization algorithms. ACM Transactions on Database Systems (TODS) 25(1), 43–82 (2000)
35. Langegger, A., Wöß, W., Blöchl, M.: A Semantic Web Middleware for Virtual Data Integration on the Web. In: Proceedings of the 5th European Semantic Web Conference, Tenerife, Canary Islands, Spain, pp. 493–507 (2008)
36. Maduko, A., Anyanwu, K., Sheth, A., Schliekelman, P.: Graph Summaries for Subgraph Frequency Estimation. In: Proceedings of the 5th European Semantic Web Conference, Tenerife, Canary Islands, Spain (2008)
37. Manola, F., Miller, E.: RDF Primer, W3C Recommendation (February 10, 2004), `http://www.w3.org/TR/rdf-primer/`
38. Miles, A., Matthews, B., Wilson, M., Brickley, D.: SKOS Core: Simple Knowledge Organisation for the Web. In: Proceedings of the 3rd European Semantic Web Conference, Budva, Montenegro, pp. 95–109 (2006)
39. Moerkotte, G., Neumann, T.: Analysis of Two Existing and One New Dynamic Programming Algorithm for the Generation of Optimal Bushy Join Trees without Cross Products. In: Proceedings of the 32nd International Conference on Very Large Data Bases, Seoul, Korea, pp. 930–941 (2006)
40. Muralikrishna, M., DeWitt, D.: Equi-Depth Histograms For Estimating Selectivity Factors For Multi-Dimensional Queries. In: Proceedings of the 1988 ACM SIGMOD International Conference on Management of Data, pp. 28–36. ACM Press, Chicago (1988)
41. Neumann, T., Weikum, G.: RDF-3X: a RISC-style Engine for RDF. In: Proceedings of the 34th International Conference on Very Large Data Bases, Auckland, New Zealand, pp. 647–659 (2008)

42. Neumann, T., Weikum, G.: Scalable Join Processing on Very Large RDF Graphs. In: Proceedings of the 35th SIGMOD International Conference on Management of Data, Providence, RI, USA, pp. 627–640 (2009)
43. Ning, X., Jin, H., Wu, H.: RSS: A framework enabling ranked search on the semantic web. Information Processing and Management 44(2), 893–909 (2007)
44. Ntarmos, N., Triantafillou, P., Weikum, G.: Counting at Large: Efficient Cardinality Estimation in Internet-Scale Data Networks. In: Proceedings of the 22nd International Conference on Data Engineering, Atlanta, Georgia, USA (2006)
45. Oren, E., Delbru, R., Catasta, M., Cyganiak, R., Stenzhorn, H., Tummarello, G.: Sindice.com: A Document-oriented Lookup Index for Open Linked Data. International Journal of Metadata, Semantics and Ontologies 3(1), 37–52 (2008)
46. Pérez, J., Arenas, M., Gutierrez, C.: Semantics and Complexity of SPARQL. ACM Transactions on Database Systems 34(3), 1–45 (2009)
47. Prud'hommeaux, E.: SPARQL Federation Extensions 1.1, Editor's Draft (March 25, 2010), `http://www.w3.org/2009/sparql/docs/fed/service`
48. Prud'hommeaux, E., Seaborne, A.: SPARQL Query Language for RDF, W3C Recommendation (January 15, 2008), http://www.w3.org/TR/rdf-sparql-query/
49. Quilitz, B., Leser, U.: Querying Distributed RDF Data Sources with SPARQL. In: Proceedings of the 5th European Semantic Web Conference, Tenerife, Canary Islands, Spain, pp. 524–538 (2008)
50. Schenk, S., Saathoff, C., Staab, S., Scherp, A.: SemaPlorer – Interactive Semantic Exploration of Data and Media based on a Federated Cloud Infrastructure. Journal on Web Semantics: Science, Services and Agents on the World Wide Web 7(4), 298–304 (2009)
51. Schenk, S., Staab, S.: Networked Graphs: A Declarative Mechanism for SPARQL Rules, SPARQL Views and RDF Data Integration on the Web. In: Proceeding of the 17th International World Wide Web Conference, Beijing, China, pp. 585–594 (2008)
52. Schmidt, M., Hornung, T., Lausen, G., Pinkel, C.: SP^2Bench: A SPARQL Performance Benchmark. In: Proceedings of the 25th International Conference on Data Engineering, Shanghai, pp. 222–233 (2009)
53. Schmidt, M., Meier, M., Lausen, G.: Foundations of SPARQL Query Optimization (2008); Arxiv preprint arXiv:0812.3788
54. Selinger, P., Astrahan, M., Chamberlin, D., Lorie, R., Price, T.: Access Path Selection in a Relational Database Management System. In: Proceedings of the ACM SIGMOD International Conference on Management of Data, Boston, MA, USA, pp. 23–34 (1979)
55. Sheth, A., Larson, J.: Federated Database Systems for Managing Distributed, Heterogeneous, and Autonomous Databases. ACM Computing Surveys 22(3), 183–236 (1990)
56. Stuckenschmidt, H., Vdovjak, R., Houben, G.J., Broekstra, J.: Index Structures and Algorithms for Querying Distributed RDF Repositories. In: Proceedings of the 13th International World Wide Web Conference, New York, NY, USA, pp. 631–639 (2004)
57. Tomasic, A., Raschid, L., Valduriez, P.: Scaling Heterogeneous Databases and the Design of Disco. In: Proceedings of the 16th International Conference on Distributed Computing Systems, Hong Kong, pp. 449–457 (1996)
58. Tran, T., Haase, P., Studer, R.: Semantic Search – Using Graph-Structured Semantic Models for Supporting the Search Process. In: Proceedings of the 17th International Conference on Conceptual Structures, Moscow, Russia, pp. 48–65 (2009)
59. Tran, T., Wang, H., Haase, P.: Hermes: Data Web search on a pay-as-you-go integration infrastructure. Web Semantics: Science, Services and Agents on the World Wide Web 7(3), 189–203 (2009)
60. Weiss, C., Karras, P., Bernstein, A.: Hexastore: Sextuple Indexing for Semantic Web Data Management. In: Proceedings of the 34th International Conference on Very Large Data Bases, Auckland, New Zealand, pp. 1008–1019 (2008)

Chapter 6
Queries over Web Services

Efthymia Tsamoura, Anastasios Gounaris, and Yannis Manolopoulos

Aristotle University of Thessaloniki,
Thessaloniki, Greece
{etsamour,gounaria,manolopo}@csd.auth.gr

1 Introduction

Nowadays, technologies such as grid and cloud computing infrastructures and service-oriented architectures have become adequately mature and have been adopted by a large number of enterprizes and organizations [2,19,36]. A Web Service (WS) is a software system designed to support interoperable machine-to-machine interaction over a network and is implemented using open standards and protocols. WSs became popular data management entities; some of their benefits are interoperability and reuseability.

Seeking to benefit from the above opportunities, the web and grid data management infrastructures are moving towards a service oriented architecture by putting their databases behind WSs, thereby providing a well-documented, interoperable method of interacting with their data (e.g., [5,32]). Furthermore, data not stored in traditional databases can be made available via WSs. As a consequence, there is a growing interest in systems that are capable of processing complex queries (i.e., tasks) spanning services deployed on remote resources. The services can perform two operations; they either perform processing of data, or they play the role of a wrapper that retrieves data from a resource.

Currently, two classes of infrastructures that employ WSs to process data have been developed, namely the WS query infrastructures and the workflow management systems (WfMSs). The former process SQL-like queries or search queries over information sources (e.g. [5,4,42]). Like traditional database management systems, they perform the following tasks in order to answer a submitted query: query translation, service selection and query optimization. In the first two steps, the appropriate services that can correctly answer the submitted query are selected (either with or without user interaction), while the final step, which is the main topic of this chapter, aims to provide an efficient service execution plan. The other category comprises WfMSs, where the workflow components are services (e.g., [33,23]). In WfMSs the user has to select the services to process the data of interest, the location of the input data (which are either extracted by a service that polls a resource or they form a data stream) and the service invocation order, which is fixed. Languages such as BPEL4WS have emerged for specifying WS composition in workflow-oriented scenarios [1].

A. Vakali & L.C. Jain (Eds.): New Directions in Web Data Management 1, SCI 331, pp. 139–169.
springerlink.com

An example of a problem of optimization queries over WSs is given below. We assume that WSs provide an interface of the form $WS : X \rightarrow Y$, where X and Y are sets of attributes, i.e., given values for attributes in X, WS returns values for the attributes in Y, as shown in the following example adapted from [42]. In the generic case, the input tuples may have more attributes than X, while attributes in Y are appended to the existing ones.

Example 1. Suppose that a company wants to obtain a list of email addresses of potential customers selecting only those who have a good payment history for at least one card and a credit rating above some threshold. The company has the right to use the following WSs that may belong to third parties, the first of which contains a database of person ids.

$WS_1 : \emptyset \rightarrow SSN\ id\ \ (ssn)$
$WS_2 : SSN\ id\ \ (ssn, threshold) \rightarrow\ credit\ rating\ \ (cr)$
$WS_3 : SSN\ id\ \ (ssn) \rightarrow\ credit\ card\ numbers\ \ (ccn)$
$WS_4 : card\ number\ \ (ccn, good) \rightarrow good\ history\ \ (gph)$
$WS_5 : SSN\ id\ \ (ssn) \rightarrow\ email\ addresses\ \ (ea)$

There are multiple valid orderings to perform this task, although several precedence constraints exist: WS_1 must always be at the beginning and WS_3 must precede WS_4. The optimization process aims at deciding on the optimal (or near optimal) ordering under given optimization goals. When there are multiple logically equivalent services for the same task (e.g., there are two services containing email addresses at distinct places) or the physical placement of a service is flexible, then the problem becomes more complex. □

In this chapter we will discuss several different flavors of queries over WSs and the corresponding optimization algorithms. Note that some of these cases can be reduced to problems that have been examined in the context of traditional database queries in a straightforward manner. Traditional database solutions for such cases can be easily transferred to our setting by replacing database operators with WSs; for this reason, throughout the text, we will use the terms operators and services interchangeably. For example, the problem of optimal ordering of centralized WSs with a view to minimizing the response time may resemble the problem of ordering commutative filters in pipelined queries with conjunctive predicates [24,22], in the sense that the calls to WSs may be treated in the same way as expensive predicates. Note that ordering some types of relational joins can be reduced to the same problem, as well [7].

However, reducing the problem of optimizing queries over WSs to the problem of optimizing traditional queries is not always feasible because there are also many substantial differences, and, as such, several optimization problems encountered in queries over WSs have not been investigated in traditional query processing. These differences stem from the fact that, in queries over WSs, there may exist precedence constraints between the WSs, selectivities may be higher than 1 (e.g., WS_3 in the example can return more than one tuple) and, typically, the execution of queries over WSs typically takes place in a both distributed and parallel manner.

1.1 Optimization Problems of Queries over WSs

In this chapter, we examine several distinct query optimization problems that can be broadly classified into four main categories, namely operator ordering, operator scheduling, tuple routing and data transfer planning. For each problem, we present some of the known solutions. Note that these solutions are not directly comparable with each other since they deal with different problems.

Operator ordering where the goal is to build an operator (or WS) execution plan that minimizes a pre-defined criterion by defining an appropriate partial or total ordering of the operators. In other words, the optimization decisions relate to the ordering of operators in the execution plan exclusively and issues such as allocation of operators to resources do not apply. Note that the ordering need not be linear. Problems that fall into this category assume that necessary metadata (e.g., operator cost per input tuple, selectivity, etc.) are available and in addition, the operators are pre-allocated on host machines. A well-known problem is the min-cost operator ordering problem. Given a set of operators, the aim is to define an ordering of the operators so that all input queries are evaluated with the minimum total execution cost of operators. Optimization criteria will be discussed in more detail in Sec. 2.

Tuple routing which is a generalization of operator ordering in the sense that not only a single operator plan to be followed by all input tuples is created. The alternative approach, advocated by tuple routing techniques, may route input tuples through different plans, which are also termed as interleaving plans [13]. A set of interleaving plans consists of multiple simultaneously active operator plans, each of which processes different partitions of the original input tuple set. When a new tuple enters the system it is routed to one of these plans, according to a probability weight.

Operator scheduling where the goal is to decide the processor on which each service is evaluated. Problems of this category appear when the system is also responsible for resource allocation. It is assumed that the system is capable of performing dynamic service deployment before the execution of the query and there are multiple choices regarding the host nodes for each service. Operator scheduling can be examined either in conjunction with operator ordering or in isolation. In the latter case, operator ordering has been fixed in a previous step.

Data transfer planning where the focus is shifted to data transmission. The aforementioned query optimization problems are operation centric, i.e., they define the operator execution order and/or the operator location. In the data transfer problems, the primary concern is to optimize data transmissions. As such, these problems emphasize more on scheduling the data transmission operations, or on the specification of the amount of data exchanged between the hosts. Obviously, operator scheduling and data transfer planning problems are met only in parallel or distributed environments, whereas operator ordering and tuple routing problems are encountered in centralized settings, as well.

1.2 Chapter Contributions and Structure

The contribution of this chapter is twofold. First, it presents a detailed overview of the problems encountered in the optimization of queries over WSs. The problems do not differ only in their nature as detailed above, but also in regard to the type of queries, type of services or operators and the exact execution environment to which they are tailored. This discussion results in the development of a taxonomy-like classification of the problems in WS queries that appears in Sec. 2. Second, this chapter discusses state-of-the-art solutions to distinct flavors of the problem of optimizing queries over WSs in Sec. 3. Especially for the problem of minimizing the response time in decentralized pipelined queries, a novel algorithm is presented. A comparison of the key properties of the different solutions appears in Sec. 3.6, while Sec. 4 concludes the chapter.

2 Different Aspects of the Problem of Optimizing WS Queries

Before probing into advanced query optimization algorithms that are relevant to queries over WSs, we must first discuss the factors of the problem, which greatly affect its complexity. These factors refer to the execution environment, the type of input queries, the type of operators involved and the query optimization criteria and are common to all kinds of problems mentioned in Sec. 1. The taxonomy presented here aims at providing a complete view of these factors in a systematic way.

2.1 Execution Environment

We are mainly interested in queries in parallel and distributed query execution environments, like those in [35], since these environments are more common in WS queries. However, a great number of algorithms originally proposed for centralized environments are still relevant. In such single-processor systems, only a central node evaluates input queries, although the queries may process data from multiple distributed resources. A distributed environment, such as the Internet and the grid [19], consists of multiple, possible heterogeneous, independent and potentially autonomous sites that are loosely connected via a wide-area network. On the other hand, a parallel environment consists of multiple, homogeneous processors and data resources spread over a local network. As such, the communication cost may dominate the query execution process in a distributed environment, which, usually, is not the case in a parallel setting. Nevertheless, the similarities between parallel and distributed systems are more significant compared to their differences; so, we prefer to distinguish between centralized and non-centralized (i.e., either parallel or distributed) systems, only.

A parallel or distributed environment may be either static or dynamic. In the latter case, the environmental characteristics, such as the number of available processors, the processor workload, the network traffic, etc., may change over time rendering the problem of query optimization more challenging. Query optimization in dynamic environments, also called adaptive query processing [15], has been a topic of investigation since late 70s [18]; however, the problem has received renewed attention in the last decade. The vast majority of works on adaptive query processing, like those mentioned above, deal with changes in the operator characteristics and the input data rather than changes in the execution environment; only a few exceptions to this are known (e.g., [20]). Centralized environments are considered to be static; of course this is not always true, e.g., the amount of available memory may be subject to unpredictable changes, but the dynamicity of the environment can be safely overlooked when the resource characteristics we are mostly interested in, such as processing cost per tuple, usually play a minor role in optimization.

In addition, a distributed or parallel environment may utilize parallelism with a view to speeding up and scaling up query execution [17]. Three types of parallelism have been identified in parallel query processing, namely independent, partitioned and pipelined parallelism. In independent parallelism, query operators none of which use data produced by the others, may run simultaneously on distinct machines. In pipelined parallelism, data already processed by an operator may be processed by a subsequent operator in the pipeline, at the same time as the sender operator processes new data. Finally, partitioned parallelism refers to running several instances of the same operator on different machines concurrently, with each instance only processing a partition of the same original data set.

The three aforementioned forms of parallelism can co-exist within a single query execution plan. For instance, in the introductory example, WS_2 and WS_3 can process in parallel output data items of WS_1; this corresponds to independent parallelism. Also, WS_1 and WS_3 can be active simultaneously, i.e., WS_3 processes output tuples of WS_1, while the latter keeps generating new tuples; this corresponds to pipelined parallelism. Finally, consider a scenario where WS_3 is physically deployed

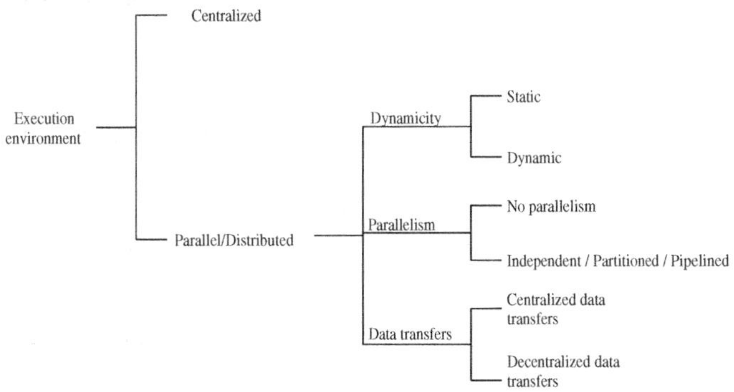

Fig. 1. Diagram of the different aspects regarding the execution environment

on two nodes, each processing half of the tuples of WS_1; in that case, partitioned parallelism is applied, as well. Obviously, parallelism can yield significant benefits only in multi-processor, parallel or distributed environments.

Distributed environments are also differentiated regarding the type of management of intermediate data transfers. In multi-processor systems, query operators can be placed and evaluated anywhere across the network. Regarding the intermediate data transfers, in a centralized data transfer approach, the intermediate data is transferred between resources via a central point. On the other hand, in a decentralized managed data transfer approach, processors exchange data directly. Since data is transferred from the source resource to the destination directly, the bottleneck problem caused in centralized approaches is ameliorated and these approaches are characterized by lower transmission times. Fig. 1 summarizes the different aspects regarding the execution environment.

2.2 Input Queries

The optimization algorithms that apply to queries over WSs either optimize each input query separately or optimize multiple queries simultaneously. In the latter case, they try to benefit from the overlap regarding the data resources they access, or even the constituent predicates. Multi-query optimization algorithms try to leverage this overlap in order to minimize the execution, communication and I/O cost. Additionally, input queries can be classified with respect to their time duration into continuous or non-continuous. Continuous queries are persistent queries that allow users to receive new results when they become available [43]. They are mainly met in streaming environments, where new data is continuously supplied and passed to WS sets for further processing. On the other hand, the non-continuous ad-hoc queries are executed on finite data. Optimization techniques that treat each tuple separately can be applied to both continuous and ad-hoc finite queries. An example of a continuous query over WSs is the following (adapted from [12]), where it is assumed that separate WSs are responsible for checking the price variations of Dell, Micron and Intel stocks:

"Notify me whenever the price of Dell or Micron stock drops by more than 5% and the price of Intel stock remains unchanged over the next three months."

Regarding their type, input queries can be expressed as traditional SQL-like database queries in the form of select-project-join (SPJ) and aggregates, or as search, information retrieval queries over information resources. Search queries are typically unstructured and often ambiguous; users submit one or more keywords to a search engine and the search engine returns approximate, i.e., incomplete answers with information that is related to the keywords provided in decreasing order of relevance. Fig. 2 provides a diagram of the different query aspects.

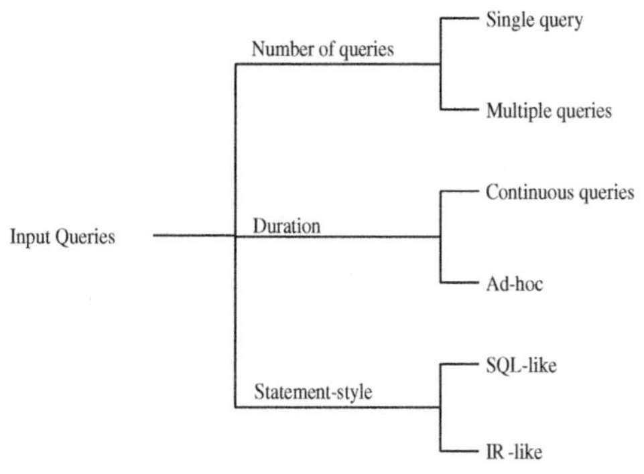

Fig. 2. Diagram of the different aspects regarding input queries

2.3 Input Operators

The type of input queries is also strongly correlated to the type of operators in the query execution plan. The operator attributes that are of interest include selectivity and precedence constraints (see Fig. 3). Selectivity is defined as the average ratio of output and input tuples. A WS that receives as input a country name and returns a list of major cities has average selectivity above one, and another service that, for the same input, returns just the capital has selectivity equal to one. Similarly, a service that may receive the name of any city in the world and returns airport codes only if the given city is nearby an airport has average selectivity below one, since, worldwide, there are fewer airports than cities. The operators in a query can be selective, i.e., their selectivity is between 0 and 1, or proliferative, i.e., their selectivity is greater than 1. IR-style services are typically characterized by high average selectivity values: given a single tuple containing a keyword, multiple data items are returned.

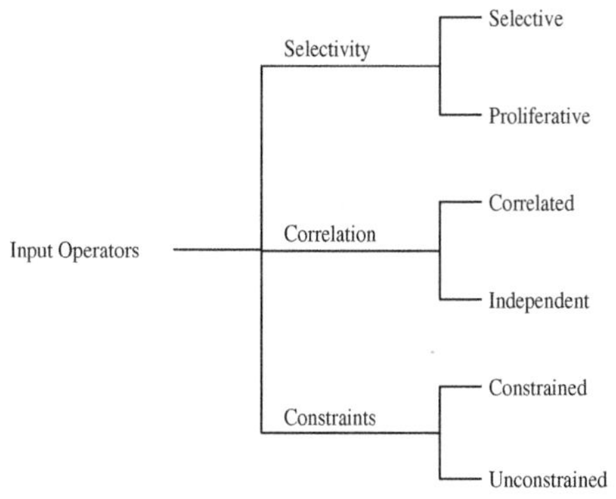

Fig. 3. Diagram of the different aspects regarding input operators

Moreover, the operators are considered to be correlated when their selectivity depends on the operators upstream in the query plan. If the selectivities are independent of the ordering, then the operators are called independent. Note that the selectivity of an independent operator may be correlated with the values of the input attributes, which is the case in [9]. A last parameter that categorizes the operators is the existence or not of prerequisite operators. More specifically, a constrained operator cannot be executed before the completion of its prerequisite operators, in contrast to an unconstrained one. The prerequisite operator O_j of an operator O_i is denoted by $O_j \prec O_i$.

2.4 Optimization Criteria

A critical factor in the optimization process is the exact optimization goal. Multiple criteria exist, such as maximizing query throughput, minimizing monetary cost, energy consumption, etc. In this chapter, we focus on two aspects, namely the minimization of the total query execution cost and the minimization of the query response time. Minimization of the total execution cost can be split in two parts. In centralized environments, execution cost encapsulates the cost for processing the operators and the disk I/Os. In distributed environments, the cost for transferring data among the operators is also considered. The minimization of execution cost aims at minimizing the sum of the processing and transmission cost for all operators in the query plan.

However, the opportunities imposed by parallelization have moved the interest to the optimization of other criteria, such as the response time, i.e., the time needed to produce the full result set. In a pipelined parallel environment, all operators process data simultaneously. As such, minimizing the query response time is equivalent to the minimization of the execution time of the longest running operator (often referred to as the bottleneck operator) instead of the sum of the execution times of all operators. When the query is evaluated with the help of interleaving plans (see Sec. 1.1), then the minimization of response time can be expressed as the maximization of the tuple flow [13]. These optimization criteria are depicted in Fig. 4.

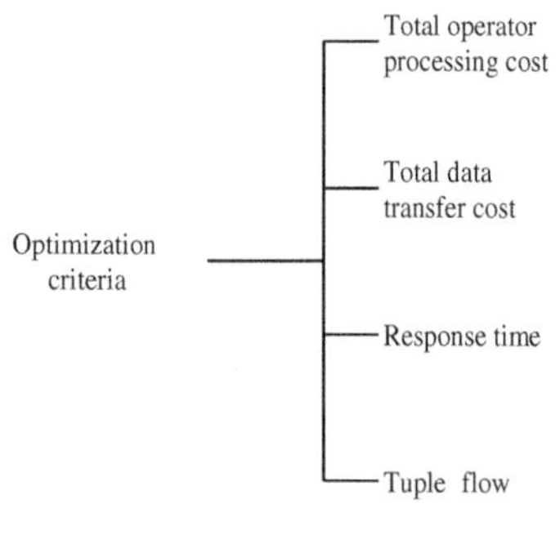

Fig. 4. Optimization criteria

3 Optimization Approaches

This section studies state-of-the-art algorithms for the problems presented in Sec. 1.1. The section starts by presenting some operator ordering problems in both static and adaptive execution environments and continues with tuple routing, scheduling and data transfer planning problems.

Independently of the execution environment, in the problems that are presented, except the data transfer planning ones, the data to be processed is either streamed by a single data resource or extracted from a database and then sent to subsequent services for processing. On the other hand, in the presented data transfer planning problems, we consider two different data resource models. In the first case, multiple data resources transfer data to a centralized processing component, while in the second case, data reside on traditional databases that are disparate across a network.

In order to answer a query involving calls to multiple services, the following actions must be performed by a query management component. First, the candidate services that may take place during the query execution phase must be selected. After that, statistics, regarding the per-tuple processing cost and the selectivity of the services, as well as the network status, must be gathered. This data is utilized by an optimization component that builds a feasible and efficient (in terms of a pre-selected optimization criterion) service ordering. It is assumed that the algorithms that deal with operator ordering and tuple routing problems exploit such query management components.

3.1 Operator Ordering Problems in a Static Environment

The operator ordering problems that are studied in the current subsection deal with static execution environments. In Sec. 3.1.1, we study problems, where the optimization objective is the query response time minimization, while in Sec. 3.1.2, we present problems, where the objective is the minimization of the per tuple total execution cost.

The services in Sec. 3.1 are considered to provide an interface of the form $WS: X \rightarrow Y$, where X and Y are sets of input and output attributes, respectively. Each WS typically performs operations such as filtering out data items that are not relevant to the query, transforming data items, or appending additional information to each input tuple.

3.1.1 Minimizing the Response Time

Srivastava *et al.* are among the pioneers that deal with query optimization when the data resources and the operators that process data are implemented as WSs. They consider a parallel and static execution environment, in which data is pipelined among services that are placed in arbitrary places. To this end, they propose a WSMS that, given an SQL-like input query, undertakes the task to produce an appropriate ordering of the services, in order to minimize the query response time. Query execution proceeds as follows. The output of one WS is returned to the WSMS and the latter redirects the received tuples to a subsequent WS, finally producing the query results. After giving a brief description of the execution environment, we present a formal problem definition utilizing the term operator instead of service.

More specifically, given an ad-hoc SPJ query Q that is defined over a set of N operators $O = \{O_1, O_2, \ldots, O_N\}$, the goal is to identify an operator ordering P for all input tuples that minimizes the response time of the query, in a parallel and static execution environment in which data is pipelined among the operators. Since the operators are executed in parallel, the maximum rate at which input tuples can be processed through the pipelined plan P is determined by the bottleneck operator (see Sec. 2.4). For every input tuple to P, the average number of tuples that an operator O_i needs to process is given by

$$R_i(P) = \prod_{k|O_k \in P_i(P)} \sigma_k \tag{1}$$

where $P_i(P)$ is the set of operators that are invoked before O_i in the plan P and σ_i is the service selectivity. The average processing time required by operator O_i per input tuple is $R_i(P)c_i$, where c_i is the per tuple processing cost of O_i. Since the cost of a plan is determined by the operator with the maximum processing time per input tuple, the bottleneck cost of a plan P is given by

$$cost(P) = \max_{1 \le i \le N} (R_i(P) \cdot c_i) \tag{2}$$

Srivastava *et al.* have proposed a greedy algorithm for the special case where the intermediate data transfers are centralized [42]. The operators are assumed to be independent, whereas arbitrary selectivity values and existence of precedence constraints are supported. In the produced plans, the output of an operator may be fed to multiple operators simultaneously. Starting from an empty operator plan, in every iteration of the algorithm, the next operator O_r to be appended to P is the one that incurs the minimum processing cost per tuple. In order to find the minimum cost of appending O_r to P, the best cut in P is found, such that on placing edges from the operators in the cut to O_r, the incurred cost is minimized. As such, the problem is reduced to a network flow problem [11]. The worst case complexity of the algorithm is $O(N^5)$ and the algorithm is provably optimal. For selective operators, the complexity is significantly lower since the optimal plan P is a linear ordering of the operators by increasing cost, ignoring their selectivity. For proliferative services, the produced plans may be parallel, i.e., a partial ordering is produced.

Example 2. Let $O = \{O_1, \ldots, O_{10}\}$ be a set of 10 operators with corresponding costs and selectivities shown in Table 1 and 2, respectively. Since all operators are selective, the proposed algorithm orders them by increasing processing cost. Thus, the optimal ordering that minimizes Eq.(2) according to [42] is $P = \{O_1 O_5 O_2 O_7 O_4 O_{10} O_8 O_9 O_3 O_6\}$. □

A drawback of this algorithm is that it does not take the potentially heterogeneous communication links between the operators into account. This is significant when the execution is decentralized, given also that the communication cost may be the

Table 1. Costs of operators in Example 2

O_i	1	2	3	4	5	6	7	8	9	10
c_i	2	7	12	8	4	16	7	10	10	9

Table 2. Selectivities of operators in Example 2

O_i	1	2	3	4	5	6	7	8	9	10
σ_i	0.8	0.7	0.9	0.3	0.5	0.6	0.4	0.1	0.6	0.7

dominant cost. In [42], it is assumed that the output of an operator is fed to the subsequent operators indirectly, through a central management component thus annihilating the need to consider the different communication costs explicitly. Tsamoura *et al.* address the afore-mentioned limitation by proposing a novel efficient algorithm for the optimal total ordering of operators, when the intermediate result transfers are decentralized and the communication costs between the operators may differ [46].

Let $t_{i,j}$ be the time needed to transfer a tuple from operator O_i to O_j. Similarly to [42], there is no limitation regarding the operator selectivities and the existence of precedence constraints; however, the selectivities are assumed to be independent, as well. The response time of a linear operator ordering S is given by the bottleneck cost metric in accordance to [42] with $t_{i,j}$ factored in:

$$cost(S) = \max_{1 \leq i \leq N} R_i(S)(c_i + \sigma_i t_{i,i+1}), \tag{3}$$

where $t_{N,N+1} = 0$. $T_{i,j} = c_i + t_{i,j}\sigma_i$ is the aggregate cost of O_i with respect to O_j. The above formula implies that in general $T_{i,j} \neq T_{j,i}$, since c_i and σ_i values may differ from c_j and σ_j values. Note that if $t_{i,j}$ is equal for all service pairs, the problem can be solved in polynomial time, as shown in [42].

The proposed algorithm is based on the branch-and-bound optimization approach. It proceeds in two phases, namely the expansion and the pruning one. During expansion, new operators are appended to a partial operator ordering C, while during the latter phase, operators are pruned from C with a view to exploring additional orderings. The decision whether to append new operators or prune existing ones from a partial plan C is guided by two cost metrics, ε and $\overline{\varepsilon}$ respectively. The former corresponds to the bottleneck cost of C, and is given by Eq. (3), while the latter is the maximum possible cost that may be incurred by operators not currently included in C:

$$\overline{\varepsilon} = \max_{l,r} \left\{ \begin{array}{ll} \left(\prod_{j|O_j \in C} \sigma_j\right) T_{l,r}, & O_l \notin C,\ O_r \notin C \\ \left(\prod_{j=0}^{l-1} \sigma_j\right) T_{l,r}, & O_l: \ \textit{last operator in } C,\ O_r \notin C \end{array} \right\} \tag{4}$$

The algorithm consists of the following simple steps. Starting with an empty plan C and an empty optimal linear plan S with infinity bottleneck cost, in every iteration of the algorithm, the parameters ε and $\overline{\varepsilon}$ are computed. If the bottleneck cost ε of

C is lower than $\overline{\varepsilon}$, then a new operator is appended to C; this operator is the one having the minimum aggregate cost with respect to the last operator in C. If the bottleneck cost ε of the current plan C is higher than or equal to the bottleneck cost ρ of the best plan found so far S, then the operators in C after the bottleneck service, including the latter, are pruned. Finally, whenever the condition $\overline{\varepsilon} \leq \varepsilon < \rho$ is met, a new solution is found. That condition implies that the ordering of the services that are not yet included in C does not affect its bottleneck cost. As a consequence, all plans with prefix the partial plan C have the same bottleneck cost. So, a candidate optimal solution S is found that consists of the current plan C followed by the rest of the services in any order. The bottleneck cost ρ of the best plan found so far S is set to $\rho = \varepsilon$. The last solution is the optimal one. The detailed description of the algorithm, along with the proofs of correctness and optimality can be found in [46]. Furthermore, detailed real-world ([45]) and simulation ([46]) evaluation has shown that the proposed algorithm can yield significant performance improvements (of an order of magnitude in many cases).

The following example demonstrates the steps of the algorithm proposed in [46] for minimizing the response time in a distributed and static environment, which employs pipelining during query execution.

Example 3. Let us assume that the operators in Example 2 are allowed to communicate directly with each other and the network connections are heterogeneous. The corresponding aggregate costs of the operators are shown in Table 3. For example, the cell $T_{1,2}$ of Table 3 is evaluated as $T_{1,2} = c_1 + t_{1,2}\sigma_1 = 2 + 65 * 0.8$, where 0.8 is the per cost to transfer an tuple from O_1 directly to O_2.

Fig. 5 shows the partial plans at the end of each iteration. Initially, the plans C and S are empty and the bottleneck cost of S is set to ∞. The algorithm starts by identifying the operator pair, which incurs the minimum bottleneck cost. The corresponding operators are O_1 and O_7. After that, $C = O_1O_7$. In the second iteration, since $\varepsilon = 8 < \overline{\varepsilon} = \sigma_1 \times T_{7,3} = 36$ and $\varepsilon < \rho = \infty$, a new operator is appended to C, the one having the minimum aggregate cost with respect to O_7; that operator is O_9.

Table 3. Aggregate cost matrix **T**

i \ j	1	2	3	4	5	6	7	8	9	10
1	-	54	35	42	14	50	8	33	17	10
2	52	-	18	33	47	40	69	37	42	43
3	49	26	-	60	68	74	98	40	66	57
4	23	19	24	-	17	46	21	9	42	27
5	11	33	35	19	-	10	40	52	14	32
6	52	44	57	91	23	-	44	22	72	46
7	10	43	45	24	36	26	-	35	17	19
8	14	15	14	11	20	11	17	-	17	16
9	21	40	46	78	22	66	25	48	-	79
10	16	45	44	53	48	44	29	47	90	-

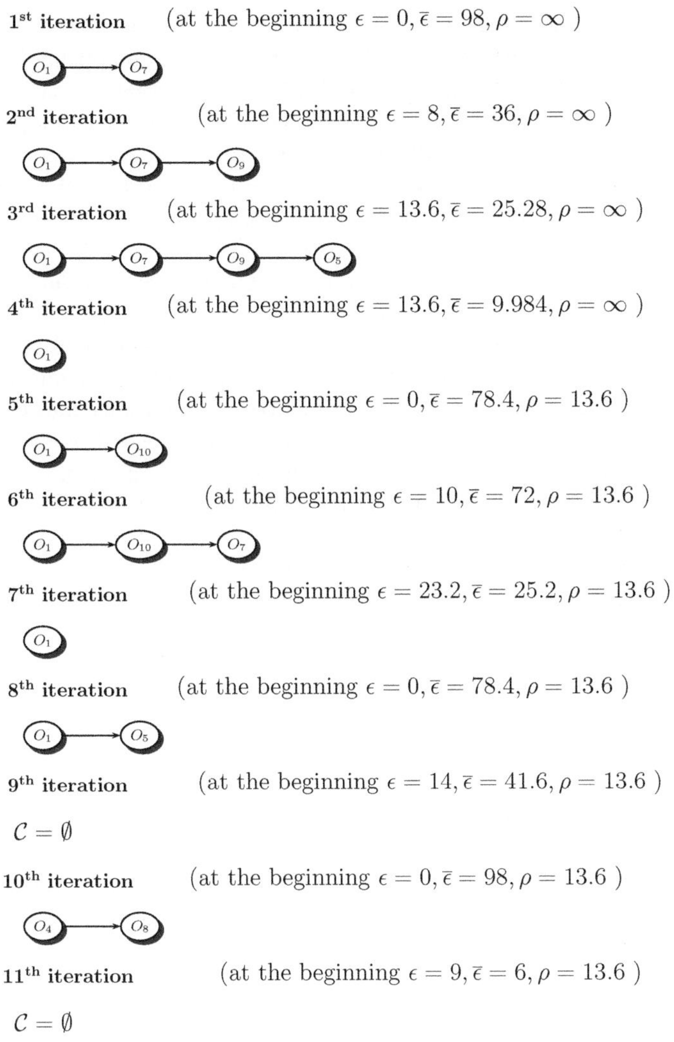

Fig. 5. The steps in Example 3

In the third iteration, since $\varepsilon = 13.6 < \overline{\varepsilon} = \sigma_1 \times \sigma_7 \times T_{9,10} = 25.28$ and $\varepsilon < \rho = \infty$ the operator O_5 is appended to C forming the partial plan $C = O_1 O_7 O_9 O_5$. Now, since $\varepsilon = 13.6 > \overline{\varepsilon} = \sigma_1 \times \sigma_7 \times \sigma_9 \times T_{5,8} = 9.984$, and $\varepsilon < \rho = \infty$, a solution is found. Thus, S is set to C, $\rho = 13.6$ and C is pruned. After the pruning, $C = O_1$ (the bottleneck operator is O_7). The termination condition, see [46], is not triggered given that there exists a two operator prefix that has not been investigated and its cost is less than ρ: $T_{4,8} = 9$.

In the fifth iteration, since $\varepsilon = 0 < \overline{\varepsilon} = 78.4$ and $\varepsilon = 0 < \rho = 13.6$, a new operator is appended to $C = O_1$; that is O_{10}. A new operator is also appended in the sixth iteration forming the partial plan $C = O_1O_{10}O_7$. In the seventh iteration, the partial plan is set to $C = O_1$, as $\varepsilon = 23.2 > \rho = 13.6$ and the bottleneck operator is the second one, i.e., O_{10}. In the eight iteration, O_5 is appended to $C = O_1$, while in the ninth iteration, the partial plan is set to $C = \emptyset$, as $\varepsilon = 14 > \rho = 13.6$ and the bottleneck operator is the first one, i.e., O_1. As a result, any other plan starting with O_1 can be safely ignored. Since the plan C is empty, the algorithm searches for the pair of operators with the minimum aggregate cost. In our example, this pair consists of O_4 and O_8. In the eleventh iteration, a new solution is found, since $\varepsilon = 9 > \overline{\varepsilon} = 6$ and $\varepsilon < \rho = 13.6$. Thus, $S = O_4O_8$, the bottleneck operator is O_4 and ρ is set to 9. After the pruning $C = \emptyset$, and the algorithm safely ignore plans starting with O_4O_8. This causes the algorithm to terminate, since the cost of the less expensive operator pair except those beginning with O_1, which is O_5O_6, is higher than ρ: $T_{5,6} = 10 > \rho = 9$. So the algorithm terminates, after having essentially explored all the 10! orderings in just 11 iterations. □

The characteristics of these two state-of -the-art algorithms for the optimization of queries over WSs in a pipelined parallel environment are summarized in Table 4.

Table 4. Operator ordering algorithms for minimizing the response time

Work	Execution environment	Input queries	Input operators
[42]	Parallel/distributed, static, centralized data transfers, pipelined parallelism	Single, ad-hoc, SQL-like	Independent, both selective and proliferative, both constrained and unconstrained
[46]	Distributed, static, decentralized data transfers, pipelined parallelism	Single, ad-hoc, SQL-like	Independent, both selective and proliferative, both constrained and unconstrained

3.1.2 Minimizing the Total Processing Time

In previous sections we saw that the query response time equals the maximum execution cost spent by an operator in order to process an input tuple and/or to send them to a subsequent operator. On the other hand, in a min-cost operator ordering problem, the goal is to minimize the total operator execution cost (processing and or transferring) that is incurred per input tuple. From now on, the term execution cost, unless clarified otherwise, encapsulates both the processing and transferring cost spent by an operator.

Ordering operators with a view to minimizing the per tuple total execution cost, a problem also commonly referred to as the min-cost operator ordering problem, is essential for achieving good system throughput. In general, solutions to the min-cost problem initially proposed for single-node settings may be applied to parallel settings characterized by resource homogeneity in a straightforward manner.

The min-cost ordering problem comes in several flavors. One of the most interesting ones refers to a parallel and static execution environment, where data is directly exchanged between the operators through pipelining; data communication can occur via a coordinator as well, without essentially modifying the problem, as long as homogeneous network links are assumed. If the operators are independent, then well-established fast solutions apply (e.g., [24,22]). However, correlated operators pose a more challenging problem. More specifically, given an ad-hoc select query Q that is defined over a set of unconstrained, correlated and selective operators $O = \{O_1, O_2, \ldots, O_N\}$ with fixed processing cost c_i and selectivity σ_i, the goal is to find an operator linear ordering S that minimizes the total execution cost of operators per input tuple. This cost encapsulates only the processing cost of tuples and is formally given by the following equation:

$$cost(S) = c_1 + \sum_{i=2}^{N} c_i D_i,\ D_i = \prod_{j=1}^{i-1}(1 - d(j|j-1)) \tag{5}$$

$d(i|j)$ is the conditional probability that the operator O_i will drop a tuple that has not been dropped by any of the operators that precede O_i in S, and $d(i|0) = 1 - \sigma_i$ is the unconditional probability that operator O_i will drop a tuple. Any drop probability is linearly related to selectivity, given that $d(i|j) = 1 - \sigma(i|j)$. Babu *et al.* have proved that this problem is equivalent to the pipelined set cover problem [7]. The pipelined set cover problem is MAX SNP-Hard [30], which implies that any polynomial operator ordering algorithm can at best provide a constant-factor approximation guarantee for this problem. In [30], a 4-times approximation algorithm is introduced to solve this problem. According to that algorithm, the operators must be ordered in a way that satisfies the following condition (termed greedy invariant):

$$\frac{d(i|i-1)}{c_i} \geq \frac{d(j|i-1)}{c_j}, 1 \leq i \leq j \leq N \tag{6}$$

Example 4. We continue Example 2, aiming now at minimizing the total execution time (only the processing time is considered). In this example, the selectivities of the operators are independent, so the algorithm in [7] is reduced to those in [24,22] and the operators are ordered in decreasing order of $(1 - \sigma_i)/c_i$. As such, first O_5 is selected, followed by O_1, O_8 and so on. If the operators were correlated, then, after selecting O_5, the conditional selectivities $\sigma(i|5)$ of all other operators would have to be estimated, in order to detect the second operator. □

Next, the min-cost operator ordering problem is studied in a multi-query setting. This problem is also known as the shared min-cost operator ordering problem. More formally, let $Q = \{Q_1, Q_2, \ldots, Q_M\}$ be a set of M, potentially continuous select queries that are evaluated over a set of N selective, unconstrained and correlated operators O. Each query is a conjunction of the operators in O. For each input tuple, operator $O_i \in O$ either returns a tuple or rejects it. The proportion of rejected tuples is defined by the operator selectivity. The goal is, given an input tuple t, to find the ordering that identifies the queries satisfied by t with the minimum cost. Note that an input tuple satisfies a query if it is not rejected by none of its constituent operators.

Obviously, if a query is satisfied, then all of its constituent operators must be evaluated. On the other hand, if an operator of a query rejects a tuple, then we do not have to evaluate the rest operators belonging to the same query (and are not evaluated so far). Thus, in a produced ordering, only a subset of operators $O' \subseteq O$ have to be evaluated, in order to determine the queries that are satisfied, and thus, the per tuple total processing cost is given by:

$$cost(S) = \sum_{i|O_i \in O'} c_i \tag{7}$$

Munagala *et al.* have proved the equivalence of this problem to the minimum set cover problem, and proposed an approximate greedy algorithm as a solution [31]. More formally, at any stage of the algorithm, the next operator to be evaluated is expected to resolve the maximum number of unresolved queries per unit cost. We say that for a given tuple t, a query is resolved if all its constituent operators return t or the currently evaluated operator rejects t. Let p_i be the number of unresolved queries the operator O_i is part of and $1-\sigma_i$ the probability that the operator O_i rejects an input tuple. Then, the expected number of queries resolved by O_i is $p_i(1-\sigma_i)$. The next operator to be evaluated is the one that minimizes the ratio $rank_i = c_i/p_i(1-\sigma_i)$. O_i is then removed independently from filtering out or not an input tuple. In addition, the queries that have been resolved due to the operator evaluation, and any other operator, which is not part of at least one not yet resolved query, is also removed. The algorithm terminates when all submitted queries are resolved.

Example 5. This example presents the steps of the algorithm proposed by Munagala *et al.* for the shared min-cost operator ordering problem. Suppose that the following queries are available $Q_1=\{O_1,O_3,O_7,O_{10}\}$, $Q_2=\{O_4,O_5,O_8\}$, $Q_3=\{O_2,O_4,O_{10}\}$, $Q_4=\{O_1,O_7,O_9,O_{10}\}$, $Q_5=\{O_3,O_6,O_7,O_9\}$. Let t be an input tuple, which is not rejected by any operator (except O_1, O_5 and O_9). As a consequence, only the query Q_3 is satisfied for t, since none of its constituent operators rejects this tuple. Figure 6 shows the results after every iteration of the algorithm. The algorithm starts by identifying the operator which minimizes the ratio $rank_i = c_i/p_i(1-\sigma_i)$, which is O_7 with $rank_7 = 3.88$ (see also Tables 1 and 2). Since O_7 does not reject the input tuple, no query is resolved. After that, operator O_1 is selected with $rank_1 = 5$. Operator O_1 rejects t, thus resolving queries Q_1 and Q_4. None of the not yet evaluated operators is removed, since they are included in at least one of the not yet resolved queries, i.e., Q_2, Q_3 and Q_5. The next two operators are O_4 and O_5 with $rank_4 = 5.71$ and $rank_5 = 8$, respectively. Since O_5 rejects t, query Q_2 is also resolved. The next operator is O_2 with $rank_2 = 23.33$. After evaluating operator O_2, operator O_9 is evaluated with $rank_9 = 25$. Since O_9 rejects input tuple t, query Q_5 is also resolved. Apart from that, operators O_3 and O_6 do not have to be evaluated, since they are not part of any unresolved query. Finally, the remaining operator, i.e., O_{10} is evaluated, and thus query Q_3 is satisfied. □

Liu *et al.* have proposed an edge-coverage-based approximate greedy algorithm for the same problem that achieves a better approximation ratio [28]. In [31], the shared min-cost operator ordering problem is viewed as the problem of covering the input

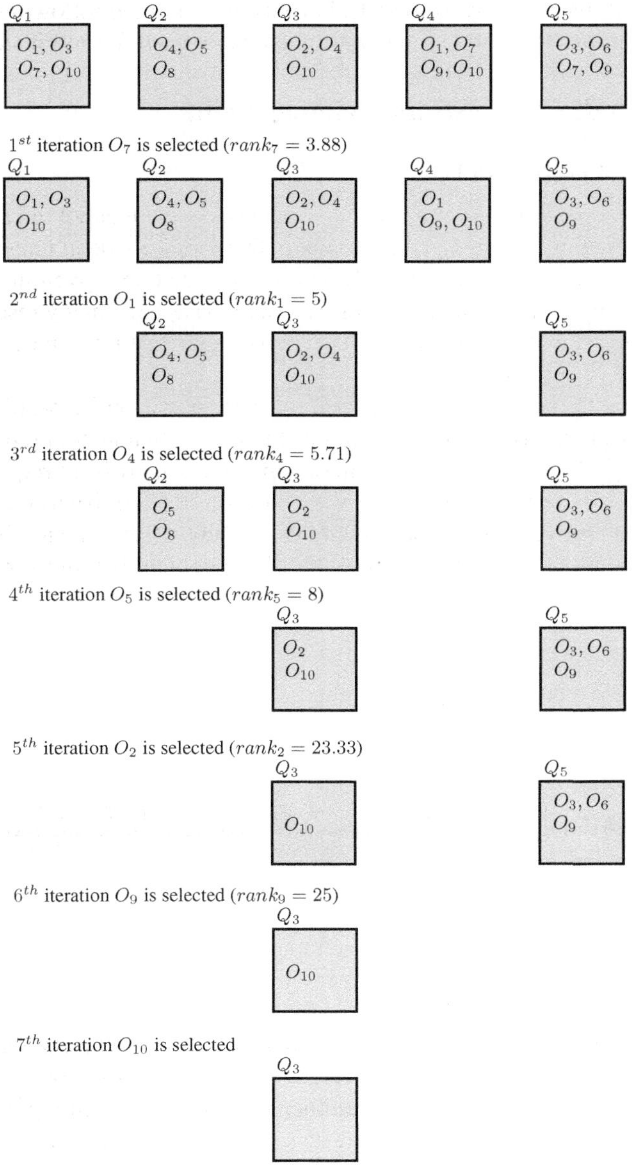

Fig. 6. The steps in Example 5, where each box corresponds to an unresolved query containing its remaining operators

queries through a suitable choice of operators. However, in [28], the same problem is viewed as the problem covering the connections between queries and operators through a suitable choice of operators, rather than covering the queries themselves.

The algorithm makes use of a bipartite graph. A bipartite graph $G = (O,Q,E)$ consists of two partitions, the set of the not yet evaluated operators (initially O) and the set of the not yet resolved queries (initially Q); an edge $e = (O_i,Q_j) \in E$ between an operator and a query indicates the fact that the operator is present in the corresponding query. Given an input tuple t, the next operator to be evaluated in each step is the one that covers the maximum number of edges in the bipartite graph with the minimum processing cost. For an operator O_i, the expected number of edges covered is the sum of the expected number of the queries that O_i evaluates to true[1] plus the expected number of the operators that do not have to be evaluated if operator O_i evaluates to false (these are the not yet evaluated operators that belong to the not yet evaluated queries, where O_i is part of). More formally, the next operator to be evaluated, given a tuple t, is the one that minimizes the ratio

$$unit - price_i = \frac{c_i}{\sigma_i \delta(O_i) + (1-\sigma_i) \sum_{\forall Q_k | (O_i,Q_k) \in E_i} \delta(Q_k)} \tag{8}$$

where E_i is the remaining set of edges in the current iteration, δ is the degree of an operator or a query respectively in the bipartite graph and Q_k is any query involving O_i. After each operator evaluation, the bipartite graph is updated with the performed actions being identical to those in [31].

Example 6. We reconsider the problem in Example 5 employing the edge-coverage based algorithm proposed in [28]. Figure 7 shows the operator-query bipartite graph after every iteration of the algorithm. Let t be the current input tuple. In the first iteration, operator O_1 is selected for evaluation with unit-price $= 2/(0.8*2+0.2*(4+4)) = 0.625$. After that, queries Q_1 and Q_4 are removed from the graph along with operator O_1, since O_1 rejects t. In the second iteration, the operator O_4 is selected with unit-price $= 8/(0.3*2+0.7*(3+3)) = 1.66$. In the third iteration, O_7 is selected, while in the fourth iteration we evaluate O_5 with unit-price $= 4/(0.5+0.5*2) = 2.66$. Since operator O_5 rejects the input tuple, query Q_2 is removed from the graph along with operator O_8. The latter is removed as it is not part of any unresolved query. In the fifth iteration, O_2 is selected for evaluation, while in the sixth iteration, O_9 is selected with unit-price$=10/(0.6+0.4*3) = 5.55$. After evaluating operator O_9, query Q_5 along with operators O_3 and O_6 are removed from the graph. Finally, operator O_{10} is evaluated, since it is the only remaining operator. □

A common problem with the performance of WSs is that they may be too slow or prohibitively expensive in some cases. In that case, if there exist some additional highly selective operators that are inexpensive and correlated to the expensive ones, it is beneficial to incorporate them early in the plan. This is the main rationale in

[1] We say that, given a tuple t, an operator O_i evaluates to true a query Q_i if $O_i \in Q_i$ and O_i returns t. Otherwise, we say that O_i evaluates Q_i to false.

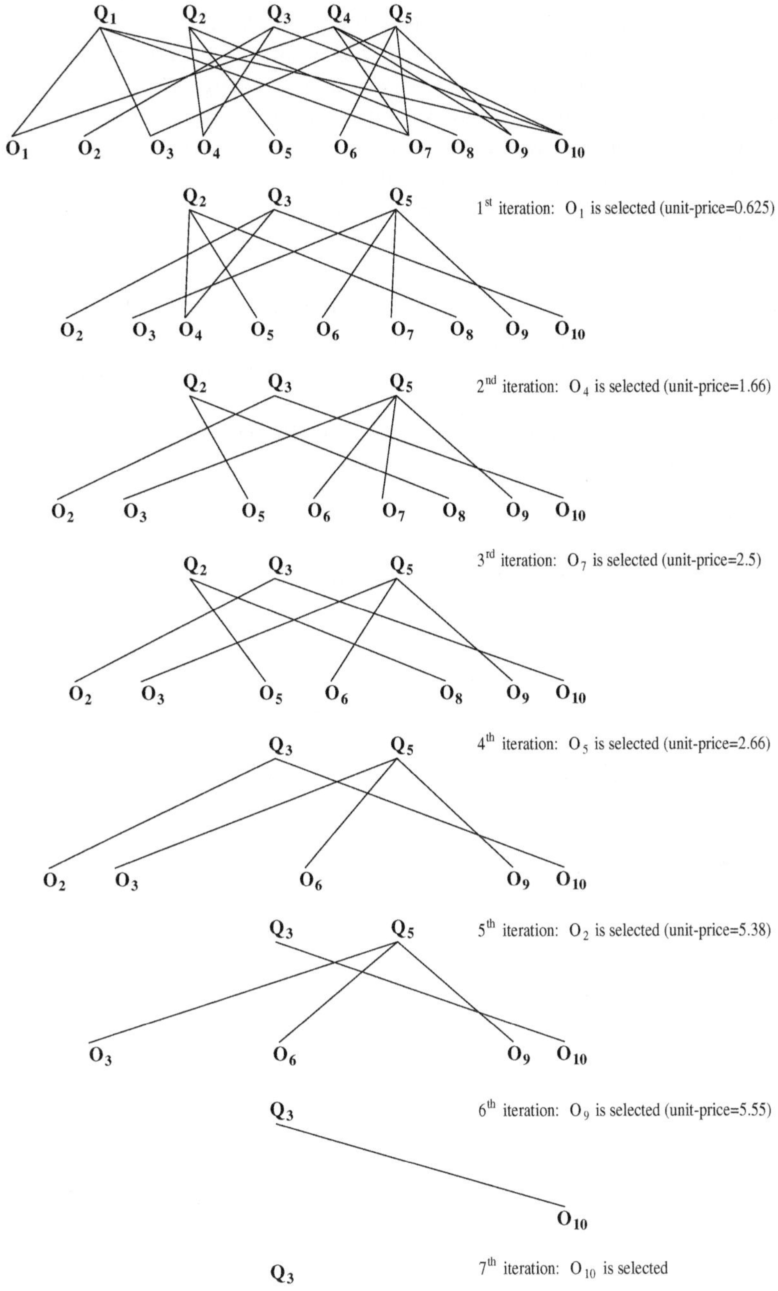

Fig. 7. The steps in Example 6

[14]. More specifically, the work in [14] exploits the fact that some additional low-cost operators $O_i \notin O$ can be evaluated so as to reject tuples at lower cost avoiding the cost of evaluating expensive operators. In this work, the proposed solution is a conditional plan. A conditional plan is a decision tree where each interior node corresponds to an operator that splits the plan into several conditional plans, in turn. During evaluation, the tree is traversed. At every node, the query processor evaluates the corresponding operator and follows one of the sub-plans depending on its output.

Lazaridis and Mehrota in [25] deal with a problem similar to [14]. Let Q be an ad-hoc, select query evaluated over a set of N independent, unconstrained and selective operators O. The goal is to find an operator ordering that minimizes the total operator processing cost per tuple, in order to decide whether an input tuple is rejected by any of the operators or not. In [25], as in [14], this is done by inserting additional lower cost operators that are not part of the original operators mentioned in the query explicitly.

A motivation example is as follows. Suppose a query that uses an expensive face identifier method, which compares input images with stored images containing faces of criminals:

*SELECT * FROM Camera, Criminals*
WHERE FaceIdentifier(Camera.image, Criminals.image)

Suppose also that the user has access to two additional, less expensive methods, ObjectDetector and FaceDetector which detect foreign objects and human faces, respectively. By using these methods we can reject some input images at lower cost: if an image does not contain an object or a face, then there is no reason to test whether it contains a particular face of a criminal. We infer a negative result for FaceIdentifier from a negative result by either of the two methods ObjectDetector and FaceDetector.

Table 5 summarizes the main characteristics of the proposed solutions to different flavors of the min-cost operator ordering problem. As already explained, although some of these solutions were originally proposed for centralized settings,

Table 5. Algorithms for flavors of the min-cost operator ordering problem

Work	Execution environment	Input queries	Input operators
[7]	Parallel, decentralized data transfers, pipelined parallelism	Single, continuous, SQL-like	Correlated, selective and unconstrained
[31]	Centralized	Multiple, continuous, SQL-like	Correlated, selective and unconstrained
[28]	Centralized	Multiple, continuous, SQL-like	Independent, selective and unconstrained
[14]	Centralized	Single, ad-hoc, SQL-like	Correlated, selective and unconstrained
[25]	Centralized	Single, ad-hoc, SQL-like	Independent, selective and unconstrained

their results can be easily transferred to parallel settings or distributed settings with centralized data transmission, and, as such, they can be employed to optimize queries over potentially remote WSs. Also, techniques proposed for continuous queries may be applicable to ad-hoc queries on finite data, too.

3.2 Operator Ordering Problems in Dynamic Environments

Wide area settings hosting WSs are typically subject to changes, which may have significant impact on queries. Babu *et al.* have extended their work in [7] to address the more general problem where the execution environment is dynamic. This is achieved by utilizing two components, a so-called "profiler" and a "re-optimizer". The profiler maintains a time-based sliding window of tuples dropped in the recent past. A profile tuple is created for every tuple in the sliding window and shows which operators have unconditionally rejected it. The re-optimizer can then compute any selectivity estimates that it requires from the profile tuples. The re-optimizer's job is to ensure that the current operator ordering satisfies Eq. (6). Similarly, the operator costs can be monitored, as well. The above render the algorithm proposed in [7] robust to environmental changes.

In the context of adaptive query processing [15], Avnur and Hellerstein have proposed the eddies execution model for minimizing the response time of ad-hoc SPJ queries at runtime [6]. The operators can be of arbitrary type, i.e., both selective and proliferative, both constrained and unconstrained, and both correlated and independent. In the eddies model, every tuple may follow a different plan. The original eddy implementation employed two main approaches to routing. The first one, called back-pressure, causes more tuples to be routed to fast operators early in query execution. The second approach augments back-pressure with a ticket scheme, whereby the eddy gives a ticket to an operator whenever it consumes a tuple and takes a ticket away whenever it sends a tuple back to the eddy. In this way, higher selectivity operators accumulate more tickets. When choosing an operator to which a new tuple should be routed, the ticket-routing policy conducts a lottery between the operators, with the chances of a particular operator winning being proportional to the number of tickets it owns. In this way, higher selectivity operators tend to receive more tuples early in their path through the eddy. The algorithm in [7] can also be incorporated into eddies routing policies. Several extensions to eddies have been proposed, including the works in [9,38,29,44]. The work of Tian and DeWitt [44] explicitly considers distributed execution environments supporting decentralized data transferring. In a distributed eddy, each operator, instead of returning processed tuples back to a central eddy, it redirects them to a subsequent operator. The operators learn query execution statistics and exchange them with other operators periodically. Based on these statistics, each operator makes its own routing decisions without consulting the central eddy or any other operator. By employing such eddies in distributed queries over WSs, the need of an optimizer that constructs an execution plan becomes obsolete.

3.3 Tuple Routing and Scheduling Problems

WSs can usually process many requests concurrently due to multi-threading. Each server hosting a WS has limited capacity though, so an optimizer has to build plans that respect the capacity constraints. Allowing multiple concurrent calls to operators is considered in tuple routing problems. These problems deal with the selection of one or more operator orderings, in order to maximize the number of tuples processed per unit time, this is why they are also called flow maximization problems. Their main rationale is to benefit from as much capacity of the processors hosting the operators as possible.

Condon *et al.* have proposed a solution for the special case where the orderings are linear, the execution environment is parallel and static, the data transfers are decentralized and pipelined parallelism is employed [13]. Let Q be an ad-hoc select query consisting of calls to N independent, unconstrained and selective operators O. r_i is the rate limit of each operator and is measured in tuples per time unit. Each tuple can be routed individually, so that different tuples can follow distinct routes. The problem is to find one or more operator orderings in order to maximize the tuple flow per unit time. More formally, suppose that a set of M different linear operator orderings are available, $\{\pi_1, \pi_2, \ldots, \pi_M\}$ that process different subsets of input tuples in parallel. Let f_i be the number of tuples sent through linear plan π_i per unit time. Then, the total number of tuples processed per unit time by the different linear orderings is given by

$$F = \sum_{\pi_i} f_i, \quad f_i > 0 \,\forall\, \pi_i \tag{9}$$

The goal is to find the set of f_i and π_i values that maximize Eq. (9) without violating the rate limits r_i of the operators.

Condon *et al.* proposed a recursive algorithm for this problem, which is detailed in [13]. Operators are initially ordered (from N to 1) in a way that satisfies the following condition:

$$r_i \sigma_i \leq r_{i+1} \forall i, 1 \leq i \leq N-1, \tag{10}$$

After that, the flow of tuples along each ordering is increased until either (i) an operator is saturated, i.e., it processes the maximum possible number of input tuples according to its rate limits, or (ii) the residual capacity of O_i, $1 \leq i \leq N$ times its selectivity σ_i becomes equal to the residual capacity of O_{i+1}. In [13], it is shown that if stopping condition (i) is satisfied, the constructed flow is optimal. On the other hand, if stopping condition (ii) is satisfied, the operator O_{i+1} is immediately placed after O_i and the operators O_{i+1} O_i are replaced by a single operator $O_{i,i+1}$ with rate limit equal to the residual capacity of O_i and selectivity equal to the product $\sigma_i \sigma_{i+1}$. The resulting smaller problem is then solved recursively.

Example 7. Let $O_2 = \{O_1, O_2, O_3, O_4, O_5\}$ be a set of five operators with rate limit and selectivity values $\{12,8,7,4,2\}$ and $\{0.9,0.3,0.7,0.5,0.8\}$, respectively. Initially, the operators are sorted in descending rate limit order, i.e., O_3 O_4 O_2 O_5 O_1; this ordering satisfies Eq. (10). The minimum flow of tuples that triggers either of the two conditions (see [13]) is $f_{3,4,2,5,1} = 4.36$. If $f_{3,4,2,5,1} = 4.36$ tuples per time unit are sent through the ordering O_3 O_4 O_2 O_5 O_1, then the residual capacity of O_2 times its selectivity is equal to the residual capacity of O_4. After that, the ordering O_3 O_4 O_2 O_5 O_1 is kept and a new operator ordering is created. To this end, operators O_2 and O_4 are merged into a single operator with residual capacity $r_{2,4} = 5.82$ (the residual capacity of O_2) and selectivity $\sigma_{2,4} = \sigma_2 \times \sigma_4$. The new smaller sub-problem is solved recursively. In the second iteration, $f_{3,2,4,5,1} = 5.25$ is the minimum flow of tuples that triggers stoping condition (2), while none of the operators becomes saturated with less flow. Thus, the ordering O_3 $O_{2,4}$ O_5 O_1 is kept and the operators $O_{2,4}$ and O_5 are merged into an operator $O_{5,2,4}$ with residual capacity equal to the residual capacity of O_5 and selectivity $\sigma_{5,2,4} = 0.105$, forming a new ordering O_3 $O_{5,2,4}$ O_1. The problem is again solved recursively. The algorithm terminates in the fifth iteration, where a single operator $O_{1,5,2,4,3}$ has been left with residual capacity $r_{1,5,2,4,3} = 0.2316$, i.e., $f_{1,5,2,4,3} = 0.2316$ tuples per time unit must be sent along this ordering, in order to saturate the single operator. Together, the flows constructed in the aforesaid five stages yield the following optimal solution to the max-throughput tuple routing problem for the given input instance: $f_{3,4,2,5,1} = 4.36$, $f_{3,2,4,5,1} = 5.25$, $f_{3,5,2,4,1} = 1.58$, $f_{3,5,2,4,1} = 1.58$, $f_{3,1,5,2,4} = 0.79$, $f_{1,5,2,4,3} = 0.2316$ and $f_\pi = 0$ for all other π orderings. The steps of the algorithm are shown in Figure 8. □

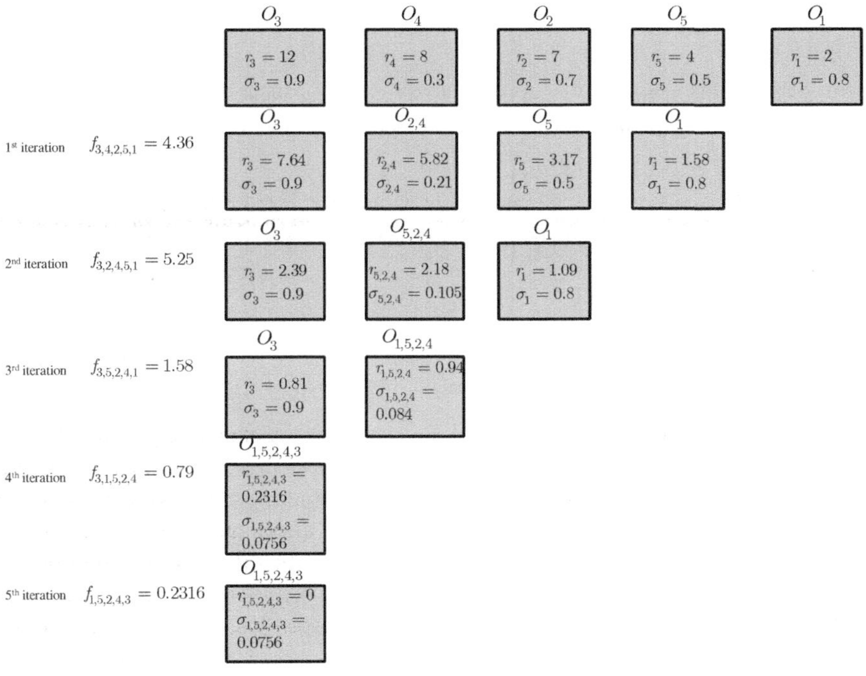

Fig. 8. The steps in Example 7

[16] extends the algorithm in [13] so that precedence constraints among operators and arbitrary selectivities are supported.

Liu *et al.* deal with the joint problem of flow maximization and scheduling in a heterogeneous, multi-query, parallel-processor environment [27]. More formally, let $Q = \{Q_1, Q_2, \ldots, Q_K\}$ be a set of K ad-hoc select queries and $O = \{O_1, O_2, \ldots, O_N\}$ be a set of N, potentially correlated, unconstrained and selective operators. Each query is a conjunction of the operators in O. For every input tuple t, operator $O_i \in O$ either returns a tuple or rejects it. Furthermore, let M be a set of heterogeneous processors and $y_i(M_k)$ be the cost of evaluating operator O_i on processor M_k. The goal is to find, for each input tuple t, one or more operator orderings and an associated allocation scheme so that the flow of tuples processed per unit time is maximized. In the proposed algorithm, each operator O_i is always evaluated on the processor M_k for which the incurred load $y_i(M_k)$ is minimized (for simplicity we denote $y_i(M_k)$ by y_i). For the operator ordering problem, Liu *et al.* leverage the fact that the flow maximization problem in heterogeneous processing environments is equivalent to the problem of min-cost operator ordering in centralized environments, where the cost of each operator O_i is given by y_i. The problem addressed is a generalized case of the shared min-cost operator problem introduced in [31] and [28]. Other operator scheduling problems are studied in [47,8,41,3,37].

The main characteristics of the algorithms are summarized in Table 6.

Table 6. Solutions to tuple routing problems (first row) and scheduling problems (second row)

Work	Execution environment	Input queries	Input operators
[13,16]	Parallel, static, decentralized data transfers, pipelined parallelism	Single, ad-hoc, SQL-like	Independent, both selective and proliferative, both constrained and unconstrained
[27]	Parallel, static, heterogeneous processors, pipelined parallelism	Multiple, ad-hoc, SQL-like	Both independent and correlated, both selective and unconstrained

3.4 Data Transfer Planning Problems

Consider a dynamic environment, where multiple data sources stream data to a central processing node through heterogeneous communication links, in order to evaluate aggregate queries. These queries combine data from multiple data sources and their answers must be re-computed as data updates arrive to the sources. It is assumed that each data source stores the values of a single data attribute. The goal is to minimize the total communication cost, in order to evaluate multiple (possibly overlapping) aggregate continuous queries. No parallelism is employed during queries execution.

Olston *et al.* have provided a solution for this problem exploiting the fact that the precise answer of a continuous query may not always be necessary [34]. In such cases, approximate answers of sufficient precision may be computed from a small fraction of the input stream items. Users need to submit quantitative precision constraints along with continuous queries, which the processing node uses to filter stream items at the remote data sources. Each query is associated with a pair of real values, L and H that define an interval $[L\ H]$ in which the precise answer is guaranteed to lie. The reasoning behind the algorithm is quite simple: a data source does not need to stream the data that does not affect the answer, according to the previously defined precision requirements. For example, if the current exact answer is 10 and the precision interval is $[7\ 13]$, then data sources holding updated data with values from 7 to 13 do not have to proceed to data transmission. The heuristic algorithm for filtering the stream items on the remote resources, called filter tuning, consists of an iterative two-step procedure. In the first step, each data resource shrinks the bound width periodically at a predefined rate. Each time the bound width of a data source shrinks, the so-called "leftover" width is reallocated to other data sources, ensuring all precision constraints are still satisfied. In the second step, the data sources that increase their bound widths are heuristically selected; the algorithm selects the ones that stream data at high rates and are connected with expensive communication links.

Li *et al.* in [26] deal with another data transfer planning problem. In this work, it is assumed that the data resources are spread across a wide-area distributed environment, and there is a single data resource per host. The links between hosts are heterogeneous, while the data resources can directly transfer data to other resources. Any SQL-like query can be submitted, and there is no limitation regarding the type of operators. For every submitted query, a query plan is provided in the form of a rooted tree. The plan specifies the operators to be evaluated on each data resource and the evaluation order. The aim is to schedule the data transfers across the queried data resources, in order to minimize the total data transferring cost when evaluating the query plans. It is proven that the problem is NP-hard for arbitrary communication networks by a reduction from the Steiner tree problem in graphs [39]. Li *et al.* proposed a polynomial time algorithm for this problem, which relies on the weighted hyper-graph minimum cut algorithm [26]. The produced data movement plan is optimal for tree-shaped communication networks, while it is an approximation to the optimal one for more general communication networks.

3.5 Other Problems Related to Queries over WSs

Thus far we have dealt with problems in which the services provide exact answers. Search queries belong to a different paradigm. In [10], Braga *et al.* deal with the joint problem of finding a WSs plan and an access pattern for each service for search query optimization. The execution environment is distributed and static, while the services can exchange data directly through pipelining. The proposed algorithm explores the space of plans using a heuristic, branch and bound strategy and it is applicable, under some modifications, for the optimization of both the total service processing cost and the response time criteria. Another common problem with WSs is that they are slow and the communication cost may well dominate the processing cost. Block-based data transmission may alleviate this problem, as proposed in [42,21]. Finally, [40] explore adaptive approaches to parallelizing calls to WSs.

3.6 Discussion and Open Issues

The purpose of the current section is to summarize the problems that have been studied for query optimization over WSs and the state-of-the-art algorithms that have been proposed.

Concerning the operator ordering problem in static execution environments, we have presented several algorithms that aim to minimize either the response time of the submitted query, or the per tuple total processing cost. Srivastava *et al.* introduced a general purpose WSMS for query optimization in a parallel environment that utilizes pipelined parallelism during query execution [42]. The major assumptions that are made are the following. The services do not exchange data directly, but a central component undertakes the intermediate data transfers. Also, the selectivity and the processing cost of the operators are constant and independent of the input attribute values. Under these assumptions, a provably optimal algorithm has been proposed that schedules parallel invocations of the operators, in order to minimize the response time of the submitted queries. The work of Tsamoura *et al.* comprises an extension of the work in [42] for distributed execution environments, where the operators exchange data directly over non-negligible and heterogeneous communication links [46]. The pipelined execution model is also applied. However, the proposed algorithm builds only linear operator orderings. A problem of significant importance that has not been addressed is the generalization of the latter algorithm for building parallel operator invocations. Furthermore, it would be very interesting to study the query response time minimization problem in a dynamic environment, where the per tuple processing and transferring costs, change significantly over time.

After that, we have presented several flavors of the min-cost ordering problem both in centralized [31,28,14,25] and parallel execution environments [7]. The above min-cost operator ordering problems deal with select queries, while the cost needed to transfer tuples from one service to another is negligible. In [7], given an input query that is evaluated through a set of correlated, unconstrained and selective operators, the goal is to build a linear operator ordering that minimizes the total cost of processing operators per input tuple.The proposed algorithm provides a 4-times approximation solution, while an heuristic technique has been introduced for the generalization of the above algorithm when the wide-area settings are subject to changes. Furthermore, Munagala *et al.* [31] and Liu *et al.* [28] deal with a multi-query flavor of the min-cost ordering problem in a centralized execution environment. In particular, given a set of one or more queries that consist of a set of independent, selective and unconstrained operators, the goal is to find an optimal operator ordering, in order to answer all input queries with the minimum total processing cost. The last two works that are studied try to minimize the per tuple total processing cost for an input query by utilizing additional, lower cost and highly selective operators [14,25]. Those operators need not be part of the initial operator set. To summarize, for operator ordering in a static execution environment, the following problems have been addressed:

- Response time minimization of single SPJ queries employing pipelined parallelism and independent operators both in a parallel and distributed execution environment. Decentralized data transfers have been considered, as well.
- Total operator execution cost minimization. Three different problem flavors that consider unconstrained operators are discussed; namely, (i) optimization of a single-query that employs parallelism and assumes correlated operators in a parallel environment, (ii) optimization of a single query with correlated operators in a centralized environment both for correlated and independent operators and (iii) optimization of multiple queries with both independent and correlated operators in a centralized environment.

Regarding the response time minimization, no work has been done for multi-query optimization or correlated operators. Furthermore, other types of parallelism (such as partitioned) have not been considered. The above works deal with SQL-like queries. The only work that deals with IR-like queries is presented in [10]. [10] deals with the joint problem of selecting the more appropriate services to invoke, when multiple services have the same functionality but different binding patterns, and of ordering the selected services in a distributed and static execution environment that employees pipelined parallelism. However, no performance guarantees have been provided.

In the context of adaptive operator ordering for minimizing the response time of a query, eddies [6] and distributed eddies [44] try to overcome the "hassle" of varying processing and communication costs in a dynamic execution environment by routing each tuple independently.

In Sec. 3.3, we have presented two throughput maximization problems [13,27]. Both of them employ inter-operator parallelism in order to maximize the tuple throughput, i.e., the number of tuples processed by the operators per unit time. The work in [16] deals with single query optimization, imposing only the independent assumption on input operators. On the other hand, the work in [27] deals with multiple select queries and unconstrained, selective operators. It also performs operator scheduling. For both proposals, the underlying execution environment is static and parallel. In general, adaptive query processing is in its infancy.

We have dedicated the last part of Sec. 3 to the description of two data transfer planning problems in a static and dynamic execution environment. Both of them deal with multiple input queries, while the communication links are heterogeneous. The work of Olston *et al.* deals with a centralized execution environment, where multiple continuous aggregate queries are evaluated in a central processing component [34]. There the data resources are disparate in a wide-area network and periodically stream data to the central component. Considering that the precise answer is not always necessary for some or all input queries, the goal is to appropriately tune the amount of data sent by each remote data resource, in order to minimize the total communication cost for answering input queries. Li *et al.* deal with another data transfer planning problem. As in [26], the data resources are disparate in a wide-area heterogeneous environment, while the latter can directly exchange data. Given a plan that specifies the operations to be performed on input data the goal is to appropriate schedule the data exchanges among the resources, in order to minimize the total data transferring cost when evaluating the input queries. Both works do not encapsulate the processing cost in order to answer the submitted queries. A limitation of the problem considered in [34] is that it handles only aggregate queries, while a limitation of [26] is that it requires a plan that specifies the operator invocation order. As both works deal with the min-cost metric, an interesting aspect would be the exploration of problems having other optimization criteria, such as the response time of the submitted queries. For example, regarding the problem in [26], the objective might be to minimize the maximum response time of the submitted queries. The characteristics of the proposed algorithms are summarized in Tables 4, 5 and 6.

The following remarks arise from the above discussion. The data transfer cost and the network heterogeneity issue are largely overlooked. In the majority of the works, the state-of-the-art operator ordering and tuple routing algorithms consider parallel and/or centralized execution environments, where the processing cost dominates. Another important issue that needs more attention is dynamicity. The presented problems mainly deal with static execution environments which is not the case in wide-area infrastructures such as the grid. The operator independence assumption must be reconsidered, since, in a real setting, the processing cost and the selectivity of utilized operators may be tightly related with the input attributes values. Another problem that should be investigated in the future is the combination of different forms of parallelism. Finally, the majority of presented problems deal with SQL-like queries. Extending current approaches or investigating new ones for IR queries optimization is crucial, since querying information sources is an important part of information management in Web and other distributed wide-area organizations.

4 Conclusion

This chapter discussed queries over WSs focusing on their optimization. Queries over WSs are becoming increasingly common due to the proliferation of publicly available WSs and remote and decentralized computing infrastructures such as grid and cloud computing. We presented a taxonomy of the problems encountered in the optimization of such queries taking into account the type of the optimization problems, the type of queries, the type of services or operators and the exact execution environment to which the queries are tailored. Some of the problems can be efficiently solved by utilizing known algorithms from the database community, whereas, for some others, novel algorithms have been proposed. This chapter discussed the state-of-the-art solutions that apply to the problem of optimizing queries over WSs, explaining their main characteristics. Especially for the problem of minimizing the response time in decentralized pipelined queries, a novel algorithm was presented.

References

1. Business process execution language for web services, `http://bpel.xml.org/tags/bpel4ws`
2. Abadi, D.J.: Data management in the cloud: Limitations and opportunities. IEEE Data Eng. Bull. 32(1), 3–12 (2009)
3. Agrawal, K., Benoit, A., Dufossé, F., Robert, Y.: Mapping filtering streaming applications with communication costs. In: Proc. of the Twenty-First Annual Symposium on Parallelism in Algorithms and Architectures (SPAA), pp. 19–28 (2009)
4. Aloisio, G., Cafaro, M., Fiore, S., Mirto, M., Vadacca, S.: Grelc data gather service: a step towards P2P production grids, pp. 561–565 (2007)
5. Alpdemir, M.N., Mukherjee, A., Gounaris, A., Paton, N.W., Watson, P., Fernandes, A.A.A., Fitzgerald, D.J.: Ogsa-dqp: A service for distributed querying on the grid. In: Proc. of the International Conference on Extended Database Technologies (EDBT) (2004)
6. Avnur, R., Hellerstein, J.M.: Eddies: continuously adaptive query processing. In: Proc. of the International Conference on Management of Data (SIGMOD), pp. 261–272 (2000)
7. Babu, S., Matwani, R., Munagala, K.: Adaptive ordering of pipelined stream filters. In: Proc. of the International Conference on Management of Data (SIGMOD), pp. 407–418 (2004)
8. Benoit, A., Dufosse, F., Robert, Y.: Filter placement on a pipelined architecture. In: International Symposium on Parallel and Distributed Processing, vol. 0, pp. 1–8 (2009)
9. Bizarro, P., Babu, S., DeWitt, D., Widom, J.: Content-based routing: different plans for different data. In: Proc. of the 31st International Conference on Very Large Data Bases (VLDB), pp. 757–768 (2005)
10. Braga, D., Ceri, S., Daniel, F., Martinenghi, D.: Optimization of multidomain queries on the web. In: Proc. of the VLDB Endowment, vol. 1, pp. 562–573 (2008)
11. Burge, J., Munagala, K., Srivastava, U.: Ordering pipelined query operators with precedence constraints. Technical Report 2005-40, Stanford InfoLab (2005), `http://ilpubs.stanford.edu:8090/705/`
12. Chen, J., DeWitt, D.J., Tian, F., Wang, Y.: Niagaracq: a scalable continuous query system for internet databases. In: Proc. of the 2000 ACM SIGMOD International Conference on Management of Data (SIGMOD), pp. 379–390 (2000)
13. Condon, A., Despande, A., Hellerstein, L., Wu, N.: Algorithms for distributional and adversarial pipelined filter ordering problems. ACM Transactions on Algorithms 5(2), 24–34 (2009)
14. Deshpande, A., Guestrin, C., Hong, W., Madden, S.: Exploiting correlated attributes in acquisitional query processing. In: Proc. of the 21st International Conference on Data Engineering (ICDE), pp. 143–154 (2005)
15. Deshpande, A., Ives, Z.G., Raman, V.: Adaptive query processing. Foundations and Trends in Databases 1(1), 1–140 (2007)
16. Despande, A., Hellerstein, L.: Flow algorithms for parallel query optimization. In: Proc. of the 24th International Conference on Data Engineering (ICDE), pp. 754–763 (2008)

17. DeWitt, D., Gray, J.: Parallel database systems: the future of high performance database systems. Communications of the ACM 35(6), 85–98 (1992)
18. Epstein, R.S., Stonebraker, M., Wong, E.: Distributed query processing in a relational data base system. In: Lowenthal, E.I., Dale, N.B. (eds.) Proc. of the 1978 ACM SIGMOD International Conference on Management of Data, June 2, pp. 169–180. ACM, New York (1978)
19. Foster, I., Kesselman, C.: The Grid: Blueprint for a New Computing Infrastructure, second edn. Morgan Kaufmann Publishers, San Francisco (2003)
20. Gounaris, A., Smith, J., Paton, N.W., Sakellariou, R., Fernandes, A.A., Watson, P.: Adaptive workload allocation in query processing in autonomous heterogeneous environments. Distrib. Parallel Databases 25(3), 125–164 (2009)
21. Gounaris, A., Yfoulis, C., Sakellariou, R., Dikaiakos, M.D.: Robust runtime optimization of data transfer in queries over web services. In: Proc. of the ACM International Conference on Data Engineering (ICDE), pp. 596–605 (2008)
22. Hellerstein, J.M., Stonebraker, M.: Predicate migration: Optimizing queries with expensive predicates. In: Proc. of the ACM SIGMOD International Conference on Management of Data SIGMOD, pp. 267–276 (1993)
23. Taylor, I., Shields, M., Wang, I.: Resource management of triana p2p services. In: Grid Resource Management (2003)
24. Krishnamurthy, R., Boral, H., Zaniolo, C.: Optimization of nonrecursive queries. In: Proc. of VLDB, pp. 128–137 (1986)
25. Lazaridis, I., Mehrotra, S.: Optimization of multi-version expensive predicates. In: Proc. of the ACM SIGMOD International Conference on Management of Data (SIGMOD), pp. 797–808 (2007)
26. Li, J., Deshpande, A., Khuller, S.: Minimizing communication cost in distributed multi-query processing. In: Proc. of the 21st International Conference on Data Engineering (ICDE), pp. 772–783 (2009)
27. Liu, Z., Parthasarathy, S., Ranganathan, A., Yang, H.: Generic flow algorithm for shared filter ordering problems. In: Proc. of the 27th ACM SIGMOD-SIGACT-SIGART Symposium on Principles of database systems (PODS), pp. 79–88 (2008)
28. Liu, Z., Parthasarathy, S., Ranganathan, A., Yang, H.: Near-optimal algorithms for shared filter evaluation in data stream systems. In: Proc. of the ACM International Conference on Management of Data (SIGMOD), pp. 133–146 (2008)
29. Madden, S., Shah, M., Hellerstein, J.M., Raman, V.: Continuously adaptive continuous queries over streams. In: Proceedings of the 2002 ACM SIGMOD International Conference on Management of Data (SIGMOD), pp. 49–60. ACM, New York (2002)
30. Munagala, K., Babu, S., Motwani, R., Widom, J.: The pipelined set cover problem. Technical Report 2003-65, Stanford InfoLab (2003)
31. Munagala, K., Srivastava, U., Widom, J.: Optimization of continuous queries with shared expensive filters. In: Proc. of 26th ACM SIGMOD-SIGACT-SIGART Symposium on Principles of Database Systems (PODS), pp. 215–224 (2007)
32. Nieto-Santisteban, M.A., Gray, J., Szalay, A.S., Annis, J., Thakar, A.R., O'Mullane, W.: When database systems meet the grid. In: CIDR, pp. 154–161 (2005)
33. Oinn, T., Addis, M., Ferris, J., Marvin, D., Senger, M., Greenwood, M., Carver, T., Glover, K., Pocock, M., Wipat, A., Li, P.: Taverna: a tool for the composition and enactment of bioinformatics workflows. Bioinformatics 20(17), 3045–3054 (2004)
34. Olston, C., Jiang, J., Widom, J.: Adaptive filters for continuous queries over distributed data streams. In: Proc. of the 2003 ACM SIGMOD International Conference on Management of Data (SIGMOD), pp. 563–574. ACM, New York (2003)

35. Ozsu, M., Valduriez, P. (eds.): Principles of Distributed Database Systems, 2nd edn. Prentice-Hall, Englewood Cliffs (1999)
36. Papazoglou, M.P., Traverso, P., Dustdar, S., Leymann, F.: Service-oriented computing: State of the art and research challenges. Computer 40(11), 38–45 (2007)
37. Pietzuch, P., Ledlie, J., Shneidman, J., Roussopoulos, M., Welsh, M., Seltzer, M.: Network-aware operator placement for stream-processing systems. In: Proc. of the 22nd International Conference on Data Engineering (ICDE), pp. 49–60 (2006)
38. Raman, V.: Interactive query processing. PhD thesis, UC Berkeley (2001)
39. Robins, Y., Zelikovski, A.: Improved steiner tree approximation in graphs. In: Proc. of the 11th ACM-SIAM Symposium on Discrete Algorithms, pp. 770–779 (2000)
40. Sabesan, M., Risch, T.: Adaptive parallelization of queries over dependent web service calls. In: Proc. of the International Conference on Data Engineering (ICDE), pp. 1725–1732 (2009)
41. Srivastava, U., Munagala, K., Widom, J.: Operator placement for in-network stream query processing. In: Proc. of the twenty-fourth ACM SIGMOD-SIGACT-SIGART Symposium on Principles of Database Systems (PODS) (2005)
42. Srivastava, U., Munagala, K., Widom, J., Motwani, R.: Query optimization over web services. In: Proc. of the 32nd Conference on Very Large Databases (VLDB), pp. 355–366 (2006)
43. Terry, D., Goldberg, D., Nichols, D., Oki, B.: Continuous queries over append-only databases. In: Proc. of the 1992 ACM SIGMOD International Conference on Management of Data (SIGMOD), pp. 321–330. ACM, New York (1992)
44. Tian, F., DeWitt, D.J.: Tuple routing strategies for distributed eddies. In: Proc. of the 29th Conference on Very Large Databases (VLDB), pp. 333–344 (2003)
45. Tsamoura, E., Gounaris, A., Manolopoulos, Y.: Decentralized execution of linear workflows over web services (submitted for publication)
46. Tsamoura, E., Gounaris, A., Manolopoulos, Y.: Optimal service ordering in decentralized queries over web services. Technical Report, `http://delab.csd.auth.gr/~tsamoura/publications.html`
47. Yu, J., Buyya, R., Tham, C.K.: Cost-based scheduling of scientific workflow application on utility grids. In: Proc. of the First International Conference on e-Science and Grid Computing (E-SCIENCE), pp. 140–147 (2005)

Chapter 7
Towards Adaptively Approximated Search in Distributed Architectures

Barbara Catania and Giovanna Guerrini

Dipartimento di Informatica e Scienze dell'Informazione
Università degli Studi di Genova
Via Dodecaneso 35, 16146 Genova, Italy
{catania,guerrini}@disi.unige.it

Abstract. Innovative applications over distributed architectures, like the Web, often require the analysis of strongly related, highly heterogeneous data, stored in remote and autonomous data sources, that can be either totally available at query processing time (stored data) or become available in a continuous stream (data stream). In these contexts, search efficiency is a key issue. However, classical processing techniques, according to which queries are executed exactly, both for what concerns the request and for what concerns the processing technique, which is set at the beginning of the execution, may not ensure adequate performance and quality (in terms of completeness and of accuracy) of the returned result. To overcome such problem, approximate and adaptive query processing techniques have been proposed. Adaptive techniques aim at ensuring an efficient query processing whenever a priori information, needed to statically select once at the beginning of the processing the most efficient processing technique, is not available. Approximation, by contrast, has been proposed for ensuring a higher result quality in presence of data heterogeneity and limited data knowledge. In highly dynamic and heterogeneous environments, these two approaches have usually been considered as orthogonal. However, we claim that applications exist that could benefit from a combined approach. An example are Web applications allowing to specify queries on heterogeneous data (streams), retrieved through mash-up from different sites. Since data are dynamically acquired, they cannot be statically reconciled, before processing queries. Moreover, adopting a single approximate search strategy, fixed a priori, could penalize the system efficiency and/or the quality of result, whenever heterogeneity only characterizes subsets of input data. The aim of this chapter is to make one step towards the integration of such approaches by introducing Approximate Search with Adaptive Processing (ASAP for short) systems. In ASAP, decisions concerning when, how, and how much to approximate are taken dynamically, with the goal of optimizing both the quality of result and the efficiency of processing.

A. Vakali & L.C. Jain (Eds.): New Directions in Web Data Management 1, SCI 331, pp. 171–212.
springerlink.com

1 Introduction

One of the main reasons for DBMS success is logical data independence, that is, the neat separation between the specification of 'what' we are searching from 'how' these searches (queries) are processed. The system is responsible for transforming declarative queries into execution plans, determined before the processing starts.

This approach ensures excellent performance for query execution on data with a completely known structure, executed in quite stable environments, with the availability of a reasonable set of statistical information on data. In the last years, however, there has been a rapid evolution of environments and applications that need to query data collections, that has radically modified the processing context. Specifically, in data integration applications, Web services, data streams, P2P systems, and hosting, data characteristics, as well as the dynamic processing conditions, are much more variable and unpredictable. The higher and higher resource sharing and the increasing interactivity in query processing make even those data properties that are traditionally conceived as static (such as relation cardinality and number of distinct values for an attribute) difficult to be known a priori and to be estimated in the above mentioned new contexts. The need thus emerged on one side to adapt the processing to dynamic conditions, thus giving up the a priori selection of a single execution strategy, fixed before processing starts. On the other side, as a consequence of data heterogeneity and limited data knowledge, we often cannot claim to get only 'precise' answers, that exactly satisfy the search condition expressed by the query, because in the above mentioned new contexts data are quite heterogeneous and exactly characterizing what we are looking for is really difficult. As a consequence, the search conditions of traditional queries, such as selections and joins, are relaxed or their evaluation is approximated, to improve result quality,[1] in terms of completeness and relevance with respect to the original query.

Up to now, query approximation and adaptive query processing have been mainly investigated as independent approaches. An exception is given by rank-aware optimization techniques (see Section 6) which deals with the detection of efficient execution plans for top-k queries (which can be seen as a particular type of approximate queries) rather than with the selection of approximation techniques in the choice of the execution plan for traditional queries, like selections and joins. There are however applications, working in highly dynamic and heterogeneous environments, that will benefit from combining the two approaches. In the Web context, a typical example is given by Web applications allowing the specification of queries on heterogeneous data (streams), retrieved through mash-up from different sites. Other examples concern interactive queries over highly heterogeneous XML data collections and multi-sensor applications. In all these examples, data are heterogeneous because collected from different sources but heterogeneity

---

[1] We remark that, under limited, insufficient resources, approximation has also been proposed as an approach to quickly provide an answer to a query. In this chapter, we do not further consider this kind of approximation.

information is not necessarily known in advance. At the same time, user queries can be imprecise since the user, due to such heterogeneity, may not exactly know the characteristics of data to be queried. As a consequence, a single execution plan of the user request may not be the more reasonable choice due to the high variability of the environment.

Based on these considerations, the aim of this chapter is to make one step towards the integration of approximate and adaptive query processing techniques by introducing Approximate Search with Adaptive Processing (ASAP for short) systems. In ASAP, decisions concerning when, how and how much to approximate are taken dynamically, with the goal of optimizing both the quality of result and the efficiency of processing. Indeed, in environments like the Web, data quality parameters are as important as classical parameters like response time. ASAP can therefore be defined as a new type of adaptive query processing which takes into account quality issues, in terms of completeness and accuracy of the result, besides performance. To this purpose, it leads to the definition of execution plans which interleave both precise and approximate evaluation in the most efficient way. The choice of approximation techniques may increase result completeness, when the original query result is stretched, or relevance, when the original query result is shrinked. This leads to the definition of QoD (Quality of Data)-oriented execution plans, in contrast with usual QoS (Quality of Service)-oriented execution plans, that are driven by efficiency or availability goals.

In the chapter, we first of all provide a survey of the main constituents of ASAP, that is, query approximation and adaptive query processing techniques. Approximation techniques will be classified with respect to their main goal and their reference approach, taking into account the characteristics of applications in distributed environments, with the Web as a special case. Adaptive query processing will be surveyed with respect to their main application contexts: local query processing, distributed query processing, and query processing on streaming data. In revising existing work, we consider several data types whose management may benefit from an ASAP approach: relational data, XML data, geo-spatial data (a specific type of structured data, nowadays quite relevant for distributed applications), both totally available at query time or made available as a continuous stream. Then, we present the notion of adaptively approximated search, through the definition of a generic query processing framework that can reveal useful in very diverse applicative scenarios, and we discuss the requirements a query processor needs to meet to support adaptively approximate processing. Different choices in addressing these requirements, possibly motivated by different application scenarios and execution contexts, lead to different instantiations of the framework and to different types of query processors relying on ASAP (called ASAP systems). Each execution plan, analyzed post-mortem and selected by an ASAP optimizer, is called ASAP technique and corresponds to a specific approach to adaptively approximate the processing of a traditional query (e.g., selection or join). The presented requirements are independent of the considered data model and summarize the several challenges that, from our point of view, should be addressed in order to make ASAP a practical technology.

We remark that a first step towards ASAP techniques has been made in [LMF+09] where an adaptive technique has been proposed to change an exact join operator into an approximate one when the presence of heterogeneity on the joining attribute prevents the identification of new matches, in the context of situational applications and mash-up [Jhi06]. Differently from [LMF+09], here the attempt is to provide an overall and reference framework for adaptively applying any kind of approximation. The ASAP approach also differs from rank-aware and quality-based query processors since in those cases the aim is to provide efficient processing for specific approximate techniques (top-k) or for traditional queries taking into account QoD parameters, respectively.

The paper is organized as follows. In Section 2, various examples of application contexts are provided which can benefit from adaptively approximated search. Query approximation techniques are surveyed in Section 3 while adaptive query processing is revised in Section 4. Requirements for ASAP systems are then discussed in Section 5. Section 6 briefly surveys work related to ASAP. Section 7 concludes the paper by providing a roadmap to be followed in order to make ASAP techniques a practical technology.

2 Examples

In this section we introduce some examples of contexts in which ASAP techniques can be profitably exploited. These examples differ on the type of data involved (i.e., relational, XML, spatial, spatio-temporal), on the processing context (i.e., stored vs streaming data), on the type of approximation (i.e., stretching vs shrinking).

Interactive queries over XML data collections. Consider the context in which several collections of XML documents, stored at autonomous sources, are interactively queried, through a Web application. Consider for instance a Web application providing help in planning evening programs, involving for instance cinema, theatre, or live music and dinner. According to her preferences, a user may formulate a query like "*Determine the cinemas or theaters accessible to user with disabilities where no later than 7 pm there is a movie/performance whose subject is fantasy with a close Indonesian ethnic restaurant*". The collections, though containing related documents, do not share a common schema thus exact approaches are impractical due to the great (and unpredictable) structural and content variations of the diverse sources. Heterogeneity may appear in the collections at different levels:

- Different tags may be employed in different collections to label the same information (e.g., *subject* instead of *type*, *hours* instead of *time*).
- The hierarchical structure of documents in different sources may be slightly different (e.g., accessibility information may be represented as an attribute of a cinema element, or as a sub-element of this elements, or as a descendant of this element, for instance as a sub-element of a *features* sub-element).

- Different strings may be employed at content level to represent the same information (e.g., *Indonesian* instead of *Indonesia*).

Given the casual and interactive nature of queries, and since the application interacts with different data sources and Web sites to dynamically crawl data, it is not reasonable to reconcile these heterogeneities. Thus, the original user queries will be solved in an approximate way, relaxing some equality constraints to similarity constraints, and returning only the most similar results to avoid returning too many results, relying on some similarity function. Once again, given the casual and interactive nature of queries, it is not reasonable to fully analyze the documents available in the source to extract enough information on their heterogeneity degrees to obtain the best similarity function to be used in approximating queries over the sources [SBM+07]. The most effective approach in this context could be to start querying the sources with some similarity function and use feedbacks on query execution to tune the approximation and similarity functions to be used in subsequent queries. Thus, in processing the subsequent queries on the same source, that refine, adjust, or completely modify the previous one, the application will rely on feedbacks from previous query executions. For instance, feedbacks can reveal that a certain tag or a certain data content were matched exactly in the exploited source, thus there is no need to approximate conditions involving them when evaluated against data in that source.

Mash-up applications. Consider the following monitoring request, submitted in the context of a Web application: *"Determine the transportation lines that may be delayed because of car accidents and display them on a map"*. This example involves at least three types of autonomous data sources, from which data is retrieved through mash-up:

- those containing relational information on car accidents: address, geographical position, and involved area (data stream);
- the one containing the road map (stored);
- those containing geo-spatial information about transportation lines (for instance stored in two different sources, one for buses and the other one for cable cars).

Since data on car accidents vary over time and can be collected by different subjects, the set of roads they refer to varies as well, and so does the associated geometry. This example exhibits various kinds of heterogeneity:

- Among strings, if the accident data source and the road map employ different formats for the alphanumeric representation of addresses.
- Among dynamic geo-spatial data, if areas related to car accidents are represented in different ways; for example, one source may represent such area just as a single point, another may represent the area as a polygon. We notice that such data are dynamic since they dynamically arrive from various sources. Therefore, information about the area representation may not be available in advance.

– Among static geo-spatial data, if transportation lines are represented in different ways in the data sources (e.g., as lines for buses and as regions for cable cars). In this case we may assume that such information are static in the sense that they are all available to the application and they do not change during computation.

As pointed out above, the dynamicity of car accidents does not allow to know in advance information on these data (specific representation of a data value – an address – or the type used to represent a geo-spatial information – the accident area –). Additionally, since data are dynamically acquired, their static reconciliation, before query processing starts, is not completely feasible (and not viable from a performance viewpoint). Finally, to cope with heterogeneity, the adoption of a priori fixed approximate search strategy is not a reasonable solution since it could penalize the system efficiency and/or the quality of result, whenever heterogeneity only characterizes subsets of input data (these information may not be available a priori because of the high dynamicity).

Multi-sensor applications. Consider a traffic operation center that produces and makes available on the Web a speed map of a metropolitan area, where the different motorway and freeway networks are broken into short segments and different color-coded speed are displayed for each segment. The center has two different sources of data: a stream of fixed sensors in the road network and a stream of readings from vehicles with on board GPS systems. Fixed sensors provide traffic speed and volume data from their location, vehicle data consists of speed and location for each vehicle. Suppose moreover that more detailed information, with different speed levels associated with different colors, of the five most congested areas closest to the city center should be provided as side bars. In this context we have to cope with different needs:

- Continuously monitor the data stream coming from the fixed sensor network, to determine the speed levels of the various segments.
- Determine the five most congested areas closest to the city centre (approximate top-k continuous query).
- For these areas, dynamically revert to rely on the more detailed stream data coming from vehicles to produce the more detailed maps.

In this example, the diverse representations of speed on the map cannot be statically decided since they depend on dynamic conditions, which can be checked through the execution of approximate continuous queries. Further approximation issues may arise due to possible different levels of details of different network maps and to mismatches and heterogeneity in data provided by the different sensor networks.

3 Query Approximation

In the following, after a brief introduction to query approximation and a classification of the main approaches, we review existing proposals for traditional, XML, and geo-spatial data.

3.1 An Introduction to Query Approximation

In the last years, there has been a rapid evolution of environments and applications that has radically modified query processing over data collections. Indeed, in data integration applications, Web services, data streams, P2P systems, and hosting, just to cite a few, data is highly heterogeneous and their characteristics are quite variable and unpredictable. In those contexts, query processing is therefore influenced by two main factors:

- *Data heterogeneity.* Data referring to the same entity may be contained in distinct datasets and represented in different ways. Processing such datasets inside the same query may therefore lead to data quality problems. In general, the degree of heterogeneity depends on the considered data types. Heterogeneity in relational/XML data arises from different value representation formats (for relational data) or structures (for XML data), which may be due to the different provenance of data on processing. Heterogeneity in geo-spatial data is often due to the employment of different resolution levels in representing data referring to the same geographical area.
- *Limited data knowledge.* The user may not always be able to specify the query in a complete and exact way since she may not know all the characteristics of data to be queried, even if data come from just one single source (possibly, because such characteristics may change during query execution, as in mash-up applications [Jhi06]).

Based on the previous considerations, we often cannot claim to get only 'precise' answers, that exactly satisfy the search condition expressed by the query, since in the above mentioned contexts it is quite common to find inconsistent or ambiguous data and it is really difficult to exactly characterize what we are looking for. As a consequence, the quality of the obtained result, in terms of completeness and accuracy, may be reduced, since interesting objects may not be returned. On the other hand, several uninteresting objects can be returned as answer, thus reducing user satisfaction. More generally, data heterogeneity and limited data knowledge may lead to the following two problems [ACDG03]:

- *Empty or few answers.* This problem arises when the query is too selective or data are quite heterogeneous (see [Luo06] for an approach detecting empty result queries without actual execution). In this case, it would be relevant for the user to stretch the query in order to get a set of approximate items, as result, ranked according to their relevance to the original query.

- *Many answers.* This problem arises when the query is low selective. In this case, ranking the obtained results and returning only a subset of them, corresponding to the more relevant ones (thus, shrinking the result), could be quite useful to the user.

In order to address the problems described above, in presence of highly heterogeneous data under highly dynamic architectures, *approximation techniques* can be used in order to *stretch* or *shrink* the result set, with the aim of providing more interesting or less unsatisfactory answers, respectively. Two main groups of approximation techniques can be devised (see Figure 1):

- *Query relaxation.* The concept of *query relaxation* has been introduced in Information Retrieval and adopted in several contexts as an approach for avoiding empty or too many answers. Query relaxation techniques allow the retrieval of a result also in very heterogeneous contexts or when the characteristics of data we are looking for are not completely known and this may lead to imprecise query specification. The main idea of query relaxation is to modify the query in order to stretch or shrink the result set. Query relaxation approaches can be classified depending on their scope into:
  - *Query rewriting approaches*: the query is rewritten using less or more strict operators, in order to get a larger or smaller answer set. Techniques of this type can be used to address both the empty or few answers and the many answers problems.
  - *Preference-based approaches*: user or system preferences are taken into account in order to generate the result, with the aim of providing best results first. Such techniques usually address the empty or few answers problem. However, they can also be thought as a shrinking approach with respect to the overall set of possible results, since they reduce the cardinality of the result dataset.
  - *Recommendation systems*: data correlation and past user history are used in order to suggest to the user further results besides those obtained by executing the specified query. Such techniques increase the number of results with respect to the original query but of course they are not suitable to solve the empty answers problem.
- *Approximate query processing.* It refers to all the techniques for executing a traditional query (e.g., a join) by using ad hoc query processing algorithms which automatically apply the minimum amount of relaxation based on the available data, in order to return the non-empty result more similar to the user request. This is possible for example by replacing equality checks with similarity checks, using a specific similarity function, depending on the domain of the attributes to be joined. Approximate query processing is suitable in all environments where data can be quite heterogeneous and may contain errors. Differently from query relaxation approaches, in this case the query is not changed; rather, its execution is modified in order to get more and approximate answers with respect to a traditional execution. Usually, approximate query processing allows one to get a larger result set, thus they are stretching techniques solving the empty or few answers problem.

Table 1 classifies the approaches described above based on: (i) the problem for which they have been defined (data heterogeneity, limited data knowledge); (ii) the underlying approach, stretching (+) or shrinking (-) the result; (iii) which information should be supplied by the user or the system in order to apply the corresponding techniques; (iv) the base operators to which they can be applied. Details about the content of the table will be provided in the remainder of this section.

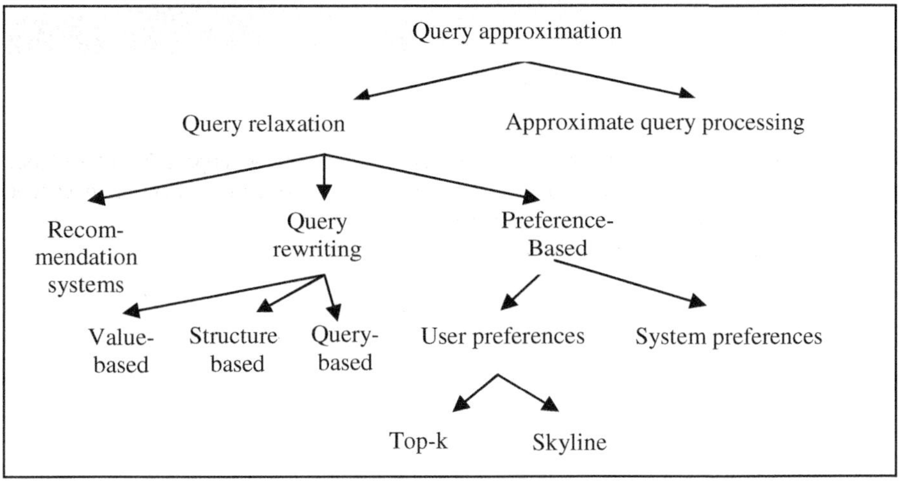

Fig. 1. Classification of query approximation techniques

Table 1. Properties of approximation techniques

Approach	Reference problem	Approach	User information	Reference queries
Query rewriting	Query spec	+ / -	Cardinality constraints	Selection, join
User preference-based	Query spec.	+/ -	Ranking function, set of relevant attributes	Selection (join)
System preference-based	Query spec.	+		Selection, join
Recommendation systems	Query spec.	+		Selection, join
Approximate query processing	Heterogeneity	+	Similarity function	Selection, join

In the following, we discuss and classify in more details query relaxation and approximate query processing approaches for traditional and XML data as well as for geo-spatial data. As pointed out in the following, query approximation constitutes nowadays one of the hot topics in traditional and XML query processing. Very recently, approximation approaches has also been proposed to deal with geo-spatial data, which represents a further fundamental data type for innovative applications in distributed architectures.

Finally, we notice that some techniques already exist that provide approximation through a summary representation of the input dataset. Such techniques have been mainly proposed for semi-structured data (as [PGI04]). Since we are interested in query approximation approaches which do not alter the input dataset, we do not further consider such approaches in the following.

3.2 Query Rewriting

The aim of *query rewriting* is to rewrite the user query into a new one when the results of the original query are too few or too many. This problem is in general different from the query rewriting issue in data integration problems, since in that case an input query, expressed over a global schema, has to be rewritten according to a possibly different local schema, in order to be locally executed. The main advantage of the query rewriting approach for query relaxation is that the generated queries can be executed using already existing query processing algorithms without the need of additional infrastructure. Of course, more efficient query processing algorithms can however be provided in order to exploit the properties of the resulting query set.

Query rewriting approaches can be used to address both the empty or few answers as well as the many answers problem. Preliminary work on query rewriting has been proposed in [Chau90], where a formal model for query stretching (called *query generalization*) has been provided. More practical approaches, instantiating that basic idea, have later been proposed. They can be classified into value-based, structure-based, and query-based depending on the used information for relaxing the query. More precisely:

- *Value-based* techniques rely on information concerning data distribution and query size estimation. This is the approach used in [MK09], for range and equality predicates on numerical and categorical attributes, and in [KWFH04], for range queries. In both cases, input information concerning the desired cardinality of the result set and information concerning data distribution are used to relax queries. The approach proposed in [MK09] is interactive, in the sense that the user can specify her preferences during the relaxation process.
- S*tructure-based* query relaxing techniques use schema or structure information (in case of semi-structured information like XML documents) during the relaxation process. This is the case of the approach proposed in [ZGBN07], for relaxing queries over data modeled using a modeling tool, called malleable schema, for modeling all the diverse and vague, both structured and semi-structured, entities of the real world. Most structure-based query rewriting

approaches have been proposed for XML documents, where structure information may refer to the type of relationships existing between nodes and order information [Lee02].
- *Query-based* approaches relax queries based on properties of the used query conditions. An example of query-based technique, which can be either directly applied by the system or requested by the user, is given by the geo-spatial relaxed topological selection operators and nearest neighbor operators proposed in [BBC+06]. Relaxation in this case is applied by considering during the execution topological relations whose similarity with respect to that specified in the query is higher than a given threshold, specified by the user in the query. Nearest neighbor operators, if the query condition is undefined for the spatial data at hand (for example, the user asks for all the rivers that cross a street, when rivers and streets are polygons and therefore relation cross is not defined) substitute the query relation with the most similar ones defined for the considered objects. Such operators can however return an empty answer when no object in the dataset satisfies the set of relaxed topological relationships.

With respect to data streams, query rewriting is mainly applied for efficiency issues, in order to limit the scope of the considered query, for example by applying a window size reduction. Such techniques are called static in [MWA+03].

3.3 Preference-Based Methods

Preference-based methods allow the user or the system to specify preferences in various ways and then use them in computing the result. Preferences can be specified in order to enlarge the result set in presence of the empty or few answers problems. When considered with respect to the overall set of data, preference-based methods can however be though as a shrinking approach, since only the 'best' results are returned, thus addressing the many answers problem. In the following, both user-based and system-based approaches will be surveyed.

User preferences. Two main operators have been proposed based on user preferences: top-k and skyline operators. The aim of the top-k operator is to restrict the number of returned results to a fixed number (k), based on some ranking. The aim of the skyline operator is to return only the best matches to the specified query. In both cases, specific query processing algorithms have been provided in order to efficiently generate the result.

While top-k operators return a small result at the price of specifying a ranking function, which is not a simple task, skyline operators avoid this specification at the price of a larger result set, which, even for two dimensional interest attributes, may be quite huge. In both cases, processing over join operations is challenging [IBMS08].

Top-k operators. The top-k operator returns the k objects that best satisfy a given condition. The answer to a top-k query is an ordered set of objects, where the ordering reflects how closely each object matches the given query condition. Specific similarity measures can be used for ordering objects [IBMS08]. The properties of the chosen ranking function impact in the design of top-k query processing techniques. Most of the existing top-k processing approaches rely on monotone ranking functions, which guarantee very good performance. They are used in several common applications, especially in the Web context [MBG04]. However, for complex applications, the ranking function can be expressed as a generic numeric expression to be optimized [ZHC+06].

A survey and a classification of existing top-k processing approaches for traditional and XML data has been proposed in [IBMS08]. Even if top-k operators can be themselves considered as a query approximation approach, they can further be executed in an approximate way, in order to improve performance. In those cases, approximate answers are associated with a probabilistic measure pointing out how far they are from the exact top-k answers [ARSZ03,YHC05]. Approximate top-k query processing has also been investigated in P2P environments [MTW05].

Concerning XML documents, top-k approaches rely on scoring functions that take into account similarity of both the content and the structure of the documents with respect to the considered query. Some of the existing approaches just extend techniques developed for relational data to the XML context [TBM+08]. Others consider specific XML scoring functions [AKM+05] and keyword search [GSBS03].

In the geo-spatial context, top-k operators return the first k objects that satisfy a given condition, based on a preference function computed over spatial and non spatial attributes. There exist top-k operators for various spatial operations, including spatial join [ZPZ+05], distance-based selection [HS99, XZK+05], preference query with respect to neighborhood objects [YDM+07]. The most extensively studied spatial query mechanism is ranking neighboring objects by the distance to a single query point [RKV95,BGRS99], which constitutes an instance of a top-1 problem. For multiple query points, Papadias et al. [PTMH05] studied ranking by the "aggregate" distance, for a class of monotone functions aggregating the distances to multiple query points.

The methods described above assume that all the relevant data are available before processing. Further, they report a single result and terminate. On the other hand, in stream environments, data are not known in advance. In those contexts, the aim is no more to compute a top-k query over all stream data; rather, the objective is to continuously monitor the top-k tuples over the stream and, accordingly, changing the result in a continuous way. One of the preliminary approach of this kind has been provided in [MBP06]. In [MAA06], an integrated approach for solving the problems of finding the top-k elements and finding frequent elements in a data stream are considered.

Skyline operators. In some applications, such as data exploration and decision making, it might be important to rank objects without using a specific ranking function, thus returning the 'best results' among all the possible ones, since defining a ranking function in those contexts could be cumbersome. This problem can be addressed in two distinct ways: (i) the system is delegated to compute the more adequate ranking function taking into account data information; (ii) preferences are still provided by the user but in terms of sets of attributes considered relevant for the ranking. Approach (i) will be described below, when we introduce techniques based on system-based preferences. The second approach relies on the definition of a partial relation among items, which can be formally specified using the concept of *dominance* and of *skyline*.

Given a set of points, each representing a list of values for the list of relevant attributes, the *skyline* contains the points that are not *dominated* by any other point. A point A dominates a point B if it is better in at least one dimension and equal or better in all the others, by considering a specific scoring function [BKS01].

Various algorithms have been proposed for skyline computation [BKS01]. Index based techniques (B-tree [BKS01], bitmap [TEO01], nearest neighbor [KRR02]) avoid scanning the overall set of data for skyline computation, improving performance with respect to basic techniques [BKS01]. The approaches cited above assume the interest is specified by a set of totally ordered attributes. This assumption does not allow the application of the existing techniques to temporal intervals and categorical data. To solve this problem, in [CET05a,CET05b] partially ordered domains are transformed into pairs of integers, over which the skyline is then computed.

The main problem in dealing with skyline is due to their often huge size. In order to address this problem, a common approach is to integrate both top-k and skyline advantages in a single technique. Three main approaches can be considered to this purpose [HB08]:

- The first approach consists in evaluating the quality of skyline objects and returning first high quality results. An approach of this type has been proposed in [YLL+05]. Starting from the observation that queries often refer to attribute subsets, the idea is that of pre-computing all possible skylines of all subspaces with all possible preferences (skycubes), which in general, for numeric attributes, are just two: the-more-the-better, the-less-the-better. When the user poses a query to a subspace, pre-computed skylines can be returned as result. A ranking is also provided in order to return 'best' skyline objects first. Other approaches for evaluating skyline relevance are proposed in: [CJT+06a], where the concept of skyline frequency is proposed as a way to measure the importance of a point based on the number of subsets of dimensions for which it is in the skyline; [CJT+06b], where the concept of k-dominance, relaxing the usual notion of dominance by considering at least k dimensions, is introduced; [LYZZ07,TDLP09], which consider finding k skyline tuples that best represent the contour of the entire skyline, called *k representative skyline*; [YM09], which proposes the *top-k dominating* query as the query scoring each tuple with the number of tuples it dominates; [XZT08, MC09, LYH09], proposing various approaches for scoring attribute importance.
- The second approach suggests to propose first to the user an overview of the skyline, which can be later refined. This is the case of the technique presented in [BZG05], for computing an approximation of the skyline object set. In this case, the result of a skyline query is just a sample of the real skyline result set.

The sample must contain only skyline objects, should be computed fast, should be small and representative.

- The third approach consists in eliciting more user preferences. An example is presented in [BGL07a, BGL07b]; the proposed approach allows users to add preference information incrementally. Each new elicited preference reduces skyline results. New preference information can be specified as new dominance relations, as equivalence relations, or as more complex trade-offs.

Concerning skyline computation over geo-spatial data, the problem of finding spatial locations that are not dominated with respect to the *network distance* to a single query point has been considered in [HJ05]. In presence of multiple query points, in [SSK09,SLAH09] various algorithms have been proposed for computing the spatial skyline, i.e., for identifying the locations such that no other location is closer than them to all query points.

Skyline queries have also been investigated for XML and stream data. Similarly to top-k queries, also for skyline it is not possible to compute the result over all stream data, due to resource limitations. Rather, the skyline is progressively generated, in an approximated way, based on data at hand, and then continuously monitored and updated, based on new object arrival or expiration. We refer the reader to [MXA04,TP06,LWLG09,CS09,ZLC09] for some recent works in those contexts.

System preferences. Preferences are not necessarily specified by the user. Rather, they can be implicitly applied by the system. This is the case of the relaxation technique proposed in [KLTV06]. Here, relaxation is applied to relational selection and join conditions over numeric attributes by redefining the semantics of such operators based on a relaxing function, quantifying the distance of each tuple (pair of tuples) with respect to the specified condition, using a numeric function (usually, the difference between numeric values appearing in the condition and in the tuple(s)). The relaxed version of the query provides a non-empty answer while being ‘close’ to the original query formulated by the user, using a skyline-based approach. Another approach for automatic ranking database result has been presented in [ACDG03], where techniques for automatically deriving ranking functions for both the empty or few and the many answers problems are investigated, adapting typical Information Retrieval approaches to the database context.

An approach similar to that presented in [KLTV06] has been provided in [Pod10] for both topological geo-spatial selection and join operators. Queries can be either syntactically correct or contain errors with respect to the used topological predicate which may not be defined for the object dimensions at hand (*non well-defined conditions*). Queries with non well-defined conditions are quite frequent in environments where the user does not know, for some application reason, the feature type dimension (e.g., in mash-up applications getting geo-spatial data from different Web-services). Non well-defined conditions are rewritten by the system using the most similar topological relations defined for the dimension of features in the available data. Two different relaxation semantics are proposed. The Best

Fit semantics applies the minimum amount of relaxation to the query condition in order to return a non empty answer. Thus, it models a variation of a top-1 query dealing with spatial relations, based on system preferences. The Threshold semantics relaxes the topological query up to a certain fixed limit, depending on system parameters. For all the considered operators and semantics, query processing algorithms, based on the usage of R-trees, have been provided.

3.4 Recommendation Systems

Recommendation systems aim at recommending to the users items not in the results of the posed queries but of potential interests. Recommendation methods can be further classified, according to [SDP09], depending on data to be considered for recommendation, into:

- *current-state*, if they use only the query result and the database content;
- *history-based*, which uses past user history;
- *external sources*, if they consider data contained in sources which are external to the database.

History-based approaches can be further classified into [AT05]:

- *content-based*, if they recommend items similar to those the user has preferred in the past;
- *collaborative*, that recommend items that similar users have liked in the past;
- *hybrid*, if they combine both approaches.

Recommendation systems have been initially proposed for Web services, very recently some approaches have also been proposed for the database context [AT05]. In particular, a framework for the declarative specification of the recommendation process over structured data is proposed in [KBG09] while the works reported in [SDP09] and [CEP09] present specific recommendation processes of current-state and content-based type, respectively. As far as we know, no current-state approaches have been still provided for geo-spatial and XML data.

3.5 Approximate Query Processing

Approximate query processing refer to all the techniques for executing a traditional query (e.g., a join) by using ad hoc query processing algorithms which automatically apply the minimum amount of relaxation based on the available data, in order to return a non-empty result more similar to the user request.

In the relational context, most approximation techniques concern the join operator [KS05] or face approximate match issues for strings [GIJ+01,CGK06] or numeric values, a quite relevant problem in case of join between tables coming from different data sources. The presence of distinct strings representing the same information may arise for human factors (incorrect data entry or ambiguity during data specification), application factors (errors in database population or not enforced constraints), or obsolescence, since data are usually dynamic.

Formally, an approximate join of two tables R1and R2 is a subset of the Cartesianproduct of R1 and R2. Specified attributes of R1 and R2 are matched and compared using a similarity function, instead of a usual equality predicate. The used similarity functions have strong analogies with those used in the context of data quality, for detecting that two values are distinct representations of the same real world entity (record linkage [KSS06] or removal of duplicate records [EIV07]).

Match can be performed by considering as matching field either a single attribute, a set of attributes, or an entire tuple. The general problem thus becomes that of, given two field values, quantifying their similarity, as a number between 0 and 1. If the field is numeric, numeric methods can be used. If fields are strings, the problem is more challenging. The existing techniques can be broadly classified into *edit-based* functions, if compare strings with respect to the single characters they contain, *token-based* if compare strings with respect to the tokens their contain, where a token is in general a substring satisfying specific properties [KS05].

The naive method for executing an approximate join consists in computing the similarity score for each pair of fields and keep only those whose similarity value is greater than a given threshold. This method is of course I/O and CPU intensive and therefore not scalable. Several algorithms have therefore been proposed with the aim of reducing the number of pairs over which similarity is computed, by taking advantage of efficient relational join methods [KS05].

For XML data, the approximate query processing problem has been deeply investigated, due to the very flexible structures and to the highly heterogeneous contexts in which XML data are used. The proposed approaches share the goal of integrating conditions over structure with the generation of approximate results. Queries are typically expressed through twigs (i.e., small trees) to which data have to conform. Approaches differ on how conditions over structure are relaxed during approximation and on how similarity is quantified. Conditions that can be relaxed could be specified with weights assigned to nodes and edges [ACS02]. The approximation degree for the structure is higher since even the ancestor/descendant relationship may not be preserved [SMG+08]. For what concerns similarity measures, used for quantifying proximity with respect to the twig, a survey can be found in [GMB06]. In [AKM+05] for example, twigs are decomposed into paths, whose correspondence with paths in the document is evaluated separately and single evaluations are then aggregated.

Some approaches have also been proposed for the approximate match of tree structures, in the context of both approximate join between XML data [GJK+02, ABD+08] and in the context of duplicate removal [WN05]. In this case, the need arises of determining whether the similarity between two tree structures is higher than a given threshold.

In querying data streams, approximation is used to deal with situations in which limited resources do not allow to produce an exact answer in a short time [DGR03,LLR07].[2] In particular, approximate techniques have been proposed for

---

[2] We remark that these techniques are employed to cope with QoS requirements (efficiency) differently from the others we discussed in this section and from the ones that we wish to face with ASAP techniques.

data reduction and synopsis construction (a survey has been presented in [ABB+02]) including: sketches [AMS96,FM83], random sampling [AGP00,AGP+99,CMN99], histograms [IP99,PG99], wavelets [CGR+00], and sliding windows [BDM02,DGI+02]. Such approximate query processing techniques are called *dynamic techniques* in [MWA+03]. Both approximate join and aggregate queries have been considered in this context [KS03].

Similar problems can be found in the spatio-temporal context, for example with continuous nearest neighbour queries [HZY05]. Also in this case, approximation is due to efficiency issues, not to data heterogeneity. Approximation issues also arise for spatial and temporal data represented at different resolutions and granularities [USU06,QQZ06] and in multi-way spatial join [PA02].

4 Adaptive Query Processing

In the following, after a brief introduction to adaptive query processing, we discuss the different styles of adaptation and review the main approaches that have been proposed in the context of local, distributed, and streaming data.

4.1 An Introduction to Adaptive Query Processing

As the database field has broadened to consider more general data management, the weaknesses of the traditional *plan-first execute-next* query processing model, according to which a query is processed following an execution plan selected on the basis of data statistics and a query optimization strategy, and then executed according to this plan with little or no run-time decision making, have begun to show themselves.

In emerging data management environments, such as data integration and data streams, as well as in most new query processing contexts, ranging from XML engines, to continuous queries engines, to Web or text engines, data statistics or other kind of information the optimizer can rely on may not be available a priori (before the query execution), or, if available, they may not be accurate, or even wrong. As a consequence, using this information to take decisions about which query plan to execute could result in a bad choice (sub-optimal plan).

Techniques addressing problems due to the lack of reliable statistics, unexpected correlations, unforeseen execution costs, and dynamic nature of data, using feedbacks for calibrating query execution, with the main goals of achieving better response time and more efficient CPU utilization, have proliferated and has resulted in a set of approaches, collectively named *adaptive query processing* (AQP, see [DIR07] for a recent survey). Adaptive query processing techniques do not rely on a priori information; rather they incrementally gather current information, that may be less complete but is up to date, in parallel with the query execution. Experimental evidence demonstrates the efficiency of the approach, resulting in several cases in the choice of an optimal or near-optimal execution plan.

The driver for the development of AQP techniques can be identified with the limits of traditional statistics-based query optimizers in coping with certain types of queries and in working in certain querying and query processing environments, characterized by missing or unreliable cardinality estimates. As the query is concerned, AQP is motivated by characteristics, such as query parameters and correlation, that statistics-based query optimizers have shown themselves unable to deal effectively with, and complex queries involving many tables, for which traditional optimizers rely on heuristic approaches to limit the plan search space. For what concerns the environments, we have from one side interactive query environments for data exploration and from the other one processing domains like data streams and wide area data sources.

The limitations of traditional query processing are quite evident in contexts where queries may be long running, and the data characteristics - and hence the optimal query plans - may change during the execution of the query. An obvious example of such data dynamicity is constituted by data stream environments, where there is also the need for multi-query optimization, since there are several continuous queries running on the same stream data, that may share some subexpressions, and the feedback from one query may be beneficial on the execution of other queries. In traditional database systems there is (almost) no inter-query state sharing. By contrast, data and computation sharing is crucial in processing queries on data streams.

The spectrum of adaptive query processing techniques has thus been quite broad:

- they may span multiple query executions or adapt within the execution of a single query;
- they may affect the query plan being executed or the scheduling of operations within the plan;
- they have been developed for improving the performance of local DBMS queries, distributed queries, queries on streaming data.

4.2 Styles of Adaptation

In the following, we present the general reference architecture for AQP and then the main styles of adaptations that correspond to instantiations of the architecture with different adaptation frequencies, namely inter-query adaptation, plan-change based adaptation, eddies.

MAPE Architecture. As introduced in [KC03] one of the methods to implement an adaptive behavior inside a system is that of using a MAPE (Monitor, Analyze, Plan, Execute) architecture. This type of architecture is formed by two main components: a managed component and an autonomic manager. The managed component is the one that carries over the real work to be done, the processing algorithm in our case. The autonomic manager on the other hand continually uses a MAPE approach in order to guarantee that the managed component is always giving the best possible effort. The components of such an architecture:

- *Monitor* the managed component actual performance and behavior in order to obtain valuable statistics and other type of information;
- *Analyze* the previously collected statistics and information to detect problems and/or opportunities;

- *Plan* a new behavior for the managed component in order to solve current problems or exploit in the best way the new opportunities;
- *Execute* the new decided behavior.

In the context of AQP, the Monitor step involves monitoring data characteristics, such as cardinality and distribution, and system characteristics, such as memory utilization and network bandwidth. Analysis is concerned with determining how well execution is proceeding, and is mainly guided by performance, or availability, goals and based on plan cost models or local heuristics. The Plan step may reconsider the query execution plan through an optimizer (in plan-change based adaptation, discussed below) or a routing policy (in eddies [AH00], discussed below). The Execute step corresponds to switching from one plan to another with careful state migration to reuse work and ensure correct answers.

The engineering effort in developing an adaptive query processor can thus be seen in defining more in detail all the four phases of the MAPE approach in order to build an efficient adaptive system that solves the specific processing problem to be faced. This general architecture can be instantiated in very different ways, leading to a wide spectrum of adaptation, depending on the frequency of adaptation.

Inter-query adaptation. The lowest level of adaptation consists in incorporating feedback from previous query executions for better selectivity/cardinality estimation. The statistics collected during the execution of a query are employed to better optimize future queries [SLM+01]. Since this approach does not reconsider nor switch plans during query execution, it is much simpler to be realized and it is not considered as fully adaptive in some surveys [DIR07].

Plan-change based adaptation. The coarser level of intra-query adaptation is plan-change based adaptation, where adaptation is inter-operator. According to this approach, the model relying on a well defined query execution plan at any time is retained, but the plan may be changed at well defined points during query processing. Mid-query re-optimization stops query processing when it detects that optimizer estimates are too different from run-time actual, and re-invokes the optimizer to pick a new plan. Adaptation is at fairly coarse granularity, typically at materialization points in query plan.

Adaptation can then be intra-operator: the query processor can use feedback and dynamic estimates to modify the query plan during execution, namely by replacing a physical operator with another that performs the same function. This idea has proven viable for pipelined[3] query plans, primarily as a dynamic optimization technique to improve the performance of a complex query, in cases where the initial query plan produced by the optimizer proves inefficient.

---

[3] A pipelined plan executes all operators in the query plan in parallel, by sending the output data of an operator directly to the next operator in the pipeline, as opposed to materialized plans in which operators are applied in sequence, computing (and materializing if needed) whole intermediate results.

[BB05] collectively refers to those approaches as plan-based since the execution of a current plan is monitored and re-optimization triggered whenever observed plan properties (e.g., intermediate result size) or system conditions (e.g., available memory) significantly differ from the estimates. The result is an optimizer that adds additional operations to track statistics (conventional as well as statistics on query sub-expressions collected during execution), detects and corrects situations where a non efficient plan is being executed. Both pipelined and non-pipelined plans can be executed.

Plan-based approaches instantiated the general MAPE architecture as follows:

- *Monitor,* through observation, statistics collected on data that passes through selected points in a plan;
- *Assess* whether an observed value is significantly different from an estimate, or outside the range of values for which the current plan is optimal; comparison may be in terms of estimated cost to completion;
- *Plan* based on the current statistics including those tracked by the current plan;
- *Execute* plan switching, with the goal of reducing the time devoted to the switch.

Plan switching involves many issues that need to be carefully addressed:

- *Correctness:* the new plan must not output results that have already been output by previous plan, nor miss results, especially in pipelined plans.
- *Reuse of work:* the current plan and plans before it may have processed a substantial part of the query; we need to consider (in a cost-based manner) whether the new plan can reuse this work instead of restart query processing from scratch.
- *Plan state:* the state captured by a plan can be taken into account as well, including: base data for the query (may be windows over stream), intermediate materialized sub-expression, in flight data in pipelined segments, temporary structures such as hash tables and sorted sub-lists.

In processing queries over stored data with non pipelined plans correctness may be ensured by producing no output data till the processing is complete (by buffering) or by keeping track of the data output so far. Work can be reused (if deemed convenient, on a cost basis) through materialized sub-expressions. Switch cost may be minimized starting the new plan on new input, combining the data partition produced by different plans after all sources are exhausted. In processing queries over streams, or with pipelined plans, there is no problem of duplicate results (each input is seen once). The state is migrated in temporary structure (e.g., hash tables) for reusing work in the new plan. To reduce the switch cost, one approach is to let the new plan process new data as it arrives, and compute later results from the 'combination' of old and new data, in an incremental fashion during processing of new data.

Though most AQP approaches cope with this correctness and reuse issues, the validity of the runtime changes proposed is rarely addressed in a rigorous manner and the adaptation undertaken is not formally characterized. One exception is constituted by [EFP06] for the replacement of join operators in pipelined query plans. Specifically, they provide a notation for describing partially evaluated operators and for each operator characterize the states, referred to as *quiescent states*, in which the result produced by the operator in the state can be precisely defined in terms of the input to the operator at this point in time. An operator in a quiescent state can thus be replaced by any other operator able to compute the remainder of the result. The complete result is then the union of that produced by the original operator with that produced by the replacement one. Note that this union may not be carried out explicitly, since replacing operators may simply resume the evaluation of a suspended plan. This establishes a safe foundation for a fine-grain replacement of operator in the middle of the execution.

Eddies. The finest-grained instantiation of the MAPE architecture is per-tuple adaptation that do not consider at all the notion of query plan, rather it views query processing as routing of tuples through operators and realizes plan changes by changing the order tuples are routed. The eddies technique, proposed in [AH00], is based on this idea. The eddy operator is a special operator that sits at the center of a tuple dataflow, intercepting the input and the output tuples of all other operators. It allows to control the execution plan at run-time at the level of each single tuple. Each tuple is considered as a message that should be sent from one operator to another. The routing scheduling can be changed when a specific asset of the query evaluation engine is reached. The eddy operator, thus, monitors the plan execution, and takes the decision concerning how routing tuples based on the asset. Most tuples exploit the route that is more efficient currently, while the rest explore other routes. [BB05] refers to the approaches relying on this idea as *routing-based* AQP approaches. The result is a greedy approach to optimization via selective tuple routing. The statistics the approach relies on are operator-level selectivities and incremental costs during execution. Routes through operators simulates pipelined plans.

Eddies instantiate the general MAPE architecture as follows:

- *Monitor and assessment* are realized through exploration and competition;
- *Plan* is implicit, and re-optimization happens automatically when statistics change;
- *Execute* enforcing routing constraints to avoid generating duplicate results and relying on fine-grained primitives for tuple router to migrate state, if not completely pipelined with no state.

4.3 Adaptive Approaches for Local Query Processing

The main motivation for AQP in local query processing is correcting optimizer mistakes, mainly due to unavailability of statistics about attribute correlations and skewed attribute distributions. Out of date statistics can be another reason for optimizer mistakes.

The first complete AQP prototype based on possible re-optimization during execution is Re-Opt [KD98], that re-optimizes the remaining query if statistics of materialized sub-plans differ significantly from optimizer estimates. A similar approach, that avoids unnecessary re-optimization and support re-optimization within pipelines is [MRS+04], that is prototyped in a commercial optimizer and shows significant benefits on real workloads.

Parametric queries are another motivations for AQP in this context. Many approaches to parametric optimization have been proposed [BBdW09] for handling parameters (mainly user inputs, but also memory size) whose values are unknown during optimization. The resulting optimization framework generates plans optimal for partitions of the parameter domains, and defers plan selection until the actual parameter values are known at run-time.

Most adaptive techniques have been proposed in the relational contexts. In the XML context, we mention [MAK+05], in which adaptation is used to allow that distinct partial matches for the same query follow distinct execution plans, considering the top-k nature of the problem. It is therefore a case of adaptive processing of an approximate query. In the spatial context, we recall the adaptive technique proposed in [SML03], for the execution of distance-based spatial join, and that in [Yu05], for range queries over multi-dimensional points.

4.4 Adaptive Approaches for Distributed Query Processing

The main motivation for AQP in distributed query processing is for coping with unknown statistics. In data integration systems, data sources available on-line and other data management systems that support queries over autonomous remote data sources, there is the need to cope with executing a query involving one or more sources for which no statistics are available. In addition to all the issues arising in AQP on local data, here another issue to adaptively consider is how to maximize CPU utilization, given the rate at which data are received from the distributed data sources.

Query scrambling [UFA98] deals with startup delay and bursty data arrival from remote data sources, with the goal of minimizing idle time during query processing, with the overall goal of reducing query response time [UF01]. Eddies [AH00], which we have already discussed in Section 4.2, and distributed eddies [TdW03] have been proposed as an approach to cope with the widely fluctuating characteristics of resources in large federated and shared-nothing databases.

In [IFF+99] query optimization for data integration systems is addressed. The work is motivated by absence of statistics and unpredictable data arrival characteristics, as well as overlap and redundancy among sources, that requires the processor to minimize the access to redundant sources and respond flexibly when some sources are unavailable. Adaptation is considered both at operator and plan level: adaptive operators such as the double pipelined hash join, also referred to as symmetric hash join, together with a collector operator realizing an efficient union of data from a large set of possibly overlapping or redundant sources have been proposed. Adaptive behavior is coordinated by a set of event-condition-action rules, where events may be raised by the execution of operators or at materialization points in the plan. Actions include modifying operator executor, re-ordering of operators, or re-optimization.

[BFM+00] is also concerned with query optimization for data integration and supports adaptation at plan and operator scheduling levels. Adaptive data partitioning [IHW04] is based on the idea of dividing source data into regions, each executed by different, complementary plans. This approach can be applied not only for correcting badly estimated cardinality and selectivity values, but also to discover and exploit order in source data as well as source data that can be effectively pre-aggregated.

Adaptive workload allocation in autonomous heterogeneous environments have been investigated in [ASP+09] and in cloud computing in [PAL+09]. An adaptive approach to query parallelization has been proposed in [PBC+09].

4.5 Adaptive Approaches for Query Processing on Streaming Data

Processing queries on streaming data requires to reconsider most of the basics of queries on stored data: not only transient, but also persistent (continuous) queries need to be processed; query answers are necessarily approximate due to the unboundedness of the stream leading to window joins to limit scope and to synopsis structures to approximate aggregates. Query operators and plans are necessarily adaptive: reacting to changes in input characteristics and system conditions is a major requirement for long-running query processing over data streams. For instance, stream arrival may be bursty, unpredictably alternating periods of slow arrival and periods of very fast arrival. The system conditions as well, e.g., the memory available to a single continuous query, may vary significantly over the query running time. In a situation where no input statistics are known initially and input characteristics as well as system conditions vary over time, all the relevant statistics are estimated during execution. Pipelined plans only can be executed, since operators in continuous query plans are non-blocking. To minimize the overhead, sampling based techniques are used for statistics tracking, which is combined with query execution whenever possible.

The optimization objectives are also different in a streaming context: the overall objective is to maximize the output rate for a query, rather than devising the least cost plan, and optimization is also rate-based for what concerns the input: the rates of the stream is taken into account during optimization rather than the input cardinality. Another optimization goal is to minimize resource (memory) consumption. Finally, it may be driven by QoS (e.g., availability) requirements [CCD+03].

Both plan-based and routing approaches have been considered for continuous queries. Plan-based approaches for continuous queries are referred to as CQ-based approaches in [BB05]. Eddies for continuous queries have been proposed in [MSH+02] and TelegraphCQ [CCD+03]. In Niagara [NdWM+01] the focus is on the adaptive sharing of common sub-expressions. In StreaMon [BW01] on integrated statistic collection (with sample-based tracking) and query re-optimization.

In CAPE [RDS+04] adaptation appears at many levels, ranging from operators, to scheduling, to distributed processing. Borealis [AAB+05] stream processing engine distributes query processing across multiple machines, monitoring run-time load and dynamically moving operators across machines to improve performance, and relies on load shedding for detecting and eliminating CPU overload from multiple machines by selectively dropping tuples in a coordinated fashion. An overall optimization approach for sensor networks have been proposed in [GBJ+09].

The proliferation of XML data produced by Web services has led to the development of specific approaches for XML data streams. Proposed approaches consider different granularities ranging from sequences of primitive tokens (e.g., start and end tag of an element) to sequences of document fragments. In the last few years, several approaches have been proposed for processing queries on XML data stream ([DAF03], [GS03], FluXQuery [KSS+04], BEA/XQRL [FHK+03], [IHW02], XSM [LMP02], XSQ [PC03]). A survey on XML stream query processing can be found in [WLR+09].

Continuous queries on spatio-temporal data streams [MXA04], supported for instance in Place [MXA+04] and Sole [MA08], are particularly meaningful in the mobile context, where the ever increasing availability of wireless networks (i.e., Wi-Fi) and GPS-equipped mobile devices, makes it easier to develop location-aware applications, such as traffic monitoring and tourist services. The specificity of this context is mainly that mobile objects are typically numerous (e.g., the vehicles in a city), volatile, and their position needs to be frequently updated (e.g., vehicles equipped with GPS).

5 Requirements for ASAP Systems

We define ASAP (Approximate Search with Adaptive Processing) as a new approach to QoD-oriented query processing, with the aim of guaranteeing a high result data quality, with respect to completeness and relevance, in the most efficient way. In the following, we call ASAP systems query processors relying on ASAP and with ASAP technique each execution plan, analyzed post-mortem, defined by the ASAP optimizer. The main characteristics of ASAP are:

- ASAP systems use an adaptive inter- or intra-query processing approach to query execution.
- Similarly to quality-based optimizers (see Section 6), ASAP systems take into account quality information, possibly provided in an interactive way, in the selection of the best query execution plan. In ASAP, quality is defined with respect to result completeness, by increasing the number of potentially interesting results, or result significance, by reducing the number of irrelevant results. Plans at the same quality level are then chosen with respect to efficiency considerations.
- In the choice of the best query execution plan, ASAP takes into account both precise and approximate query processing techniques, in order to guarantee a high result quality in the most efficient way.

Since ASAP relies on both precise and approximate processing techniques, the space of the execution plans for a given query may in general exponentially increase with respect to a traditional query processor. Of course, depending on the application context, the levels of freedom can be reduced and we believe that, in concrete cases, ASAP will compete with respect to traditional processing in terms of performance. Possible worst performance are traded-off with an increased quality of the result set, as explained above.

In order to develop a general framework for ASAP processing there are several issues to be taken into account, which are listed below. Some of them (1-2) refers to the interactions of an ASAP system with the working environments. Others (3-7) concern the ASAP system itself and lead to a specific instantiation of the MAPE architecture introduced in Section 4.2. Among them, requirements 6 and 7 deal with specific issues, to take into account when considering intra-query adaptiveness.

1. *Application contexts.* Suitable application contexts for the usage of ASAP systems have to be identified.
2. *User participation.* A characterization of the type of interactions between the user and ASAP techniques.
3. *Frequency of adaptation.* A choice of the most appropriate granularity of implementation of the MAPE architecture for ASAP systems.
4. *Properties monitored.* A characterization of the properties to monitor and of the conditions to assess, that are the basis on which a decision in terms of the processing techniques can be taken.
5. *Re-optimization.* A characterization of the overall goal of planning, in terms of the quality-efficiency trade-off and of the general principles that can guide it.
6. *Correctness.* A characterization of the conditions that ensure that a correct result is produced upon switch, which also entails determining processing states in which a switch between different techniques can be made.
7. *Reusability.* Upon a change in the processing techniques, whether some information on the processing/result on the already processed data can be reused, or whether the processing restarts from scratch with no information available.

In the following, for each of the previous topics, a list of challenges will be provided and discussed in details, providing some concrete examples. We notice that, though we discuss different challenges independently, the arising issues are not orthogonal and may impact one another. For instance, if the adaptation is inter-query, then plan switching is not an issue.

5.1 Application Contexts

Challenge 1: *Identify the application contexts where ASAP systems can be successfully used.*

The first challenge in the design of an ASAP system concerns the identification of the application contexts in which ASAP could be effectively used. This requires the identification of reference architectures, data models, and queries.

Sub-challenge 1.1: *Identify the reference architectures and data models suitable for ASAP.*

Based on what stated in Section 1 and the examples presented in Section 3, management of stored data under distributed architectures and data streams are the more natural contexts in which applying ASAP. Additionally, based on what stated in Sections 4 and 5, at the state of the art, a large number of approximation techniques have already been defined for relational, XML, geo-spatial data, either stored or in streams. Therefore, we claim that ASAP should be considered as a framework for all such data models.

Sub-challenge 1.2: *Identify the types of queries suitable for ASAP.*

ASAP increases the degree of freedom of the optimizer since it takes into account quality issues besides performance in the selection of the query plan. As a consequence, the size of the plan space increases with respect to a traditional optimizer. In order to keep the problem tractable, we claim that the analysis of ASAP should start from simple selection and join queries, followed by conjunctive and Select-Project-Join queries.

5.2 User Participation

Challenge 2: *Formalize the types of interactions between the user and an ASAP system.*

An important issue related to user participation concerns whether the usage of an ASAP system is transparent to the user or not and, in case the answer is yes, in which measure. Two main sub-challenges can therefore be considered, as pointed out below.

Sub-challenge 2.1: *Identify the level of user awareness in using an ASAP system.*

ASAP query optimizers should be designed in order to apply ASAP techniques each time a request is posed to the system in order to always guarantee a good trade-off between response time and result quality. At each switch state (see Section 5.6), the optimizer will choose the execution plan, taking into account result quality and efficiency. Therefore, it seems reasonable to assume that the user should trust the query processor in applying the approach which guarantees the best compromise between efficiency and result quality, based on a specific contract with it [Cha90]. Thus, we claim that she has not to be informed each time an ASAP technique is applied. On the contrary, we point out that, when using ASAP techniques inside a traditional optimizer, the user should be informed of the fact that an ASAP technique is going to be used for the execution of her request, since, in this case, this is just a choice of the optimizer which may alter the expected user result.

Sub-challenge 2.2: *Develop a model for the specification of user preferences.*

Most query approximation techniques rely on user or system information, such as cardinality constraints (for some query rewriting approaches), ranking (for top-k operators) or similarity (for approximate join) functions, sets of interesting attributes (for skyline operators) (see Table 1). In order for an ASAP system to consider for optimization one approximation method, the required information for its application has therefore to be available to the query processor. While some information can be directly chosen by the system (e.g., similarity functions), other requires a user specification. A model has therefore to be developed in order to specify all user and system preferences related to a specific domain.

Sub-challenge 2.3: *Identify when user preferences have to be specified to the system.*

Different approaches can be followed for user-preference specification. They mainly depend on the user type:

- *Frequent user.* In this case, we can assume the user knows quite well the application domain. Therefore, it seems reasonable to ask her to specify preferences. This can be done a priori or at the time the user asks for the usage of an ASAP technique (in case a traditional optimizer is used).
- *Non frequent user.* In this case, the user may not know the domain in depth. Therefore, imposing the specification of preference information, that are not necessarily used during the processing, seems a too strong requirement, especially in the context of traditional optimizers. Only system-defined preferences should be therefore used in this case.

In both cases, preferences can be specified una-tantum or may change, through user-interaction, during processing. This situation seems reasonable for long-running transactions or data streams. In those cases, information upon which interaction has to be based have to be carefully defined.

5.3 Frequency of Adaptation

Challenge 3: *Choose the most appropriate granularity of implementation of the MAPE architecture for ASAP systems.*

ASAP systems as adaptive query processors will rely on the general architecture discussed in Section 4.2, however appropriate frequency of adaptation needs to be chosen. Given the more ambitious goals of the ASAP framework, that does not try to maximize a single objective function in processing and rather tries to determine the most advantageous trade-off between possibly conflicting goals (such as efficiency of processing and data quality), and on the approximation techniques considered in the framework, coarser grained plan-change based adaptation seem more adequate than techniques relying on eddies. Moreover, both intra-query and inter-query adaptation can be considered.

5.4 Properties Monitored

Challenge 4: *Characterization of the properties to monitor and of the conditions to assess.*

To properly instantiate the MAPE framework, the properties monitored during the processing need to be devised, depending on the general goals of approximation, as well as on the processing contexts. These properties may refer both to QoS (Quality of Service) and QoD (Quality of Data) properties. An appropriate model for the assessment based on the monitored properties needs to be developed, which leads to the following sub-challenges.

Sub-challenge 4.1: *Characterize properties for QoS monitoring.*

QoS factors vary mainly depending on the application and processing contexts, but they mainly refer to efficiency and availability. Thus, QoS-related properties may refer to processing time, but also to CPU usage, network bandwith, and other typical measures used in AQP approaches.

Sub-challenge 4.2: *Characterize properties for QoD monitoring.*

QoD factors vary mainly depending on the motivations that lead to introduce approximation in our processing. Thus, for stretching techniques QoD is characterized in terms of completeness, while for shrinking techniques QoD is characterized in terms of relevance. QoD-related properties surely include the cardinality of result, but other relevant properties could be devised as well, such as heterogeneity or entropy in the result, which may be reasonable to monitor and assess for inter-query adaptation.

Sub-challenge 4.3: *Develop suitable cost and estimation models for assessment.*

Depending on the properties monitored, an appropriate model for assessment needs to be defined. The basic ideas would be to rely on the ratio between the resources employed up at a certain point and the expected/affordable one for QoS and on the ratio between the cardinality (which is much easier to dynamically evaluate than completeness or significance) of the result produced up to a certain point and the expected one.
This entails many challenging issues, both for what concerns developing realistic estimation models, that will likely be probabilistic models, and combining different components into a single model. Obviously, assessment relying on different components entails all the challenges of components assessment and the further issues related to combining models.

5.5 Re-optimization

Challenge 5: *Characterize the overall goal of planning, in terms of the quality-efficiency trade-off and of the general principles that can guide it.*

To keep re-planning feasible, in such a multi-faceted approach such as ASAP, some basic choices that limit the space to search for the most convenient solution must be made. A basic choice of the approach we propose, intrinsic to the motivations that lead to its definition, is to consider QoD first. Thus, the first aspect considered during re-optimization is whether the query as it is addresses the QoD requirements and, in case it does not, it is approximated/refined accordingly. To this purpose, a new query may have to be generated and, for it, the most appropriate processing technique needs to be selected. This entails revisiting the notion of query execution plan in the ASAP context. We may assume that each instantiation of the framework specifies a pool of techniques to be used for approximation. The reasonable sets of approximation techniques which can be used together in ASAP and how they can be interleaved, as well as appropriate heuristics to limit the plan search space, need to be defined. These considerations lead to the following sub-challenges.

Sub-challenge 5.1: *Define the notion of ASAP query execution plan.*

The classical notion of query plan needs to be revisited in the context of ASAP, thus establishing the building block of QoD-driven query processing. The notion of ASAP execution plan will provide the basis on which QoD-oriented optimization will be developed and thus needs to be formally defined.

Sub-challenge 5.2: *Identify the type of interplays between techniques used by ASAP.*

We claim that three distinct types of interplays can be applied by ASAP.

1. *From exact processing to approximate processing.* The aim of this interplay is to increase result quality by choosing an approximate evaluation of the query.
2. *From approximate processing to precise processing.* The aim of this interplay is to increase efficiency by choosing a precise evaluation of the query, usually computationally less expensive than the approximate one. This kind of interplay can also be useful in stretching or shrinking the result, when the approximate technique is shrinking or stretching, respectively.
3. *From one approximate processing to another.* The aim of the interplay in this case is either:
   - To change the type of the used approximation approach when data characteristics change. As an example, this may allow the processor to switch from a relaxation technique to an approximate processing when data heterogeneity increases. This fact can be detected by checking integrity constraints, e.g., foreign keys, which, in presence of high data heterogeneity, may not be satisfied any more. In this case, result quality increases while efficiency is not taken into account.
   - To change the specific approximate technique used (but not the type of the approximation used) for efficiency reasons. This approach seems reasonable when the considered techniques are instances of the same leaf in the tree presented in Figure 1 (e.g., two different top-k evaluations, two different approximate join evaluations using different similarity measures, threshold, or

algorithms). When the techniques are instances of two distinct query relaxation approaches (e.g., query rewriting and preference queries, or top-k and skyline), the switch can be considered relevant assuming that user-preferences may change during the computation (from a cardinality constraint to a ranking function to a set of important attributes) through user interaction or system decision. In this last case, result quality is improved while efficiency is not taken into account.

Sub-challenge 5.3: *Identify the approximation techniques that can be used alone or together in ASAP.*

Based on the classification reported in Section 3.1, it seems reasonable to assume that, from a theoretical point of view, all approximation techniques can be used, alone (i.e., only involved in switches of the first two types) in ASAP. A consideration is however required for all approaches based on a global execution, i.e., that need to access all the items before output the result. We claim that, in an ASAP context, similarly to what happens in streaming processing, such techniques should be applied, possibly in a continuous way, locally to a window. Such window should be dynamically defined and may correspond to the number of items analyzed before the next switch.

For the usage of a specific approximation technique, adequate preferences should be selected based on what specified by the user and possibly refined by the system. For example, if a skyline execution is scheduled, the set of attributes over which performing the execution has to be chosen. In case a top-k operator is selected, besides choosing the ranking function, parameter k should be selected. Its value may depend on some statistics on the average window size.

Concerning groups of approximation techniques to be used together, inside the same plan, we claim that they should pursue the same goal and follow the same approach. In particular, based on Figure 1 and Table 1, the following situations may arise:

- *Heterogeneous data, stretching approach*: in this context, approximate query processing techniques should be used.
- *Query specification problems, stretching approach*: all stretching query relaxation approaches can be used together in this context.
- *Query specification problems, shrinking approach*: all shrinking query relaxation approaches can be used in this case.

We notice that, in presence of heterogeneous data, shrinking approaches do not seem to be useful (see Table 1). This is because, as pointed out in Section 3, in presence of heterogeneous data, the typical approach is to relax equality checks into similarity-based checks. As a consequence, more results can be returned (stretching approach).

We finally remark that, in case ASAP is used to solve problems coming from both heterogeneous data and query specification, we claim that the approaches listed above for each problem can be used together inside the same query plan.

Sub-challenge 5.4: *Develop suitable heuristics to prune the plan search space.*

Based on the techniques considered in the instantiation of the framework, it is crucial, to limit the complexity of ASAP optimization, to rely on heuristics to limit the size of the plan search space. Thus, well-founded heuristics to determine which alternatives to explore and which ones to disregard, thus realizing a pruning of the plan search space, need to be developed.

5.6 Correctness

Challenge 6: *Characterize the conditions that ensure that a correct result is produced upon switch.*

In AQP, correctness upon plan switch means that the new plan must not output results that have already been output by previous plan, nor miss results. Difficulties are in ensuring this, but, since the result set is fixed, the definition is rather obvious to state. In ASAP, by contrast, even defining correctness and characterize it is not trivial. First of all, approximation by itself shifts the characterization of the result from exact to similar, thus some effort is needed in this correctness characterization. Moreover, the result set itself is different, depending on the approximation introduced. The switch, therefore, may change the result set, both in a larger and in a smaller set, depending on whether we switch from an approximate to an exact technique, or vice-versa. The situation is even more complex since approximation, as discussed in Section 3, can be either stretching or shrinking. Thus a switch from exact to approximate stretching, as well as a switch from approximate shrinking to exact, causes the expected result set to enlarge, though in different ways.

As a general consideration, we claim however that each single ASAP query plan should rely on either stretching or shrinking techniques. The motivation is that, using stretching *and* shrinking techniques inside the same plan may reduce the control over the overall result and may lead to an unbound number of possible switches at each step. Indeed, we claim that, as pointed out in [MK09], by mixing stretching and shrinking approaches one could transform any query to any other query.

Sub-challenge 6.1: *Devise properties for correctness of the switch in terms of no duplicate results.*

What we aim at ensuring is that, applying ASAP with a technique T1 switched to a technique T2 at a point where a portion P of input data has been processed does not cause the production of results that would not have been produced by T1 and T2 applied separately on the portions of data processed before and after P, respectively. In other words, the same input data item does not contribute twice to the result as a consequence of the switch. While this property is particularly meaningful for materialized results, we also have the converse property, very relevant for pipelined plans, corresponding to the following sub-challenge.

Sub-challenge 6.2: *Devise properties for correctness of the switch in terms of no misses.*

What we aim at ensuring is that, applying ASAP with a technique T1 switched to a technique T2 at a point where a portion P of input data has been processed does not cause the miss of results that would have been produced by T1 and T2 applied separately on the portions of data processed before and after P, respectively. In other words, an input data item does not fail to contribute to the result as a consequence of the switch.

Once these properties have been properly characterized, we also need to characterize the processing states in which we can safely switch.

Sub-challenge 6.3: *Define safe switch states.*

This means to characterize, in dependence of the specific switch we may want to realize, in which plan states a switch can be realized so that the correctness of the switch can be guaranteed. The challenge is even more hard for fine-grain inter-operator adaptation, where the switch is realized in the middle of an operator processing, and thus a notion of *quiescent* state needs to be devised. For instance, in [LMF+09] where the switch is between an exact and an approximate join algorithm, both implemented through symmetric hash join, safe switch states are those in which a tuple from one of the two data sources have been probed against the hash table containing the set of tuples from the other data source processed so far.

5.7 Reusability

Challenge 7: *Characterize the amount of processing work that can be reused upon a switch.*

One of the main issues of adaptive and thus of ASAP techniques is to minimize switch costs and thus to avoid thrashing [DIR07]. The most obvious way to achieve this goal is to reuse as much as possible the processing work done so far, both in terms of results already produced and in terms of auxiliary structures (e.g., hash tables) already available. We refer to these auxiliary structures built by the processing technique as *plan state*. Upon a change in the processing technique, we need thus to determine whether the result on the already processed data, or some other information on the processing, can be reused, otherwise the processing restarts from scratch with no information available. However, it is not always convenient to reuse the work already done, if this requires some extra work to become usable by the new technique, thus the choice of whether or not to reuse should be made in a cost-based manner. This leads to the following sub-challenges.

Sub-challenge 7.1: *Define properties for reuse of work.*

Depending on the type of the switch, on the techniques involved, and on the processing context (e.g., stored or streamed data), we need to determine whether upon a switch the processor needs to restart query processing from scratch or it can reuse (and how) the part of the query and the portions of data already being processed before the switch.

Sub-challenge 7.2: *Develop cost models for reuse of work.*

Depending again mostly on the type of the switch and on the techniques involved, if the reuse of work is possible, but some transformation work is needed, it may be the case that this extra work is more costly than restart query processing from scratch. Thus an appropriate cost model allowing to determine whether reuse is convenient should be defined to guide this decision.

Sub-challenge 7.3: *Define properties for plan state reuse.*

The various information collected by a processing technique on data, which we collectively refer to as *plan state*, and that may include base data for the query (in streaming contexts), intermediate materialized sub-expressions, temporary structures such as hash tables and sorted sub-lists, can be taken into account for reuse upon switching. For these information as well, most of the decisions depend on the specific techniques and information involved, and should be taken on cost basis. For instance, referring again to [LMF+09], exploiting a symmetric hash join both for exact and approximate join, the hash tables built when processing tuple in one case can be exploited when switching to another technique. Specifically, maintaining hash tables both on attribute values and their q-grams, allows a technique to reuse the auxiliary structures developed by the previous one. This would have not been possible, for instance, if the switch between an approximate join based on a q-gram based hash table and an approximate join based on trie indexes interplayed.

6 Related Work

The main existing proposals closed to ASAP refer to the following research areas: query approximation, adaptive query processing, rank-aware query processing, and quality-based query processing. Query approximation and adaptive query processing have been surveyed in details in Sections 3 and 4. The others areas will be briefly described in the following.

Rank-aware Query Optimization. Rank-aware query optimization deals with the detection of efficient execution plans for top-k queries, where the ranking function is user specified. The idea is to integrate rank-aware operators in query engines through the definition of an extended rank-aware relational algebra. In [ISA+04,LCIS05], such an approach has been followed for top-k selection and join queries. An adaptive extension of this approach has then been proposed in [IAE+06]. Those approaches consider top-k queries as a first-class query type which has therefore to be optimized, as any other relational operator. Optimization chooses between the usage of specific top-k algorithms and of a traditional join-then-sort plan in generating the ranked result. Other rank-aware operators have been proposed in [BCG02,CH02,NCSLV01].

Similarly to ASAP, rank-aware query processing deals with the execution of approximated queries (top-k in this case), possibly in an adaptive way. However, there are two main differences between the approaches. Rank-aware optimization considers only performance criteria in optimization and top-k queries are seen as first-class query types to be optimized. On the other hand, in ASAP, optimization is performed taking into account quality issues first. ASAP applies on traditional queries, such as selection and join, and approximation is seen as just an opportunity to increase result quality by the optimizer. Of course, ASAP optimizers can then rely on rank-aware optimizer in order to efficiently execute a top-k query, when selected for execution.

Quality-Based Query Processing. Quality-based query processing refers to processing techniques which take into account data quality parameters in estimating the cost of a query plan. Quality-based query processing has been mainly considered in data integration systems, where the quality of the various data sources has a strong impact on query result. Various QoS and QoD criteria have been considered. For example in [BKK+01], the user can specify quality constraints on, among the others, results (e.g., size of the result) and time (e.g., time to get the first results). Such constraints are then used in all the phases of query processing. If they cannot be fulfilled, the query plan is dynamically adapted or the query is aborted. On the other hand, in [NL99], quality criteria may concern the data source, specific queries computable by a source, or the ability of a source to provide attribute values for a certain query. A QoS-oriented multi-query scheduling over data streams has then been proposed in [WTZ09].

Similarly to ASAP, quality-based query processors take into account quality parameters in choosing the best execution plan. However, they consider a broad spectrum of quality criteria while ASAP deals with specific QoD constraints related to the empty answer and the many answer problems. Additionally, ASAP uses quality criteria in choosing the best approximate query to be used for stretching or shrinking the user query. Rather, current quality-based query processing approaches use them only in the selection of traditional query execution plans.

7 Concluding Remarks

The advent of new application environments, such as data integration systems, Web services, data streams, P2P systems, has greatly increased variability and unpredictability of data characteristics and dynamic processing conditions. In order to effectively query data in such new contexts, adaptive query processing has been proposed as an approach to adapt the processing to varying dynamic conditions while approximation methods have been introduced with the aim of relaxing queries or approximating their evaluation in order to improve result quality.

In this chapter, we claimed that applications exist that could benefit from a combined approach of Approximate Search and Adaptive Processing, leading to the definition of ASAP systems. In an ASAP system, decisions concerning when, how, and how much to approximate a standard query are taken dynamically, with the goal of optimizing both the quality of result, in terms of completeness and relevance with respect to the original query, and the efficiency of processing. Differently from other quality-based query processors, ASAP systems pursue the quality goals by proposing execution plans which interleave both precise and approximate evaluation in the most efficient way.

After reviewing existing work on approximate and adaptive query processing, several requirements for setting up ASAP systems have been discussed in the chapter. The overall problem is of course quite vast. In order to make the first steps towards the design of ASAP systems, we believe that the proposed challenges should be addressed in bottom-up and example driven way. This may help in addressing challenge 1. In particular, we propose the following road map, composed of three main research directions:

- *General theoretical foundations of ASAP systems.* This research direction includes the definition of a preference model (sub-challenges 2.1 and 2.2), QoS- and QoD-oriented parameters to be monitored and used in query optimization (sub-challenges 4.1 and 4.2), an estimation model for the assessment of QoD parameters (sub-challenge 4.3), formal definitions of correctness, completeness, and safety of switches (challenge 7), formal definition of ASAP execution plan (sub-challenges 5.1 and 5.2).
- *Optimization at each switch state.* This research direction includes the definition of techniques for pruning ASAP execution-plans based on the considered parameters, at a switch point (sub-challenge 5.4). We suggest to start by first considering only QoD-oriented parameters and consider QoS-oriented parameters, together with QoD ones, as a second step, in order to first investigate issues concerning data quality, which represent a key feature of ASAP systems. Specific groups of optimization techniques should be considered in order to limit the problem (sub-challenge 5.3).
- *Adaptive optimization.* This research direction concerns the definition of the overall ASAP optimization framework. We suggest to start by considering inter-query adaptation and investigate intra-query adaptation as a second step (challenge 3) and deal with specific groups of approximation techniques (sub-challenge 5.3). Issues to be taken into account are possible user interactions (sub-challenge 2.3) and principles concerning reusability of the processing work (challenge 8).

The results coming out from such research will be the base for the development of ASAP prototype systems, to be used for a concrete evaluation of their efficiency and effectiveness when used in distributed architectures.

References

[ABB+02] Arasu, A., Babcock, B., Babu, S., McAlister, J., Widom, J.: Characterizing Memory Requirements for Queries over Continuous Data Streams. In: PODS, pp. 221–232 (2002)

[AAB+05] Abadi, D.J., Ahmad, Y., Balazinska, M., Çetintemel, U., Cherniack, M., Hwang, J.-H., Lindner, W., Maskey, A., Rasin, A., Ryvkina, E., Tatbul, N., Xing, Y., Zdonik, S.B.: The Design of the Borealis Stream Processing Engine. In: CIDR, pp. 277–289 (2005)

[ABD+08] Augsten, N., Böhlen, M.H., Dyreson, C.E., Gamper, J.: Approximate Joins for Data-Centric XML. In: ICDE, pp. 814–823 (2008)

[ACDG03] Agrawal, S., Chaudhuri, S., Das, G., Gionis, A.: Automated Ranking of Database Query Results. In: CIDR (2003)

[ACS02] Amer-Yahia, S., Cho, S., Srivastava, D.: Tree Pattern Relaxation. In: EDBT, pp. 496–513 (2002)

[AGP00] Acharya, S., Gibbons, P.B., Poosala, V.: Congressional Samples for Approximate Answering of Group-by Queries. In: SIGMOD, pp. 487–498 (2000)

[AGP+99] Acharya, S., Gibbons, P.B., Poosala, V., Ramaswamy, S.: Join Synopses for Approximate Query Answering. In: SIGMOD, pp. 275–286 (1999)

[AH00] Avnur, R., Hellerstein, J.M.: Eddies: Continuously Adaptive Query Processing. In: SIGMOD, pp. 261–272 (2000)

[AKM+05] Amer-Yahia, S., Koudas, N., Marian, A., et al.: Structure and Content Scoring for XML. In: VLDB, pp. 361–372 (2005)

[AMS96] Alon, N., Matias, Y., Szegedy, M.: The Space Complexity of Approximating the Frequency Moments. In: ACM Symp. on Theory of Computing, pp. 20–29 (1996)

[ARSZ03] Amato, G., Rabitti, F., Savino, P., Zezula, P.: Region Proximity in Metric Spaces and its Use for Approximate Similarity Search. ACM Trans.on Information Systems 21(2), 192–227 (2003)

[ASP+09] Gounaris, A., Smith, J., Paton, N.W., Sakellariou, R., Fernandes, A.A.A., Watson, P.: Adaptive Workload Allocation in Query Processing in Autonomous Heterogeneous Environments: Distributed and Parallel Databases, vol. 25(3), pp. 125–164 (2009)

[AT05] Adomavicius, G., Tuzhilin, A.: Toward the Next Generation of Recommender Systems: A Survey of the State-of-the-art and Possible Extensions. IEEE Trans. Knowl. Data Eng. 17(6), 734–749 (2005)

[BB05] Babu, S., Bizarro, P.: Adaptive Query Processing in the Looking Glass. In: CIDR, pp. 238–249 (2005)

[BBC+06] Belussi, A., et al.: Towards Similarity-Based Topological Query Languages. In: EDBT Workshops, pp. 675–686 (2006)

[BBdW09] Bizarro, P., Bruno, N., DeWitt, D.J.: Progressive Parametric Query Optimization. IEEE Trans. Knowl. Data Eng. 21(4), 582–594 (2009)

[BCG02] Bruno, N., Chaudhuri, S., Gravano, L.: Top-k Selection Queries over Relational Databases: Mapping Strategies and Performance Evaluation. ACM Trans. Database Syst. 27(2), 153–187 (2002)

[BDM02] Babcock, B., Datar, M., Motwani, R.: Sampling from a Moving Window over Streaming Data. In: ACM-SIAM Symp. on Discrete Algorithms, pp. 633–634 (2002)

[BFM+00] Bouganim, L., Fabret, F., Mohan, C., Valduriez, P.: Dynamic Query Scheduling in Data Integration Systems. In: ICDE, pp. 425–434 (2000)

[BGL07a] Balke, W.-T., Güntzer, U., Lofi, C.: Eliciting Matters – Controlling Skyline Sizes by Incremental Integration of User Preferences. In: Kotagiri, R., Radha Krishna, P., Mohania, M., Nantajeewarawat, E. (eds.) DASFAA 2007. LNCS, vol. 4443, pp. 551–562. Springer, Heidelberg (2007)

[BGL07b] Balke, W.-T., Güntzer, U., Lofi, C.: User Interaction Support for Incremental Refinement of Preference-based Queries. In: RCIS, pp. 209–220 (2007)

[BGRS99] Beyer, K.S., Golstein, J., Ramakrishnan, R., Shaft, U.: When is " Nearest Neighbor" Meaningful? In: EDBT 1999. LNCS, vol. 1540, pp. 217–235. Springer, Heidelberg (1999)

[BKK+01] Braumandl, R., Keidl, M., Kemper, A., Kossmann, D., Kreutz, A., Seltzsam, S., Stocker, K.: ObjectGlobe: Ubiquitous Query Processing on the Internet. VLDB Journal 10(1), 48–71 (2001)

[BKS01] Börzsönyi, S., Kossmann, D., Stocker, K.: The Skyline Operator. In: ICDE, pp. 421–430 (2001)

[BW01] Babu, S., Widom, J.: Continuous Queries over Data Streams. SIGMOD Record 30(3), 109–120 (2001)

[BZG05] Balke, W.-T., Zheng, J.X., Güntzer, U.: Approaching the Efficient Frontier: Cooperative Database Retrieval Using High-Dimensional Skylines. In: Zhou, L.-z., Ooi, B.-C., Meng, X. (eds.) DASFAA 2005. LNCS, vol. 3453, pp. 410–421. Springer, Heidelberg (2005)

[CCD+03] Chandrasekaran, S., Cooper, O., Deshpande, A., Franklin, M.J., Hellerstein, J.M., Hong, W., Krishnamurthy, S., Madden, S., Raman, V., Reiss, F., Shah, M.A.: TelegraphCQ: Continuous Dataflow Processing for an Uncertain World. In: CIDR (2003)

[CEP09] Chatzopoulou, G., Eirinaki, M., Polyzotis, N.: Query Recommendations for Interactive Database Exploration. In: SSDBM, pp. 3–18 (2009)

[CET05a] Chan, C.Y., Eng, P.-K., Tan, K.-L.: Efficient Processing of Skyline Queries with Partially-Ordered Domains. In: ICDE, pp. 190–191 (2005)

[CET05b] Chan, C.Y., Eng, P.-K., Tan, K.-L.: Stratified Computation of Skylines with Partially Ordered Domains. In: SIGMOD, pp. 203–214 (2005)

[CH02] Chang, K.C.-C., Hwang, S.: Minimal Probing: Supporting Expensive Predicates for Top-k Queries. In: SIGMOD, pp. 346–357 (2002)

[Cha90] Chaudhuri, S.: Generalization and a Framework for Query Modification. In: ICDE, pp. 138–145 (1990)

[CGK06] Chaudhuri, S., Ganti, V., Kaushik, R.: A Primitive Operator for Similarity Joins in Data Cleaning. In: ICDE (2006)

[CGR+00] Chakrabarti, K., Garofalakis, M.N., Rastogi, R., Shim, K.: Approximate Query Processing using Wavelets. In: VLDB, pp. 111–122 (2000)

[CJT+06a] Chan, C.-Y., Jagadish, H.V., Tan, K.-L., Tung, A.K.H., Zhang, Z.: On High Dimensional Skylines. In: EDBT, pp. 478–495 (2006)

[CJT+06b] Chan, C.-Y., Jagadish, H.V., Tan, K.-L., Tung, A.K.H., Zhang, Z.: Finding k-dominant Skyline in High Dimensional Space. In: SIGMOD, pp. 503–514 (2006)

[CMN99] Chaudhuri, S., Motwani, R., Narasayya, V.: On Random Sampling over Joins. In: SIGMOD, pp. 263–274 (1999)

[CS09] Cohen, S., Shiloach, M.: Flexible XML Querying Using Skyline Semantics. In: ICDE, pp. 553–564 (2009)

[DAF03] Diao, Y., Diao, Y., Altinel, M., Franklin, M.J., Zhang, H., Fischer, P.M.: Path Sharing and Predicate Evaluation for High-Performance XML Filtering. ACM Transaction on Database Systems 28(4), 467–516 (2003)

[DGI+02] Datar, M., Gionis, A., Indyk, P., Motwani, R.: Maintaining Stream Statistics over Sliding Windows. In: ACM-SIAM Symp. on Discrete Algorithms, pp. 635–644 (2002)

[DGR03] Das, A., Gehrke, J., Riedewald, M.: Approximate Join Processing Over Data Streams. In: SIGMOD, pp. 40–51 (2003)

[DIR07] Deshpande, A., Ives, Z., Raman, V.: Adaptive Query Processing. Foundations and Trends in Databases 1(1), 1–140 (2007)

[EFP06] Eurviriyanukul, K., Fernandes, A.A.A., Paton, N.W.: A Foundation for the Replacement of Pipelined Physical Join Operators in Adaptive Query Processing. In: Grust, T., Höpfner, H., Illarramendi, A., Jablonski, S., Fischer, F., Müller, S., Patranjan, P.-L., Sattler, K.-U., Spiliopoulou, M., Wijsen, J. (eds.) EDBT 2006. LNCS, vol. 4254, pp. 589–600. Springer, Heidelberg (2006)

[EIV07] Elmagarmid, A.K., Ipeirotis, P.G., Verykios, V.: Duplicate Record Detection: A Survey. IEEE Trans. Knowl. Data Eng. 19(1), 1–16 (2007)

[FHK+03] Florescu, D., Hillery, C., Kossmann, D., Lucas, P., Riccardi, F., Westmann, T., Carey, M.J., Sundararajan, A., Agrawal, G.: The BEA/XQRL Streaming XQuery Processor. In: VLDB, pp. 997–100 (2003)

[FM83] Flajolet, P., Martin, G.: Probabilistic Counting. In: IEEE Symp. on Foundations of Computer Science (1983)

[GBJ+09] Galpin, I., Brenninkmeijer, C.Y.A., Jabeen, F., Fernandes, A.A.A., Paton, N.W.: Comprehensive Optimization of Declarative Sensor Network Queries. In: Winslett, M. (ed.) SSDBM 2009. LNCS, vol. 5566, pp. 339–360. Springer, Heidelberg (2009)

[GIJ+01] Gravano, L., Ipeirotis, P.G., Jagadish, H.V., Koudas, N., Muthukrishnan, S., Srivastava, D.: Approximate String Joins in a Database (Almost) for Free. In: VLDB, pp. 491–500 (2001)

[GMB06] Guerrini, G., Mesiti, M., Bertino, E.: Structural Similarity Measures in Sources of XML Documents. In: Darmont, J., Boussaid, O. (eds.) Processing and Managing Complex Data for Decision Support, pp. 247–279. IDEA Group, USA (2006)

[GJK+02] Guha, S., Jagadish, H.V., Koudas, N., Srivastava, D., Yu, T.: Approximate XML Joins. In: SIGMOD, pp. 287–298 (2002)

[GS03] Gupta, A.K., Suciu, D.: Stream Processing of XPath Queries with Predicates. In: SIGMOD, pp. 419–430 (2003)

[GSBS03] Guo, L., Shao, F., Botev, C., Shanmugasundaram, J.: XRANK: Ranked Keyword Search over XML Documents. In: SIGMOD, pp. 16–27 (2003)

[HB08] Hwang, S., Balke, W.: Preference Query Formulation and Processing: Ranking and Skyline Query Approaches. In: DASFAA Tutorial (2008)

[HJ05] Huang, X., Jensen, C.S.: In-Route Skyline Querying for Location-Based Services. In: Kwon, Y.-J., Bouju, A., Claramunt, C. (eds.) W2GIS 2004. LNCS, vol. 3428, pp. 120–135. Springer, Heidelberg (2005)

[HS99] Hjaltason, G.R., Samet, H.: Distance Browsing in Spatial Databases. ACM Trans. Database Syst. 24(2), 265–318 (1999)

[HZY05] Hsueh, Y.-L., Zimmermann, R., Yang, M.H.: Approximate Continuous k Nearest Neighbor Queries for Continuous Moving Objects with Pre-defined Paths. In: ER Workshops, pp. 270–279 (2005)

[KBG09] Koutrika, G., Bercovitz, B., Garcia-Molina, H.: Flexrecs: Expressing and Combining Flexible Recommendations. In: SIGMOD, pp. 745–758 (2009)

[KC03] Kephart, J.O., Chess, D.M.: The Vision of Autonomic Computing. IEEE Computer 36(1), 41–50 (2003)

[KD98] Kabra, N., DeWitt, D.J.: Efficient Mid-Query Re-optimization of Suboptimal Query Execution Plans. In: SIGMOD, pp. 106–117 (1998)

[KLTV06] Koudas, N., Li, C., Tung, A.K.H., Vernica, R.: Relaxing Join and Selection Queries. In: VLDB, pp. 199–210 (2006)

[KRR02] Kossmann, D., Ramsak, F., Rost, S.: Shooting Stars in the Sky: an Online Algorithm for Skyline Queries. In: VLDB, pp. 275–286 (2002)

[KS03] Koudas, N., Srivastava, D.: Data Stream Query Processing: A Tutorial. In: VLDB Tutorial (2003)

[KS05] Koudas, N., Srivastava, D.: Approximate Joins: Concepts and Techniques. In: VLDB Tutorial (2005)

[KSS06] Koudas, N., Sarawagi, S., Srivastava, D.: Record Linkage: Similarity Measures and Algorithms. In: SIGMOD Tutorial (2006)

[KSS+04] Koch, C., Scherzinger, S., Schweikardt, N., Stegmaier, B.: FluXQuery: An Optimizing XQuery Processor for Streaming XML Data. In: VLDB, pp. 1309–1312 (2004)

[KWFH04] Kadlag, A., Wanjari, A.V., Freire, J., Haritsa, J.R.: Supporting Exploratory Queries in Databases. In: Lee, Y., Li, J., Whang, K.-Y., Lee, D. (eds.) DASFAA 2004. LNCS, vol. 2973, pp. 594–605. Springer, Heidelberg (2004)

[IAE+06] Ilyas, I.F., Aref, W.G., Elmagarmid, A.K., Elmongui, H.G., Shah, R., Vitter, J.S.: Adaptive Rank-aware Query Optimization in Relational Database. ACM Transactions on Database Systems 31(4), 1257–1304 (2006)

[IBMS08] Ilyas, I.F., Beskales, G., Soliman, M.A.: A Survey of Top-k Query Processing Techniques in Relational Database Systems. ACM Comput. Surveys 40(4) (2008)

[IFF+99] Ives, Z.G., Florescu, D., Friedman, M., Levy, A.Y., Weld, D.S.: An Adaptive Query Execution System for Data Integration. In: SIGMOD, pp. 299–310 (1999)

[IHW02] Ives, Z.G., Halevy, A.Y., Weld, D.S.: An XML Query Engine for Network-bound data. VLDB Journal 11(4), 380–402 (2002)

[IHW04] Ives, Z.G., Halevy, A.Y., Weld, D.S.: Adapting to Source Properties in Processing Data Integration Queries. In: SIGMOD, pp. 395–406 (2004)

[IP99] Ioannidis, Y., Poosala, V.: Histogram-Based Approximation of Set-Valued Query-Answers. In: VLDB, pp. 174–185 (1999)

[ISA+04] Ilyas, I.F., Shah, R., Aref, W.G., Vitter, J.S., Elmagarmid, A.K.: Rank-aware Query Optimization. In: SIGMOD, pp. 203–214 (2004)

[Jhi06] Jhingran, A.: Enterprise Information Mash-ups: Integrating Information, Simply. In: VLDB, pp. 3–4 (2006)

[Lee02] Lee, D.: Query Relaxation for XML Model. Phd Thesis, University of California (2002)

[LCIS05] Li, C., Chang, K.C.-C., Ilyas, I.F., Song, S.: RankSQL: Query Algebra and Optimization for Relational Top-k Queries. In: SIGMOD, pp. 131–142 (2005)

[LLR07] Li, Y., Lokey, S.W., Ramakrishna, M.V.: Performance Study of Data Stream Approximation Algorithms in Wireless Sensor Networks. In: ICPADS, pp. 1–8 (2007)

[LMF+09] Lengu, R., Missier, P., Fernandes, A.A.A., Guerrini, G., Mesiti, M.: Time-completeness Trade-offs in Record Linkage using Adaptive Query Processing. In: EDBT, pp. 851–861 (2009)

[LMP02] Ludäscher, B., Mukhopadhyay, P., Papakonstantinou, Y.: A Transducer-Based XML Query Processor. In: VLDB, pp. 227–238 (2002)

[Luo06] Luo, G.: Efficient Detection of Empty-Result Queries. In: VLDB, pp. 1015–1025 (2006)

[LYH09] Lee, J., won You, G., won Hwang, S.: Personalized Top-k Skyline Queries in High-Dimensional Space. Inf. Syst. 34(1), 45–61 (2009)

[LYZZ07] Lin, X., Yuan, Y., Zhang, Q., Zhang, Y.: Selecting Stars: The k Most Representative Skyline Operator. In: ICDE, pp. 86–95 (2007)

[LWLG09] Li, L., Wang, H., Li, J., Gao, H.: Efficient Algorithms for Skyline Top-K Keyword Queries on XML Streams. In: Zhou, X., Yokota, H., Deng, K., Liu, Q. (eds.) DASFAA 2009. LNCS, vol. 5463, pp. 283–287. Springer, Heidelberg (2009)

[MAK+05] Marian, A., Amer-Yahia, S., Koudas, N., Srivastava, D.: Adaptive Processing of Top-k Queries in XML. In: ICDE, pp. 162–173 (2005)

[MA08] Mokbel, M.F., Aref, W.G.: SOLE: Scalable Online Execution of Continuous Queries on Spatiotemporal Data Streams. VLDB Journal 17(5), 971–995 (2008)

[MAA06] Metwally, A., Agrawal, D., Abbadi, A.E.: An Integrated Efficient Solution for Computing Frequent and Top-*k* Elements in Data Streams. ACM Trans. Database Syst. 31(3), 1095–1133 (2006)

[MBG04] Marian, A., Bruno, N., Gravano, L.: Evaluating top-k Queries over Web-Accessible Databases. ACM Trans. on Database Systems 29(2), 319–362 (2004)

[MBP06] Mouraditis, K., Bakiras, S., Papadias, D.: Continuous Monitoring of Top-k Queries over Sliding Windows. In: SIGMOD, pp. 635–646 (2006)

[MC09] Mindolin, D., Chomicki, J.: Discovering Relative Importance of Skyline Attributes. PVLDB 2(1), 610–621 (2009)

[MK09] Mishra, C., Koudas, N.: Interactive Query Refinement. In: EDBT, pp. 862–873 (2009)

[MRS+04] Markl, V., Raman, V., Simmen, D.E., Lohman, G.M., Pirahesh, H.: Robust Query Processing through Progressive Optimization. In: SIGMOD, pp. 659–670 (2004)

[MSH+02] Madden, S., Shah, M.A., Hellerstein, J.M., Raman, V.: Continuously Adaptive Continuous Queries over Streams. In: SIGMOD, pp. 49–60 (2002)

[MTW05] Michel, S., Triantafillou, P., Weikum, G.: KLEE: A Framework for Distributed Top-k Algorithms. In: VLDB, pp. 637–648 (2005)

[MWA+03] Motwani, R., Widom, J., Arasu, A., Babcock, B., Babu, S., Datar, M., Manku, G.S., Olston, C., Rosenstein, J., Varma, R.: Query Processing, Resource Management, and Approximation in a Data Stream Management System. In: CIDR (2003)

[MXA04] Mokbel, M.F., Xiong, X., Aref, W.G.: SINA: Scalable Incremental Processing of Countinuous Queries in Spatio-Temporal Database. In: SIGMOD, pp. 623–634 (2004)

[MXA+04] Mokbel, M.F., Xiong, X., Aref, W.G., Hambrusch, S.E., Prabhakar, S., Hammad, M.A.: PLACE: A Query Processor for Handling Real-time Spatio-temporal Data Streams. In: VLDB, pp. 1377–1380 (2004)

[NCSLV01] Natsev, A., Chang, Y.-C., Smith, J.R., Li, C.-S., Vitter, J.S.: Supporting Incremental Join Queries on Ranked Inputs. In: VLDB, pp. 281–290 (2001)

[NdWM+01] Naughton, J.F., DeWitt, D.J., Maier, D., Aboulnaga, A., Chen, J., Galanis, L., Kang, J., Krishnamurthy, R., Luo, Q., Prakash, N., Ramamurthy, R., Shanmugasundaram, J., Tian, F., Tufte, K., Viglas, S., Wang, Y., Zhang, C., Jackson, B., Gupta, A., Chen, R.: The Niagara Internet Query System. IEEE Data Eng. Bull. 24(2), 27–33 (2001)

[NL99] Naumann, F., Lesser, U.: Quality–driven Integration of Heterogeneous Information Systems. In: VLDB, 447–458 (1999)

[PA02] Papadias, D., Arkoumanis, D.: Approximate Processing of Multiway Spatial Joins in Very Large Databases. In: Jensen, C.S., Jeffery, K., Pokorný, J., Šaltenis, S., Hwang, J., Böhm, K., Jarke, M. (eds.) EDBT 2002. LNCS, vol. 2287, pp. 179–196. Springer, Heidelberg (2002)

[PAL+09] Paton, N.W., Aragão, M.A.T., Lee, K., Fernandes, A.A.A., Sakellariou, R.: Optimizing Utility in Cloud Computing through Autonomic Workload Execution. IEEE Data Eng. Bull. 32(1), 51–58 (2009)

[PBC+09] Paton, N.W., Chávez, J.B., Chen, M., Raman, V., Swart, G., Narang, I., Yellin, D.M., Fernandes, A.A.A.: Autonomic Query Parallelization using non-Dedicated Computers: an Evaluation of Adaptivity Options. VLDB J 18(1), 119–140 (2009)

[PC03] Peng, F., Chawathe, S.S.: XPath Queries on Streaming Data. In: SIGMOD, pp. 431–442 (2003)

[PG99] Poosala, V., Ganti, V.: Fast Approximate Answers to Aggregate Queries on a Data Cube. In: SSDBM, pp. 24–33 (1999)

[PGI04] Polyzotis, N., Garofalakis, N., Yoannidis, Y.E.: Approximate XML Query Answers. In: SIGMOD, pp. 263–274 (2004)

[Pod10] Podesta', P.: Query Processing and Analysis of Multi-resolution Spatial Data in Distributed Architectures. PhD Thesis, University of Genoa, Italy (2010)

[PTMH05] Papadias, D., Tao, Y., Fu, G., Seeger, B.: Progressive Skyline Computation in Database Systems. ACM Trans. Database Syst. 30(1), 41–82 (2005)

[QQZ06] Qin, S., Qian, W., Zhou, A.: Approximately Processing Multi-granularity Aggregate Queries over Data Streams. In: ICDE (2006)

[RDS+04] Rundensteiner, E.A., Ding, L., Sutherland, T.M., Zhu, Y., Pielech, B., Mehta, N.: CAPE: Continuous Query Engine with Heterogeneous-Grained Adaptivity. In: VLDB, pp. 1353–1356 (2004)

[RKV95] Roussopoulos, N., Kelley, S., Vincent, F.: Nearest Neighbor Queries. In: SIGMOD, pp. 71–79 (1995)

[SBM+07] Sanz, I., Berlanga, R., Mesiti, M., Guerrini, G.: ArHeX: Flexible Composition of Indexes and Similarity Measures for XML. In: ICDE Workshops, pp. 281–284 (2007)

[SDP09] Stefanidis, K., Drosou, M., Pitoura, E.: "You Also Like" Results in Relational Databases. In: PersDB (2009)

[SLM+01] Stillger, M., Lohman, G.M., Markl, V., Kandil, M.: LEO - DB2's LEarning Optimizer. In: VLDB, pp. 19–28 (2001)

[SMG+08] Sanz, I., Mesiti, M., Guerrini, G., Berlanga, R.: Fragment-based Approximate Retrieval in Highly Heterogeneous XML Collections. Data Knowl. Eng. 64(1), 266–293 (2008)

[SML03] Shin, H., Moon, B., Lee, S.: Adaptive and Incremental Processing for Distance Join Queries. IEEE Trans. Knowl. Data Eng. 15(6), 1561–1578 (2003)

[SSK09] Sharifzadeh, E., Shahabi, C., Kazemi, L.: Processing Spatial Skyline Queries in both Vector Spaces and Spatial Network Databases. ACM Trans. Database Syst. 34(3) (2009)

[SSLAH09] Son, W., Lee, M.-W., Ahn, H.K., Hwang, S.-W.: Spatial Skyline Queries: An Efficient Geometric Algorithm. In: Mamoulis, N., Seidl, T., Pedersen, T.B., Torp, K., Assent, I. (eds.) SSTD 2009. LNCS, vol. 5644, pp. 247–264. Springer, Heidelberg (2009)

[TBM+08] Theobald, M., Bast, H., Majumdar, D., Schenkel, R., Weikum, G.: TopX: Efficient and Versatile Top- k Query Processing for Semistructured Data. VLDB J 17(1), 81–115 (2008)

[TDLP09] Tao, Y., Ding, L., Lin, X., Pei, J.: Distance-based Representative Skyline. In: ICDE, pp. 892–903 (2009)

[TdW03] Tian, F., DeWitt, D.J.: Tuple Routing Strategies for Distributed Eddies. In: VLDB, pp. 333–344 (2003)

[TEO01] Tan, K.-L., Eng, P.-K., Ooi, B.C.: Efficient Progressive Skyline Computation. In: VLDB, pp. 301–310 (2001)

[TP06] Tao, Y., Papadias, D.: Maintaining Sliding Window Skylines on Data Streams. TKDE 18(3), 377–391 (2006)

[UF01] Urhan, T., Franklin, M.J.: Dynamic Pipeline Scheduling for Improving Interactive Query Performance. In: VLDB, pp. 501–510 (2001)

[UFA98] Urhan, T., Franklin, M.J., Amsaleg, L.: Cost Based Query Scrambling for Initial Delays. In: SIGMOD, pp. 130–141 (1998)

[USU06] Ünal, A., Saygin, Y., Ulusoy, Ö.: Processing Count Queries over Event Streams at Multiple Time Granularities. Inf. Sci. 176(14), 2066–2096 (2006)

[XZK+05] Xia, T., Zhang, D., Kanoulas, E., Du, Y.: On Computing Top-t Most Influential Spatial Sites. In: VLDB, pp. 946–957 (2005)

[XZT08] Xia, T., Zhang, D., Tao, Y.: On Skylining with Flexible Dominance Relation. In: ICDE, pp. 1397–1399 (2008)

[YLL+05] Yuan, Y., Lin, X., Liu, Q., Wang, W., Yu, J.X., Zhang, Q.: Efficient Computation of the Skyline Cube. In: VLDB, pp. 241–252 (2005)

[YM09] Yiu, M.L., Mamoulis, N.: Multi-dimensional Top Dominating Queries. VLDB J. 18(3), 695–718 (2009)

[YDM+07] Yiu, M.L., Dai, X., Mamoulis, N., Vaitis, M.: Top-k Spatial Preference Queries. In: ICDE, pp. 1076–1085 (2007)

[YHC05] Yu, H., Hwang, S., Chang, K.: RankFP: A Framework for Supporting Rank Formulation and Processing. In: ICDE, pp. 514–515 (2005)

[Yu05] Yu, B.: Adaptive Query Processing in Point-Transformation Schemes. In: Andersen, K.V., Debenham, J., Wagner, R. (eds.) DEXA 2005. LNCS, vol. 3588, pp. 197–206. Springer, Heidelberg (2005)

[WN05] Weis, M., Naumann, F.: DogmatiX Tracks down Duplicates in XML. In: SIGMOD, pp. 431–442 (2005)

[WLR+09] Wei, M., et al.: XML Stream Query Processing: Current Technologies and Open Challenges. Open and Novel Issues in XML Database Applications: Future Directions and Advanced Technologies. IGI Publishing (2009)

[WTZ09] Wu, J., Tan, K.L., Zhou, Y.: QoS-Oriented Multi-query Scheduling over Data Streams. In: Zhou, X., Yokota, H., Deng, K., Liu, Q. (eds.) DASFAA 2009. LNCS, vol. 5463, pp. 215–229. Springer, Heidelberg (2009)

[ZPZ+05] Zhu, M., Papadias, D., Zhang, J., Lun Lee, D.: Top-k Spatial Joins. IEEE Trans. on Knowl. and Data Eng. 17(4), 567–579 (2005)

[ZGBN07] Zhou, X., Gaugaz, J., Balke, W.-T., Nejdl, W.: Query Relaxation using Malleable Schemas. In: SIGMOD, pp. 545–556 (2007)

[ZHC+06] Zhang, Z., Hwang, S., Chang, K., et al.: Boolean + Ranking: Querying a Database by k-Constrained Optimization. In: SIGMOD, pp. 359–370 (2006)

[ZLC09] Zhu, L., Li, C., Chen, H.: Efficient Computation of Reverse Skyline on Data Stream. In: CSO, pp. 735–739 (2009)

Chapter 8
Online Social Networks: Status and Trends

George Pallis, Demetrios Zeinalipour-Yazti, and Marios D. Dikaiakos

Department of Computer Science
University of Cyprus
1678, Nicosia, Cyprus
{gpallis,dzeina,mdd}@cs.ucy.ac.cy

Abstract. The rapid proliferation of *Online Social Network (OSN)* sites has made a profound impact on the WWW, which tends to reshape its structure, design, and utility. Industry experts believe that OSNs create a potentially transformational change in consumer behavior and will bring a far-reaching impact on traditional industries of content, media, and communications. This chapter starts out by presenting the current status of OSNs through a taxonomy which delineates the spectrum of attributes that relate to these systems. It also presents an overall reference system architecture that aims at capturing the building blocks of prominent OSNs. Additionally, it provides a state-of-the-art survey of popular OSN systems, examining their architectural designs and business models. Finally, the chapter explores the future trends of OSN systems, presents significant research challenges and discusses their societal and business impact.

1 Introduction

With the emergence of Web 2.0, end-users are placed at the heart of various Web technologies, which tend to reshape the future of the WWW in terms of its structure, design, and utility. In this context, *Online Social Networks (OSNs)* are emerging as a new type of "killer application" on the Internet, which can be considered as a natural extension of Web applications that establishes and manages explicit relationships between users. Specifically, an OSN consists of users who communicate with each other in an online setting in diverse ways. Nowadays, we have been witnessing the rapid rise of a large variety of OSN sites, which publish user-generated or aggregated content, allow users to annotate published content with tags, reviews, comments and recommendations, and provide mechanisms that enable the establishment of user communities based on shared interests [2], [11].

The first well-known OSN site, called SixDegrees.com, was launched in 1997; its name originates from the six degrees of separation concept. Six degrees of separation is the theory that anyone can be connected to any other person through a chain of acquaintances that has no more than five intermediaries. Through SixDegrees.com, users could create their profiles, have a list of friends and

A. Vakali & L.C. Jain (Eds.): New Directions in Web Data Management 1, SCI 331, pp. 213–234.
springerlink.com

contribute information to their community. Although this site attracted million of users, it could not evolve into a sustainable business and closed down in 2000. The founder of SixDegrees.com believes that it was ahead of its time. From 2003, we witnessed a revolution and uptake of OSN sites that established most of nowadays most popular OSN sites. This revolution has brought a dramatic shift on the business, the cultural and the research landscape of the WWW [11]. Figure 1, presents a timeline that shows the evolution of OSN sites during the last decade.

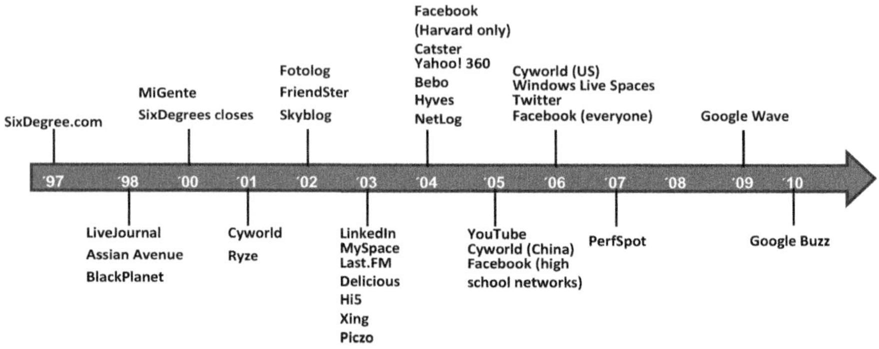

Fig. 1. Timeline of Online Social Network Sites

According to Nielsen Online's latest research[1], social network and blogging sites are nowadays the fourth most popular activity on the Internet; this means that more than two-thirds of the global on-line population visit and participate in social networks and blogs. In fact, social media have pulled ahead of e-mail in the rank of the most popular online activities. Another interesting finding is that social networking and blogging accounts for nearly 10% of all time spent on the Internet. These statistics suggest that OSNs have become a fundamental part of the online experience on the WWW throughout the world.

The key breakthrough brought by OSN sites like Facebook, Myspace, Flickr, LinkedIn, and YouTube, and the main driving force behind their success, is that OSN sites promote the vision of a Human-centric Web, where the network of people and their interests become the primary source of information, which resides entirely on social networking services. Consequently, the main objective of OSN systems is to provide social networking functionality as a core service to a variety of high-level applications and services. In addition, online social networking opens new interesting problems and creates challenges for research in an environment that becomes increasingly complex, and less structured [1], [43]. Nowadays, OSNs have become the subject of numerous startup companies, offering users the ability to create, search and manage their own OSN communities.

---

[1] Nielsen Company: `http://blog.nielsen.com/nielsenwire/wp-content/uploads/2009/03/nielsen_globalfaces_mar09.pdf`

Currently, the main technical underpinnings of OSN infrastructures and services include Web 2.0 technologies, service-oriented software, caching, database and content distribution technologies. From a technical point of view, OSN sites provide APIs, software frameworks and open-source platforms that enable application developers to build applications and manipulate their content. The core of OSN platforms is the social graph, where nodes represent individual actors within a social network and edges represent the interdependencies between the actors, which is integrated with new applications [11]. Social networking services make the social graph an integral element of their backend infrastructure, and provide direct access to parts of the graph through their end-user interfaces. In addition, the advent of the Linked Open Data W3C project opens new perspectives to OSNs allowing user contributed content to become even more open and accessible. Specifically, Linked Open Data project[2] aims to connect data sets using semantic Web technologies such as RDF (Resource Description Framework) or RDFa (a simpler variation). RDF based descriptions of social data provide a rich typed graph and offer a much more powerful and significant way to represent online social networks than social graphs. In this context, several ontologies are used to represent social networks. For instance, FOAF[3] is used for describing people profiles, their relationships and their activities online and SIOC[4] ontology provides the basis for defining the users.

The goal of this chapter is to discuss OSN platforms from different perspectives. The main contributions of this chapter can be summarized as follows:

- We present a reference architecture for OSNs in order to establish some common terminology and for ease of exposition. Such an architecture facilitates the process of identifying the technical challenges that arise in constructing OSNs.

- We develop a comprehensive taxonomy of OSNs that provides an in-depth coverage of this field in terms of OSN organizational structure and service types. The main aim of our taxonomy is to explore the unique features of OSNs and to provide a basis for categorizing present and future development in this area.

- We present a state-of-the-art survey of prominent OSN platforms that provides a basis for instantiating the blocks of our taxonomy and for understanding the current social networking landscape. It also presents the underlying Web technologies that are currently exploited in the social networking field.

- We discuss significant open problems and research challenges that need to be addressed in order to develop efficient OSN infrastructures and services. Finally, we identify the strengths, weaknesses, and opportunities as well as the business and societal impact of OSNs.

---

[2] Linked Open Data project: `http://linkeddata.org/`

[3] The Friend of a Friend Project: `http://www.foaf-project.org/`

[4] SIOC-project: `http://sioc-project.org/`

The rest of the chapter is structured as follows: Section 2 describes the architecture of OSNs, providing an insight into the technological characteristics of these platforms. Section 3 presents the taxonomy of OSNs and Section 4 performs a detailed survey of prominent OSN platforms. Section 5 outlines the research challenges of OSNs, focusing on Decentralized OSNs and on the business and societal impact of this technology. Finally, Section 6 concludes this chapter.

2 Architecture of OSNs

An online social networking site is a Web site that:

- Acts as a hub for individuals to establish relationships with other persons (friends, colleagues, etc.). Each user articulates a list of other users with whom a connection is shared.
- Includes a wide range of tools for people to build a sense of community in an informal and voluntary way. Online users interact with each other, contribute information to the common information space, and participate in different interactive activities (e.g., photo uploading, tagging, etc.).
- Contains specific components that allow people to: define an online profile, list their connections (e.g., friends, colleagues), receive notifications on the activities of those connections, participate in group or community activities, and control permission, preference and privacy settings.

A reference architecture of OSNs is depicted in Figure 1. The entire system is formed by the following layers:

- **Data Storage layer:** This layer consists of two components: The *Storage Manager*, which is responsible for efficiently storing the information of social graphs and for handling increased database loads. This is usually achieved by adopting distributed memory caching. The other component, called *Data Store*, comprises the storage elements that store information items of a social networking service. Data Stores can be Multimedia Databases, User Profiles Databases etc.
- **Content Management layer:** This layer is responsible for three main tasks. Firstly, it facilitates the incorporation of social information from remote OSN sites through a Content Aggregator that gathers and organizes content from social media but also distributes it to other OSN platforms. Secondly, it facilitates the maintenance and the retrieval of the social content graph through the Data Manager. Thirdly, it controls the access of users by creating and maintaining an access control scheme.

- **Application layer:** Each OSN site supports numerous services such as search, news feeds, mobile access, etc. The services communicate with the data manager and the access control manager in order to analyze and manage the social content graph. The applications are provided to users through an application manager. The application manager facilitates the user interaction via a set of APIs. This component also includes a service framework for scalable cross-language services development. Such a framework allows users to deploy applications by abstracting the portions of each language that tend to require the most customization into a common library that is implemented in each programming language.

The users interact with an OSN platform through HTTP requests. Each user can either be registered as an authorised user or be anonymous. Registered users usually have more rights than the anonymous ones (e.g., comment published articles, upload figures, etc.). The access control module is responsible for addressing

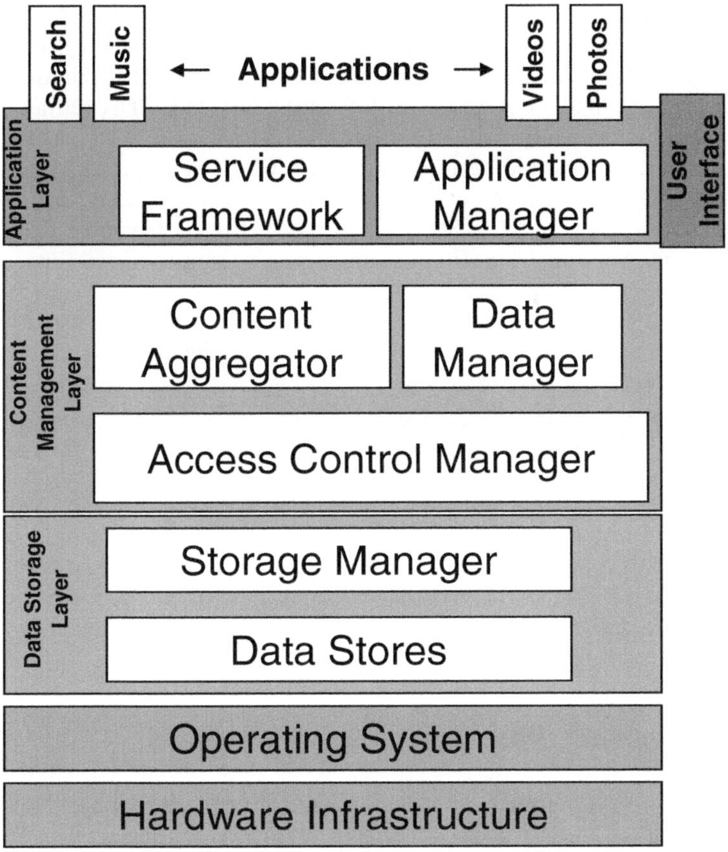

Fig. 2. Reference Architecture of an Online Social Network Platform

privacy and security settings of OSN users. Specifically, most of the research in the field of OSN security and privacy has focused on the development of privacy-preserving techniques providing answers to issues such as the ownership of personal information and the protection of privacy [8], [15], [17], [38], [41].

Each OSN platform consists of multiple application servers, which provide a set of services and APIs. A user forwards a request to the OSN platform, which forwards it to the appropriate application server. A load balancer is responsible for monitoring the application servers of an OSN platform, balancing the load of requests, handling failover, and forwarding the requests to the application servers. The graph servers track and manage the connection relationships among users.

In terms of content distribution, cache servers speed up dynamic Web applications by alleviating the load of application servers. Specifically, OSNs involve a significantly different set of requirements compared to traditional Web applications. One of the major difference among OSNs platforms, are the relations and trust among users, which change over time. This behavior dictates the design of data placement, replication and distribution algorithms in OSNs. Another significant difference is that OSNs involve a large number of small files that need to be frequently accessed and updated by a large number of users and the propagation of file updates to guarantee data coherency. To further improve the system performance, OSN platforms distribute their content over Content Distribution Networks (CDNs) (e.g., Akamai, etc.) or Cloud Infrastructures. In a CDN setting [30], user requests are automatically routed to the nearest edge location, so content is delivered with the best possible performance. In a Cloud setting, the system is built over large clusters of processors. For instance, Facebook users can build their applications on Amazon Web Services (Amazon Elastic Compute Cloud (Amazon EC2) and Amazon CloudFront), improving reliability, flexibility, and cost-effectiveness.

Finally, the Data Stores of OSN platforms can be either centralized or distributed across multiple administrative domains. Centralized OSNs raise concerns regarding the protection of privacy and scalability. To overcome these limitations, data can be stored in a peer-to-peer infrastructure [12], [44], [45]. To support such a scheme, the reference architecture of Figure 1, should be extended by i) an overlay network layer on top of the operating system and network subsystem; and ii) an overlay management layer. The overlay network layer would provide mechanisms for node identity management, topology construction and maintenance, message routing, node search services, interfacing with local resources and the underlying fabric. The overlay management layer will provide mechanisms for authentication and authorization, decentralized monitoring, management, and adaptive control of peer-to-peer resources.

3 Taxonomy of OSNs

This section proposes a taxonomy that covers the key aspects of OSNs. This taxonomy is split into four branches. The first branch covers the scope of OSN systems in terms of activities. The second branch deals with the data model of OSNs, since the data model is the way in which data sources are stored in a system. The third branch, called system model, categorizes the OSNs regarding the hosting and content distribution of application servers. The fourth branch is from the point of view of the formation of users' network within OSN platforms. Figure 3 depicts the OSNs taxonomy.

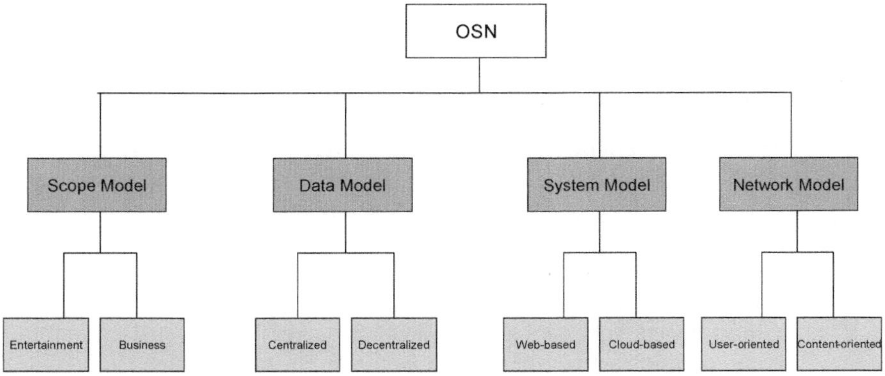

Fig. 3. OSN Taxonomy

In terms of their scope model, OSNs can be classified into the following two categories:

- **Entertainment:** Most OSNs are dedicated to entertainment. Their focus is on delivering fun and interactive social experience online to registered users. Popular OSN sites that are mainly entertainment-oriented are Facebook, Myspace, Hi5 and Flickr.

- **Business:** In this category, the focus of OSNs is to connect the world's professionals to make them more productive and successful. Through business OSNs, registered users create profiles that summarize their professional expertise and accomplishments. Indicative OSN sites in this category are LinkedIn and Xing.

The next dimension of our taxonomy is the data model followed by OSN providers. Here we identify the following categories:

- **Centralized:** the data of centralized OSNs are stored entirely in physical proximity (e.g., within a cluster or data center), concentrating the data of all their users under a single administrative domain. Today, most OSNs rely on centralized storage and functionality. However, centralized OSNs raise concerns regarding the protection of privacy and their scalability in the context of an expanding base of users and applications.

- **Decentralized:** the data of decentralized OSNs is distributed across multiple administrative domains [13], [35]. Application servers run on desktop machines (i.e., peers) owned by users. In general, hosting personal data on peers is more privacy-preserving than delegating control to a third-party service provider. In addition, this model is cheaper than acquiring dedicated centralized equipment. The main drawback of this approach is that peers might not be available continuously. Peers are prone to failures, reboots, power-offs, and network disconnections [12]. More discussion about the decentralized OSNs is given to section 5.

Our next dimension under discussion is the system model of OSNs. In particular, we identify the following two categories:

- **Web-based scheme:** the application servers are hosted by Web sites that provide a set of services and APIs. In such a scheme, the load balancer balances the load of requests, handles any failover, and forwards the requests to the appropriate application servers. In Web-based schemes, most OSN services are free to users.

- **Cloud-based scheme:** the application servers are hosted by a utility computing infrastructure such as Amazon Elastic Compute Cloud (EC2). In such a scheme, each user stores its own data on a personal virtual machine instance, called Virtual Server. The main advantages of this scheme are its high availability and its improved privacy since each user keeps its personal data in a Virtual Server residing in a Cloud computing environment. Also, Cloud-based schemes are usually integrated with CDN infrastructures (e.g., Amazon Cloud Front integrates with Amazon EC2). This results in distributing content to end users with low latency and high data transfer speeds. On the other hand, hosting data in a Cloud increases the costs due to utilizing a commercial infrastructure.

Finally, our taxonomy addresses the network model of OSNs that can be classified into the following two categories:

- **User-oriented OSNs:** The OSNs of this category emphasize on the social relationships. In such OSNs, content sharing is mainly among users who belong to the same community. Indicative OSNs of this category are the Facebook, MySpace and LinkedIn.

- **Content-oriented OSNs:** The users' network is not determined by the underlying social relationships but by their common interests. Indicative OSNs of this category are blog networks, question answering networks and video networks (e.g., YouTube).

4 Case Studies

This section presents various case studies of the most popular OSN platforms. For each OSN platform, we present a brief history, its scope, some technical issues and its business model. Table 1 summarizes the key characteristics of OSN platforms taking into account the taxonomy described in the previous section.

4.1 Facebook

History & Scope. Facebook (www.facebook.com) was founded by Mark Zuckerberg while he was a psychology student at Harvard University. In February 2004, Zuckerberg launched "The facebook"; the name was taken from the sheets of paper distributed to freshmen, profiling students and staff. Within 24 hours, 1200 Harvard students signed up, and after one month, over one half of the undergraduate population had a profile. In August 2005, it became Facebook.com and the address was purchased for $200,000. In September 2006, the network was extended beyond educational institutions to anyone with a registered email address. Nowadays, Facebook is one of the most popular online destinations with more than 300 million subscribers, whospend an average of 20 minutes on the site daily, contribute 4 billion pieces of information, 850 million photos, and 8 million videos every month. The main scope of Facebook is *entertainment*. Facebook is both a *user-oriented* and a *content-oriented* OSN site. Not only do the users create their networks among their friends, but also they can create networks based on common interests (e.g., Princeton alumni Network). Each registered user has a personal profile, adds friends and sends them messages, uploads photos, videos, links, and updates her personal profile to notify friends about herself.

Architectural Design. The data model architecture of Facebook *is centralized* and based on a typical hierarchical PHP Web application model with a layer of data caching. The caching layer is provided via the memcached open source software. Regarding the *application layer*, Facebook supports an Apache open source RPC mechanism called Thrift, which is used for both low-latency real-time RPC and persistent structured data storage across a variety of applications, such as Search, News Feed etc. The RPC language is influenced by CORBA's Interface Definition Language (IDL). Facebook supports various backend services that use the Hadoop, Scribe and Hive frameworks. Regarding the *data storage layer*, data (e.g., photos, statuses, and comments) are periodically stored to central repositories and relational (MySQL) databases. As far as the system model is concerned, Facebook application servers run on *Web-based* and *Cloud-based infrastructures* (i.e., Amazon EC2).

Business Model. Since the subscription to Facebook is free, the business model of Facebook depends on advertising. This business model is based on the principle that free subscriptions build audiences with distinct interests and expressed needs that advertisers will pay to reach. Facebook supports two advertising polices; the Pay for Clicks (also called cost per click or CPC) advertising allows advertisers to specify a certain amount that they are willing to pay each time a user actually clicks on their ad, whereas, the Pay for Views (also called cost per thousand impressions or CPM) advertising allows customers to specify how much they are willing to pay for 1000 views of their ad. Also, Facebook provides the option to target the ads to specific groups of users. This allows advertisers to write ad text that is more personalized, making their ad more appealing to the users they are reaching. Facebook supports several targeting options such as location, birthdays, likes and interests. An indicative success story of Facebook advertising is the following: *"Over 12 months, CM Photographics generated nearly $40,000 in revenue directly from a $600 advertising investment on Facebook. Of the Facebook users who were directed to CM Photographics' website from the ads, 60% became qualified leads and actively expressed interest in more information."* The worth of Facebook's social network has been estimated to $15 billion[5]. It is remarkable to note that two-thirds of comScore's U.S. top 100 websites and half of comScore's Global top 100 websites have implemented *Facebook Connect*, a Facebook API that enables individuals and organizations to link their applications to Facebook. comScore is a global leader in measuring the digital world and the preferred source of digital marketing intelligence.

4.2 MySpace

History & Scope. MySpace (www.myspace.com) was developed in August 2003 by eUniverse employees who sought to mimic the more popular features of Friendster - a social networking website. The first MySpace users were eUniverse employees. In July 2005, MySpace was sold for US$580 million to Rupert Murdoch's News Corporation. MySpace became the most popular social networking site in the US in June 2006. Nowadays, MySpace is one of the fastest growing OSN sites with 300 million users. Its main scope is *entertainment*. MySpace is a *user-oriented* OSN site. Each registered user can share among friends a personal profile, photos, videos, links, etc. The user can also play games and update her personal profile to notify friends about herself. MySpace also runs event management, such as product launches and album launches for major labels which provides additional revenue streams.

Architectural Design. MySpace architecture is *centralized* and based on ASP.Net 2.0 Web application model with a layer of data caching and extracted services components. Specifically, MySpace consists of more than 500 Web servers running windows 2003/IIS 6.0/APS.NET. The *Data Stores* of MySpace consist of 1200 cache servers running 64-bit Windows 2003 with 16GB of objects cached in RAM, and 500 database servers running 64-bit Windows and SQL Server 2005. The caching layer is provided via the memcached open source software. Within the *data storage layer*, data is stored in central repositories and relational databases. As far as the system model is concerned, MySpace application servers run on *Web-based* and *Cloud-based infrastructures* (Amazon EC2).

Business Model. The subscription to MySpace is also free. Therefore, the business model of MySpace is also dependent on advertising. Similarly to Facebook, MySpace supports two advertising polices: the Pay for Clicks and the Pay for Views. With a rich set of demographic and behavioral data, MySpace enables advertisers to target their precise audience based on attributes, interests and activities. All the services are offered free to users. Users can create new content to attract, on their turn, new users. This has led several companies to advertise their products through the MySpace platform.

---

[5] `http://www.facebook.com/press/releases.php?p=8084`

4.3 Hi5

History & Scope. Hi5 (hi5.com) was founded in 2003 by entrepreneur Ramu Yalamanchi, who is also its current Chief Executive Officer. According to comScore, in 2008 Hi5 was the third most popular social networking site in terms of monthly unique visitors. Nowadays, Hi5 is available in 50 languages and is one of the most popular online destinations for *entertainment* with more than over 60 million active users. Hi5 provides a robust platform for third-party developers to integrate games, content, and other applications. Hi5 is a *user-oriented* OSN site where each registered user can upload photos, videos, songs, personal information and share them with friends. Also, the users can join groups with semantically common interests.

Architectural Design. Hi5 architecture is based on an N-tiered Java architecture model. The *Data Stores* of Hi5 consists of PostgreSQL database servers. Hi5 runs using open-source software on a Linux platform. Specifically, Hi5 uses Linux Servers running SuSE Enterprise Linux, Apache and Lighttpd Web servers, the Squid proxy server for Web acceleration, Resin and Tomcat Java Application Servers, Struts for Model-View-Controller (MVC), Spring for Java Application Framework, iBatis for Object Relational Mapping, the Lucene library for indexing, and the Enunciate as Web service deployment framework. The caching layer is provided via the memcached open source software. Regarding the *data storage layer*, data is stored in a centralized configuration of the PostgreSQL object-relational database system. All queries are centrally maintained in XML files. Regarding the system model, application servers run on a *Web-based infrastructure*.

Business Model. Like previous OSNs, the business model of Hi5 also depends on advertising. Hi5 is collaborating with SponsorSelect - a premium advertising network that is reinventing behavioral targeting –in order to allow its users to choose the advertising they wish to see. By allowing users to self-select advertising, advertising is more relevant and performs better for advertisers, publishers (a publisher displays ads, text links, or product links on its Web site) make more money and can provide better content and services to users for free.

4.4 Flickr

History & Scope. Flickr (www.flickr.com) is the most popular OSN dedicated to photo and video sharing, where millions of photos are uploaded, tagged and organized by more than 8.5 million registered Web-users. Flickr was founded by Stewart Butterfield and Caterina Fake. In February 2004, Flickr was launched by Ludicorp, a Vancouver-based company. In March 2005, Flickr was acquired by Yahoo!. Flickr is a *user-oriented* OSN site and its main scope is *entertainment*. In Flickr, every user enters/selects new tags for a particular photo/video and the system suggests related tags to user, based on the tags that the user or other people have used in the past along with (some of) the tags already entered.

Architectural Design. Flickr architecture is based on a typical hierarchical PHP Web application model with a layer of data caching. The *Data Store* of Flickr consists of 62 MySQL databases across 124 servers, with about 800,000 user accounts per pair of servers. The MySQL databases are hosted on servers that are Linux-based, with a software platform that includes Apache, PHP, shards, Memcached, Squid, Perl and Java. The system administration tools include anglia for distributed system monitoring and Cvsup for distributing and updating collections of files across a network. Flickr supports tools for image processing (ImageMagick) and deployment (SystemImager). Data (e.g., photos, videos and tags) are stored in *central repositories* and relational MySQL databases. Regarding the system model of Flickr, application servers run on *Web-based* and *Cloud-based infrastructures* (Amazon EC2).

Business Model. With 3 billion pictures, Flickr is the biggest repository of digital images and videos on the Web. The basic account is for free, but Flickr charges users that want a professional and more sophisticated account. Flickr follows a consolidated business model, the "freemium" model, in which the registrations fees are varied from free to a premium "pro" version, with the latter featuring more capabilities.

4.5 LinkedIn

History & Scope. LinkedIn (www.linkedin.com) was founded in 2003 by Reid Hoffman. LinkedIn is an interconnected network of experienced professionals from around the world, representing 150 industries and 200 countries. In December 2009, LinkedIn had more than 55 million registered users and it was available in four languages. The scope of LinkedIn is mainly for *business*, allowing users to maintain a list of contact details of people they know and trust in business. Through this network people can find jobs and business opportunities, whereas employers can post and distribute job listings for potential candidates. LinkedIn is a *user-oriented* OSN site where registered users create their networks by sending personal invitations. A key feature of LinkedIn is that registered users can be recommended by someone in one's contact network.

Architectural Design. The data model of LinkedIn is *centralized*. It is an open architecture that consists of one monolithic Web application. Its *Data Store* includes a set of databases and a social network graph. LinkedIn runs on the Solaris operating system, uses Tomcat and Jetty as application servers, Oracle and MySQL as Databases, Spring for Java Application Framework, the Lucene library for indexing and ActiveMQ for Java Message Services (JMS). Web applications provide the GUI to the user and update the databases directly. Regarding the system model, application servers run on a *Web-based infrastructure*.

Business Model. LinkedIn is free to join. In addition, LinkedIn offers a premium version providing more tools for finding and reaching the "right" people. Specifically, with a premium account users can send messages directly to people and search for profiles that do not belong in their network. An indicative success story is when LinkedIn drove highly users to the MAZDA6 site and delivered some of the highest KPI ratings of all lifestyle sites on the plan.

4.6 Twitter

History & Scope. Twitter (www.twitter.com), founded by Jack Dorsey, Biz Stone and Evan Williams in March 2006 and launched publicly in July 2006, is a social networking and micro-blogging service that allows users to post their latest updates. An update is limited to 140 characters (called tweets) and can be posted through a Web form, a text message, or an instant message. Tweets delivered to the author's subscribers who are known as “followers”. Senders can restrict their posts to specific friends or, by default, allow open access. Registered users can also follow lists of authors instead of following individual authors. The scope of Twitter is twofold: *business* and *entertainment*. For instance, Twitter has been used for campaigning (2008 US presidential campaign), educational purposes, public relations etc. Twitter is a *content-oriented* OSN site since a user’s network is determined by the underlying social relationships; users create their networks by becoming “followers”.

Architectural Design. The data model of Twitter *is centralized* and based on the Ruby on Rails Web application framework with a layer of data caching. The caching layer is provided via the memcached open source software. Regarding the *data storage layer*, data is periodically stored to central repositories and relational (MySQL) databases. Also, it supports networked resource monitoring tool (Munin and Nagios) for analyzing resource trends. As far as the system model is concerned, Twitter application servers run on *Web-based infrastructures*.

Business Model. Twitter is free for all registered users. Contrary to most OSN sites, Twitter does not provide any advertising policy. In addition, it does not support any premium account. Nowadays, Twitter is in beta test with providing enterprise subscriptions that would target corporate customers. The idea to provide enterprise subscription is based on the assumption that the more businesses use Twitter, the more ways the company will find to monetize their traffic.

4.7 YouTube

History & Scope. YouTube (www.youtube.com), founded by Steve Chen, Chad Hurley and Jawed Harim in February 2005, is a social networking that allows users to post their videos. In November 2006, YouTube was bought by Google for $1.65 billion. Recently, YouTube has been ranked as the fourth most visited Website on the Internet According to comScore, YouTube is the dominant provider of online video in the US. It is estimated that 20 hours of new videos are uploaded to the site every minute. In March 2008, YouTube's bandwidth costs were estimated at approximately US$1 million a day. The scope of YouTube is for *entertainment*. All users can watch open videos, while registered users are permitted to upload an unlimited number of videos. YouTube is a *content-oriented* OSN site since users’ network is determined by users’ common interests.

Architectural Design. The data model of YouTube is *decentralized* and based on the distributed storage system, called BigTable[6]. BigTable developed at Google and its scope is to store efficiently large-scale structured data. To further improve its performance, BigTable is used with MapReduce[7], a framework for running parallel

---

[6] BigTable: `http://labs.google.com/papers/bigtable.html`
7 MapReduce: `http://labs.google.com/papers/mapreduce.html`

computations. Regarding the content delivery, most popular content is moved to a CDN provider (Akamai), whereas, less popular content is delivered through YouTube servers. A distributed multilevel cache is used for decreasing latency. NetScaler Web application controller is used for load balancing and caching static content. Regarding the *data storage layer*, data is periodically stored to MySQL databases. As far as the system model is concerned, YouTube application servers run on *Web-based* and *Cloud-based infrastructures* (Amazon EC2).

Business Model. YouTube is free for all the users (subscribers or not). The business model of YouTube depends on advertising (i.e., Google AdSense). Google AdSense uses its Internet search technology to serve advertisements based on Website content, the user's geographical location, users' tags, and other factors. Those wanting to advertise with Google's targeted advertisement system may enroll through AdWords. AdWords offers Pay for Clicks advertising, and site-targeted advertising. Thus, the business model of YouTube is based on mass collaboration. Providing free access to its users results to an increased number of users and, hence, to increased profits through increased advertisement rates and more advertisers.

Table 1. Main Characteristics of OSN sites

OSN Platform	Scope		Data Model		System Model		Network Model	
	Entertainment	Business	Centralized	Decentralized	Web-based	Cloud-based	User-oriented	Content-oriented
Facebook	V		V		V	V	V	V
MySpace	V		V		V	V	V	
Hi5	V		V		V		V	
Flickr	V		V		V	V	V	
LinkedIn		V	V		V		V	
Twitter	V	V	V		V			V
YouTube	V			V	V	V		V

5 Future Research Challenges

Despite the fact that most OSN sites are centralized, nowadays, we observe a paradigm shift from centralized to distributed infrastructures [12], [35]. In general, decentralization can provide answers to issues that have raised controversy in the context of centralized OSNs, such as the ownership of personal information and the protection of privacy; problems in cross-platform service provision and user lock-in [15], [17]. Decentralization promises higher performance, fault-tolerance and scalability in the presence of an expanding base of users and applications. OSN decentralization has been identified as a key research challenge [12]. However, the transition to a fully distributed architecture so as to scale up OSNs is nontrivial. It is often a costly endeavor and specially challenging for OSNs that were not designed to be fully distributed from day one [12], [35]. In this context, this paradigm shift gives rise to many research questions intersecting networking, security, distributed systems and social network analysis, leading to a better understanding of how technology can support social interactions.

In the following sub-sections we present requirements and challenges towards developing efficient *Decentralized Online Social Networking (DOSN)* infrastructures and services. Finally, we conclude this section by presenting the business and social impact of OSNs.

5.1 Overlay Networking

Researchers have analyzed different properties of OSNs, mainly focusing on their formation and evolution as well as their information propagation over the network [1], [29], [34]. The advent of DOSNs creates new perspectives and challenges in networking [12], [43]. The objective is to build fundamental overlay services that will form the foundation for DOSN services and applications. This requires the investigation of novel architectures, algorithms and protocols for overlay networks of peers so as to provide the run-time environment and the basic communication functionality required by DOSN services and applications.

Therefore, self-management techniques, P2P publish/subscribe mechanisms and software component models should be developed. Thus, it is important to devise algorithms for self-management of the existing network infrastructure as well as for efficient group-based communications and information sharing (publish/subscribe) among the participants of DOSNs. In addition, a publish/subscribe solution must be able to effectively deal with large populations of dynamic users, large numbers of topics, arbitrary subscription patterns and be robust to malicious peer behaviour. Finally, adaptive component models for DOSNs will enable the underlying infrastructure to continuously operate under ever changing network conditions, enabling the safe manipulation and configuration of existing overlay resources at runtime.

Finally, it is also important to understand how the workload of DOSNs is reshaping the Internet traffic as this is valuable in designing the next-generation Internet infrastructure and content distribution systems (e.g., CDNs). Although current DOSNs contribute a lot less than peer-to-peer applications in terms of bytes, DOSNs might add features that increase the per-user bandwidth demand [12]. Given this potential of traffic explosion (e.g., when video becomes popular within a OSN site), it is crucial to explore the network-level dynamics of DOSNs [34].

5.2 Privacy and Trust

The amount of digital content today circulated in OSNs is enormous providing personal information about their users (i.e., profiles, friend relationships, daily activities, photos, videos, etc.). Although security and privacy concerns can prevent such efforts in practice, the problem remains since the data is located on a single server. Up to now, many researchers have started to work on improving the access control systems provided by OSNs. Specifically, most of the research in the field of OSN security and privacy has focused on the development of privacy-preserving techniques to mine OSN data [14]. In addition, several research efforts [8], [15], [17], [38], [41] focus on addressing the restrictions of protection mechanisms provided by current OSNs. Decentralized OSNs address many privacy concerns since the personal data of users is distributed across multiple administrative domains. This provides users more control over their content, reducing the system's vulnerability. Therefore, novel privacy-aware access control solutions should also be developed, providing users as much control as possible over their data and the way they are protected. In [35], the authors compare the privacy, cost, and availability tradeoffs for DOSN schemes.

Considering that OSNs host a variety of personal data, it is also crucial to identify possible threats arising from distributed malicious data and content in DOSNs. In the context of Web, it has been observed a recent increase of exploits, such as drive-by downloads [32] and malicious documents [23], or other related Web threats such as Cross-Site Scripting (XSS) [6], [10] and Cross-Site Request Forgery (CSRF) [9]. OSNs are considered as ideal targets for this kind of threats, since they are usually composed by complex AJAX interfaces that serve millions of users who trust each other. For instance, a social application may be used to launch DoS attacks against third parties [5] or a fake user account may be used for SPAM distribution. Regarding the DOSNs, data replication takes place in user's desktops. This further augments the probability of hosting malicious content in a user's computer machine. Therefore, a DOSN must provide the required mechanisms for the identification and further prevention of malicious content distribution. Thus, it is important to provide practical workarounds and directed guideline for detecting and preventing further distribution of malicious content, where this is possible. This can be achieved by applying data encryption policies. Such policies ensure that data are accessible to users who have the right private keys. In [42] a secret sharing protocol for DOSNs is presented.

5.3 Knowledge Discovery and Search

The role of network structure on the Web has grown in significance in the field of knowledge discovery and information retrieval, stimulated to a great extent by the importance of link analysis in the development of Web search techniques. But the Web has always contained a second network, less explicit but equally important, and this is the social network on its users, with latent person-to-person links encoding a variety of relationships including friendship, information exchange, and influence. Developments over the past few years - including the emergence of OSN systems and rich social media, as well as the availability of large-scale e-mail and instant messaging datasets - have highlighted the crucial role played by OSNs, and at the same time have made them much easier to uncover and analyze [1].

The study of evolution of OSNs and their properties have been one of the central areas of social network analysis. In particular, it is inherently related with the problem of predicting particular attributes of OSNs [18]. A large body of work has focused on the study of global evolution of networks and the identification of communities [7], [21]. Other recent research work has focused on developing algorithmic tools for the analysis of evolving networks [34], [37] as well as inferring users' profiles based on social graphs [27].

There is now a considerable opportunity to exploit the structure, evolution and information content inherent in DOSNs since the prospect of decentralization raises a number of interesting research challenges. However, the complexity of DOSNs requires devising novel algorithms and mechanisms in order to study and model the evolution of DOSNs in both macroscopic and microscopic level. At the macroscopic level mining patterns that characterize the evolution of the network when it is viewed as a global structure should be investigated. At a microscopic level algorithms that identify primitive "patterns" of evolution will be developed [21]. A specific prediction problem that has drawn an amount of interest in the research community is the link-prediction problem [24]. The question is to infer which new interactions among the members of a DOSN are likely to occur in the near future. In addition, it is crucial to study the problem of link formation in DOSNs. New models for the evolution and the formation of links should be built by integrating all available information: topological structure of the graph, content information of the users of the network, as well mined patterns of evolution. Considering that DOSNs are very large and complex, methodologies and tools from the field of Complex Networks should be used in order to exploit them [31]. Also, methodologies and mechanisms should be investigated that allow identifying interesting patterns in the network and create/select features for characterizing the various components of the system and predicting their evolution. Such features can be the popularity of certain items or the existence of a link between two items, and they can be used for designing improved ranking models, providing early characterization of spam, early characterization of being member of a community (strongly connected components), leader, reputation of authors, etc.

In terms of searching, the current evolution of OSN sites is characterized by an increasing availability of online services and novel search facilities (e.g., services for searching scientific literature, photos, videos, vacation offers, travels, restaurants, online shops, and so on). A recent survey [28] has shown that people turn to OSNs rather than typical search engines or Q&A sites for certain question types and topics. However, the quality of their answers goes much beyond what can be achieved via conventional, general purpose search engines. Authors in [16] describe a social model of user activities before, during, and after search. Recently, Horowittz and Kamvar [20] presented a large-scale social search engine, called

Aardvark. Contrary to traditional Web search engines, the scope of Aardvark is not to find the documents that satisfy the user's information need, but to find the person that can satisfy the user's information need. This person belongs to the user's social network.

The shift in search paradigm that we observe in OSNs opens up a number of interesting research questions in information retrieval. The main challenge posed by content in DOSN sites is the fact that the distribution of quality has high variance: from very high-quality items to low-quality, sometimes abusive content [4], [13]. This makes the tasks of filtering and ranking in such systems more complex than in other domains [40]. Also, trust in a traditional search engine is based on authority, whereas in a social search engine, trust is based on intimacy. However, social media systems present inherent advantages over traditional collections of documents: their rich structure offers more available data than in other domains. In addition to document content and link structure, DOSNs exhibit a wide variety of user-to-document relation types, and user-to-user interactions [2], [3], [22], [39].

5.4 Business and Social Impact

Online social networking is a complex, large and rapidly expanding sector of the information economy whose impact is expected to be far-reaching. User-generated content is causing changes in the traditional content/media industry structure. In the future, community features will be an integral part of all digital experiences - from information/publishing to business and entertainment. Companies providing services for social networking and media (e.g., sysomos[8] - a product suite that provides customers with tools for measuring, monitoring, understanding and engaging with the social media landscape) or adding social networking features to existing services must anticipate significant growth.

From technological perspectives, the success of OSN sites represents a growing threat to the monopoly of Google. For instance, Facebook has become one of the most popular online platforms with more than 300 million registered users (one fifth of all Internet users – circa August 2009) who spend an average of 20 minutes on the site every day, contributing 4 billion pieces of information, 850 million photos, and 8 million videos every month. This means that Facebook not only has it become perhaps the largest source of personal data online, but also it has embarked on an ambitious effort to challenge Google's position as the dominant driver of Web traffic and the dominant power in Web advertising. This challenge has led to the introduction by Google of the OpenSocial API, which promises to provide functionality similar to that of Facebook's platform, even though relying on open-standard technologies. Also, to the recent introduction by Google of the Buzz social networking services, which is integrated with Gmail.

The key added ingredient of OSN platforms is their social dimension with the aim of linking users together to facilitate their interaction and make it richer and

---

[8] Sysomos: `http://www.sysomos.com/`

more productive. The power of people interacting with people in an online setting has driven the success or failure of many companies in the Internet space. Kees Winkel (http://futurecase.wordpress.com/) argues that 2% of nodes in a social network are all that need to be reached to ensure an idea or marketing initiative successfully. A white paper[9] from AT&T discusses the business impact of social networking. According to this paper, the impact of social networking in business is catalytic, driving several companies to change their vision and organization.

Except of business, OSNs have also social impact. Authors in [36] studied the impact of OSNs in marketplaces and observed that social networking improves e-commerce presenting a very positive impact. Information from OSNs can also be exploited to improve Internet search engines [26], while others have applied this information to increase profits from viral marketing [33]. OSNs have also applications in Vehicular ad hoc Networks (VANETs) where groups of vehicles' drivers socialize and communicate with each other in order to inform for the roads' conditions [19]. As far as the education is concerned, OSNs can be helpful to a students learning environment, as long as it is used correctly and responsibly [25].

6 Conclusion

OSNs are popular infrastructures for information sharing, communication and interaction on the Internet. With over half a billion users, OSNs are nowadays a mainstream research topic of interest for computer scientists, economists, sociologists etc.

In this chapter, we have analyzed and categorized the infrastructural and technical attributes of OSNs. We have developed a comprehensive taxonomy for OSNs based on their scope, data model, system model and network model. We have also provided a detailed survey of the most popular OSNs and identified the underlying Web technologies that are currently in use in social networking domain. In doing so, the readers can gain an insight into the technologies, services and business models that are currently followed in this field. It is also presented the system architecture for OSNs, where the main components of an OSN platform are described.

However, and despite their impressive success, OSN services face significant challenges that need to be addressed in order to improve the end-user experience and to allow for a healthy market expansion in the future. Consequently, content distribution, scalability and privacy issues are gaining more attention in order to meet up the new technical and infrastructure requirements of the next generation OSNs. This has led to a paradigm shift from a centralized infrastructure to a decentralized one. Thus, a section of this chapter is devoted to discuss the open problems and research challenges for OSNs. Finally, we explore the societal and business impacts of social networking.

---

[9] `http://www.bligoo.com/media/users/1/50369/files/Business - Social Networking Impact.pdf`

References

[1] Ahn, Y., Han, S., Kwak, H., Moon, S., Jeong, H.: Analysis of Topological Characteristics of huge Online Social Networking Services. In: Proceedings of the 16th International Conference on World Wide Web (WWW 2007), Banff, Alberta, Canada (May 2007)

[2] Amer-Yahia, S., Lakshmanan, L., Yu, C.: SocialScope: Enabling Information Discovery on Social Content Sites. In: Proceedings of the CIDR 2009, Asilomar, CA, USA (2009)

[3] Amer-Yahia, S., Benedikt, M., Lakshmanan, L., Stoyanovic, J.: Efficient Network-aware Search in Collaborative Tagging Sites. In: Proceedings of VLDB Endow, vol. 1(1), pp. 710–721 (2008)

[4] Anderson, C.: The Long Tail: Why the Future of Business Is Selling Less of More. Hyperion Publisher (July 2006)

[5] Athanasopoulos, E., Makridakis, A., Antonatos, S., Antoniades, D., Ioannidis, S., Anagnostakis, K.G., Markatos, E.P.: Antisocial Networks: Turning a Social Network into a Botnet. In: Proceedings of the 11th International Conference on Information Security, Taipei, Taiwan, September 2008, pp. 146–160. Springer, Heidelberg (2008)

[6] Athanasopoulos, E., Pappas, V., Markatos, E.P.: Code-Injection Attacks in Browsers Supporting Policies. In: Proceedings of the 3rd Workshop on Web 2.0 Security & Privacy (W2SP), Oakland, California, USA (May 2009)

[7] Backstrom, L., Huttenlocher, D., Kleinberg, J., Lan, X.: Group Formation in Large Social Networks: Membership, Growth, and Evolution. In: Proceedings of the 12th ACM SIGKDD International Conference on Knowledge Discovery and Data mining, New York, NY, USA, August 2006, pp. 44–54 (2006)

[8] Baden, R., Bender, A., Spring, N., Bhattacharjee, B., Starin, D.: Persona: an Online Social Network with User-defined Privacy. In: Proceedings of the ACM SIGCOMM 2009 Conference on Data Communication, Barcelona, Spain, August 2009, pp. 135–146 (2009)

[9] Barth, A., Jackson, C., Mitchell, J.C.: Robust Defenses for Cross-Site Request Forgery. In: Proceedings of the 15th ACM Conference on Computer and Communications Security (CCS 2008), Alexandria, USA (October 2008)

[10] Barth, A., Weinberger, J., Song, D.: Cross-Origin JavaScript Capability Leaks: Detection, Exploitation, and Defense. In: Proceedings of the 18th USENIX Security Symposium (USENIX Security 2009), Montreal, Canada (August 2009)

[11] Boyd, D.M., Ellison, N.B.: Social Network Sites: Definition, History, and Scholarship. Journal of Computer-Mediated Communication 13(1) (2007)

[12] Buchegger, S., Datta, A.: A Case for P2P Infrastructure for Social Networks - Opportunities and Challenges. In: Proceedings of the 6th International Conference on Wireless On-demand Network Systems and Services, Snowbird, Utah, USA (February 2009)

[13] Buchegger, S., Schioberg, D., Vu, L.-H., Datta., A.: PeerSoN: P2P Social Networking – Early Experiences and Insights. In: Proceedings of the 2nd Workshop on Social Network Systems (SocialNets 2009), Nuremberg, Germany (March 2009)

[14] Carminati, B., Ferrari, E.: Access control and Privacy in Web-based Social Networks. Journal of Web Information Systems 4(4), 395–415 (2008)

[15] Carminati, B., Ferrari, E., Perego, A.: Enforcing Access Control in Web-based Social Networks. ACM Transactions on Information & System Security 13(1), 6 (2009)

[16] Evans, B.M., Chi, E.H.: Towards a Model of Understanding Social Search. In: Proceedings of 2008 ACM Conference on Computer Supported Cooperative Work (CSCW 2008), San Diego, California, USA (November 2008)
[17] Fong, P.W.L., Anwar, M.M., Zhao, Z.: A Privacy Preservation Model for Facebook-Style Social Network Systems. In: Backes, M., Ning, P. (eds.) ESORICS 2009. LNCS, vol. 5789, pp. 303–320. Springer, Heidelberg (2009)
[18] Guo, L., Tan, E., Chen, S., Zhang, X., Zhao, Y.: Analyzing Patterns of User Content Generation in Online Social Networks. In: Proceedings of the 15th ACM SIGKDD International Conference on Knowledge Discovery and Data Mining, Paris, France (June 2009)
[19] Gupte, M.T., Hajiaghayi, M., Han, L., Iftode, L., Shankar, P., Ursu, R.M.: News Posting by Strategic Users in a Social Network. In: Leonardi, S. (ed.) WINE 2009. LNCS, vol. 5929, pp. 632–639. Springer, Heidelberg (2009)
[20] Horowittz, D., Kamvar, S.D.: The Anatomy of a Large-scale Social Search Engine. In: Proceedings of the 19th International Conference on World Wide Web (WWW 2010), Raleigh, North Carolina, USA (April 2010)
[21] Leskovec, J., Backstrom, L., Kumar, R., Tomkins, A.: Microscopic Evolution of Social Networks. In: Proceedings of the 14th ACM SIGKDD International Conference on Knowledge Discovery and Data mining, Las Vegas, USA (August 2008)
[22] Leskovec, J., Hattenlocher, D., Kleinberg, J.: Predicting Positive and Negative Links in Online Social Networks. In: Proceedings of the 19th International Conference on World Wide Web (WWW 2010), Raleigh, North Carolina, USA (April 2010)
[23] Li, W.-J., Stolfo, S., Stavrou, A., Androulaki, E., Keromytis, A.D.: A Study of Malcode-Bearing Documents. In: Proceedings of the 4th International Conference on Detection of Intrusions and Malware, and Vulnerability Assessment (2007)
[24] Liben-Nowell, D., Kleinberg, J.: The Link Prediction Problem for Social Networks. In: Proceedings of the 12th International Conference on Information and knowledge management (CIKM), New Orleans, Louisiana, USA (November 2003)
[25] Liccardi, I., Ounnas, A., Pau, R., Massey, E., Kinnunen, P., Lewthwaite, S., Midy, M.-A., Sarkar, C.: The role of Social Networks in Students' Learning Experiences. In: Working Group Reports on ITiCSE on Innovation and Technology in Computer Science Education, Dundee, Scotland (December 2007)
[26] Mislove, A., Gummadi, K.P., Druschel, P.: Exploiting Social Networks for Internet Search. In: Proceedings of the 5th Workshop on Hot Topics in Networks (HotNets 2006), Irvine, CA, USA (November 2006)
[27] Mislove, A., Viswanath, B., Gummadi, K.P., Druschel, P.: You are Who you Know: Inferring User Profiles in Online Social Networks. In: Proceedings of the 3rd ACM International Conference of Web Search and Data Mining (WSDM 2010), New York (February 2010)
[28] Morris, M.R., Teevan, J., Panovich, K.: What do people ask their social networks, and why? A Survey study of status message Q&A behavior. In: Proceedings of CHI 2010, Atlanta, Usa (April 2010)
[29] Nazir, A., Raza, S., Gupta, D., Chuah, C.-N., Krishnamurthy, B.: Network Level Footprints of Facebook Applications. In: Proceedings of Internet Measurement Conference (IMC 2009), Chicago, Illinois, USA (November 2009)
[30] Pallis, G., Vakali, A.: Insight and Perspectives for Content Delivery Networks. Communications of ACM 49(1), 101–106 (2006)

[31] Papadopoulos, F., Krioukov, D., Boguna, M., Vahdat, A.: Greedy Forwarding in Dynamic Scale-Free Networks Embedded in Hyperbolic Metric Spaces. In: Proceedings of IEEE INFOCOM, San Diego, CA, USA (March 2010)
[32] Provos, N., Mavrommatis, P., Abu Rajab, M., Monrose, F.: All your iFRAMEs point to Us. In: Proceedings of the 17th USENIX Security Symposium, San Jose, California, USA (July 2008)
[33] Richardson, M., Domingos, P.: Mining Knowledge-sharing Sites for Viral Marketing. In: Proceedings of the 8th ACM SIGKDD International Conference on Knowledge Discovery and Data Mining, Edmonton, Alberta, Canada (July 2002)
[34] Schneider, F., Feldmann, A., Krishnamurthy, B., Willinger, W.: Understanding Online Social Network Usage from a Network Perspective. In: Proceedings of Internet Measurement Conference (IMC 2009), Chicago, Illinois, USA (November 2009)
[35] Shakimov, A., Varshavsky, A., Cox, L.P., Caceres, R.: Privacy, Cost, and Availability Tradeoffs in Decentralized OSNs. In: Proceedings of the 2nd ACM Workshop on Online Social Networks (WOSN 2009), Barcelona, Spain (August 2009)
[36] Swamynathan, G., Wilson, C., Boe, B., Almeroth, K., Zhao, B.Y.: Do Social Networks improve e-commerce?: a Study on Social Marketplaces. In: Proceedings of the 1st Workshop on online Social Networks, Seattle, WA, USA (August 2008)
[37] Tantipathananandh, C., Berger-Wolf, T., Kempe, T.D.: A framework for Community Identification in Dynamic Social Networks. In: Proceedings of the 13th ACM SIGKDD International conference on Knowledge Discovery and Data mining, New York, NY, USA, August 2007, pp. 717–726 (2007)
[38] Tootoonchian, A., Gollu, K.K., Saroiu, S., Ganjali, Y., Wolman, A.: Lockr: Social Access Control for Web 2.0. In: Proceedings of the 1st Workshop on Online Social Networks (WOSN 2008), Seattle, WA, USA (August 2008)
[39] Ukkonen, A., Castillo, C., Donato, D., Gionis, A.: Searching the Wikipedia with Contextual Information. In: Proceedings of the 17th International Conference on Information and Knowledge Management (CIKM), Napa Valley, California, USA (November 2008)
[40] Vieira, M.V., Fonseca, B., Damazio, R., Golgher, P., Davi, B., Ribeiro-Neto, B.: Efficient Search Ranking in Social Networks. In: Proceedings of the 16th International Conference on Information and Knowledge Management, Lisbon, Portugal, November 2007, pp. 563–572 (2007)
[41] Villegas, W., Ali, B., Maheswaran, M.: An Access Control Scheme for Protecting Personal Data. In: Proceedings of the 6th Annual Conference on Privacy, Security and Trust (PST 2008), Fredericton, New Brunswick, Canada (October 2008)
[42] Vu, L.H., Aberer, K., Buchegger, S., Datta, A.: Enabling Secure Secret Sharing in Distributed Online Social Networks. In: Proceedings of Annual Computer Security Applications Conference (ACSAC) 2009, Hawaii, USA (December 2009)
[43] Willinger, W., Rejaie, R., Torkjazi, M., Valafar, M., Maggioni, M.: Research on Online Social Networks: Time to Face the Real Challenges. In: Proceedings of the 2nd Workshop on Hot Topics in Measurement and Modeling of Computer (2009)
[44] Zeinalipour-Yazti, D., Kalogeraki, V., Gunopulos, D.: Exploiting Locality for Scalable Information Retrieval in Peer-to-Peer Systems. Information Systems (InfoSys) 30(4), 277–298 (2005)
[45] Zeinalipour-Yazti, D., Kalogeraki, V., Gunopulos, D.: pFusion: An Architecture for Internet-Scale Content-Based Search and Retrieval. IEEE Transactions on Parallel and Distributed Systems (TPDS) 18(6), 804–817 (2007)

Chapter 9
Enhancing Computer Vision Using the Collective Intelligence of Social Media

Elisavet Chatzilari[1,3], Spiros Nikolopoulos[1,2], Ioannis Patras[2], and Ioannis Kompatsiaris[1]

[1] Centre for Research & Technology Hellas, Informatics and Telematics Institute, 6th km Charilaou-Thermi Road, Thermi-Thessaloniki, GR-57001 Thessaloniki, Greece
Tel.: +30-2311.257701-3; Fax.+30-2310-474128
ehatzi@iti.gr, nikolopo@iti.gr, ikom@iti.gr

[2] School of Electronic Engineering and Computer Science, Queen Mary University of London, E1 4NS, London, UK
Tel.: +44 20 7882 7523; Fax: +44 20 7882 7997
i.patras@eecs.qmul.ac.uk

[3] Centre for Vision, Speech and Signal Processing University of Surrey Guildford, GU2 7XH, UK
e.chatzilari@surrey.ac.uk

Abstract. Teaching the machine has been a great challenge for computer vision scientists since the very first steps of artificial intelligence. Throughout the decades there have been remarkable achievements that drastically enhanced the capabilities of the machines both from the perspective of infrastructure (i.e., computer networks, processing power, storage capabilities), as well as from the perspective of processing and understanding of the data. Nevertheless, computer vision scientists are still confronted with the problem of designing techniques and frameworks that will be able to facilitate effortless learning and allow analysis methods to easily scale in many different domains and disciplines. It is true that state of the art approaches cannot produce highly effective models, unless there is dedicated, and thus costly, human supervision in the process of learning that dictates the relation between the content and its meaning (i.e., annotation). Recently, we have been witnessing the rapid growth of Social Media that emerged as the result of users' willingness to communicate, socialize, collaborate and share content. The outcome of this massive activity was the generation of a tremendous volume of user contributed data that have been made available on the Web, usually along with an indication of their meaning (i.e., tags). This has motivated the research objective of investigating whether the Collective Intelligence that emerges from the users' contributions inside a Web 2.0 application, can be used to remove the need for dedicated human supervision during the process of learning. In this chapter we deal with a very demanding learning problem in computer vision that consists of detecting and localizing an object within the image content. We present a method that exploits the Collective Intelligence that is fostered inside an image Social Tagging System in order to facilitate the automatic generation of training data and therefore object detection models.

A. Vakali & L.C. Jain (Eds.): New Directions in Web Data Management 1, SCI 331, pp. 235–271.
springerlink.com

The experimental results shows that although there are still many issues to be addressed, computer vision technology can definitely benefit from Social Media.

1 Introduction

The recent advances of Web technologies have effectively turned ordinary people into active members of the Web, that generate, share, contribute and exchange various types of information. Web users act as co-developers and their actions and collaborations with one another have added a new social dimension on Web data. This social dimension of information was fostered by the next generation of the Web, namely Web 2.0, the applications of which have generated (and still generate) a remarkable volume of multimedia content. Based on this huge repository of content, various services have evolved [55], ranging from the field of eCommerce, to emergency response [56] and consumer collective applications such as realtravel.com [14]. The intelligence provided by single users organized in communities, takes a radical new shape in the context of Web 2.0, that of Collective Intelligence. Collective Intelligence emerges from the collaboration, communication and sharing among the users of social networks.

Although Collective Intelligence is at least as old as humans and appears in a wide variety of forms e.g., bacteria, animals, computer networks, it is now occurring in dramatically new forms. For example, Google[1] uses the knowledge millions of people have stored in the World Wide Web to provide useful answers to users' queries and Wikipedia[2] motivates thousands of volunteers around the world to create the world's largest encyclopedia. With new communication technologies and using the Internet as host, a large number of people all over the planet can now work together in ways that were never before possible in the history of humanity. But what exactly is Collective Intelligence and how can we benefit from it; The MIT Center for Collective Intelligence[3] frames the research question as *"How can people and computers be connected so that-collectively-they act more intelligently than any individuals, groups, or computers have ever done before?"*. It is now more important than ever for us to understand Collective Intelligence at a deep level so as to take advantage of these new possibilities.

In the field of multimedia data management, Collective Intelligence provides added value to the shared content and enables the accomplishment of tasks that are not possible otherwise. The acquisition of valuable knowledge is a big departure from traditional methods for information sharing, since managing Collective Intelligence poses new requirements. For example, semantic analysis has to fuse information coming both from the content itself, the social context and the emergent social dynamics. This fact has motivated increasing interest in discovering the different layers of Collective Intelligence, as well as in using these layers to empower new forms of

[1] http://www.google.com

[2] http://en.wikipedia.org

[3] http://cci.mit.edu

Web Data Management. Important progress towards this objective has been achieved in the context of the WeKnowIt[4] project were Collective Intelligence is considered to be the synthesis of 5 different layers namely, Personal Intelligence, Media Intelligence, Mass Intelligence, Social Intelligence and Organizational Intelligence.

In this chapter we investigate whether the Collective Intelligence derived from the user contributed content can be used to guide a learning process that will teach the machine how to recognize objects from visual content, the way a human does. We examine the problem both from the perspective of the teacher, which consists of knowledge that is build incrementally in an evolutionary and decentralized manner and therefore is characterized by questionable reliability, lack of structure, ambiguity and redundancy; as well as from the perspective of the learner that consists of models that apply learning algorithms on training data to capture the diversity of an object's form and appearance, and therefore demand for close supervision.

The rest of the chapter is structured as follows. Section 2 elaborates on the role of learning in computer vision and provides a description of Social Tagging Systems in the context of Web Multimedia Data. Section 3 emphasizes on the key aspect of multimedia analysis and provides an overview of the basic mechanisms that are used for learning. Section 4 presents an approach for training object detection models using data from collaborative tagging environments, that exploits the Collective Intelligence derived from the massive users' contribution. Concluding remarks are drawn in Section 6.

2 Learning and Web 2.0 Multimedia

2.1 Learning in Computer Vision

Learning has always been of primary importance for computer vision scientists. If we wish to construct a visual system that is able to scale on an arbitrary large number of concepts, effortless learning is crucial. Humans learn to recognize materials, objects and scenes from very few examples and without much effort. A 3-year old child is capable of building models for a substantial number of concepts and recognizing them using these models. By age of six humans recognize more than 10^4 categories of objects [7] and keep learning more throughout their life. Can a computer program learn how to recognize semantic concepts from images? This is the general question addressed by the computer vision scientists. But what is the process of learning; what is the mechanism that allows humans to initially require many examples to learn, as performed by little babies, and after they have learned how to learn, they can learn from just a few examples; and most importantly what is the role of the teacher in this process and what is the minimum amount of supervision that is absolutely necessary for facilitating efficient learning?

In [38] the authors make the hypothesis that, once a few categories have been learned with significant effort, some information may be abstracted from the process to make learning further categories more efficient. Similarly in [41] when images

[4] http://www.weknowit.eu/

of new concepts are added to the visual analysis model, the computer only needs to learn from the new images. What has been learned about previous concepts is stored in the form of profiling models, and the computer needs no re-training. On the other hand in [67] the authors claim that with the availability of overwhelming amounts of data, many problems can be solved without the need for sophisticated algorithms. The authors mention the example of Google's "Did you mean" tool, which corrects errors in search queries by memorizing billions of query-answer pairs and suggesting the one closest to the user query. In their paper the authors present a visual analog to this tool using a large dataset of 79 million images and a non-parametric approach for image annotation that is based on nearest neighbor matching.

However, the need for effortless learning coupled with the fact that the images archived on the Internet are growing at a phenomenal rate, has motivated other researchers to turn their interest in weakly (i.e. image level) annotated instead of strongly (i.e. region or pixel level) annotated images. Fig. 1 shows an example image with both strong and weak annotations. Photo sharing through the Internet has become a common practice and according to the reports released in 2007, flickr.com has 40 million monthly visitors and hosts two billion photos, with new photos in the order of millions being added on a daily basis. In this context, the authors of [11] use multiple instance learning to learn models from images labeled as containing the semantic concept of interest, but without indication of which image regions are observations of that concept. Similarly in [18] object recognition is viewed as machine translation by learning how to map visual objects (blobs) to concept labels. In [15] models are learned from ambiguously labeled examples, where each example is supplied with multiple potential labels, only one of which is correct. Approaches that learn an object category from just its name include [21], where the authors obtain images from the web using Google or Yahoo search engines and takes the returned results to be pseudo-positively labeled training images.

The key trade-off between the annotation-based models (which use labels provided by human annotators) and search-based models (which use models automatically obtained from the Web), is from the one side the amount of human effort that is required in annotation-based models and on the other side the expected decrease

Fig. 1. An example image annotated both strongly (i.e., each of the identified image regions is assigned with a label) and weakly (i.e., a set of labels is provided to describe the image content)

in classification performance that will result from search-based methods [34]. Recently, and driven by the widespread appeal of social sites, we have been witnessing an increasing research interest in using social sites and in particular Social Tagging Systems (STS), instead of the search engines, to obtain the necessary labels.

2.2 Social Tagging Systems and Web 2.0 Multimedia

An STS is a web-based application, where users, either as individuals or more commonly as members of a community (i.e., social networks), assign labels (i.e., arbitrary textual descriptions) to digital resources. Their motivation for tagging is information organization and sharing. Social tagging systems tend to form rich knowledge repositories that enable the extraction of patterns reflecting the way content sementics is perceived by the web users. In [27] the authors show that the tag proportions each resource receives crystallizes after about 100 annotations attributing this behavior to the users' common background and their tendency for imitation on other users' tagging habits. The availability of such content in the Web is high and the exploitation of the Collective Intelligence that is fostered by this type of content still remains a challenge.

In order to extract the knowledge that is stored and often "hidden" in social data, researches have employed various approaches: a) Clustering techniques that are based on tagging information and tag co-occurrence to derive semantically-related groups of tags and resources [4], [28], [30], b) Ontology driven tagging organization and mining, by combining Web 2.0 and Semantic Web, [29], [60], [54], c) content-based analysis of tagging-related sources that explore both tags and visual features (in a supplementary manner) for browsing and retrieving semantically related images [2], [57], [25]. Despite the active research efforts in this area, the full potential of Web 2.0 data has not been exploited yet. Few approaches exploit the fact that the tag and visual information space are highly correlated and the Collective Intelligence that will emerge from the massive participation of users in contributing and tagging multimedia content can be used to facilitate the learning process of computer vision systems.

However, while the Collective Intelligence derived from social data seems a very promising source of information, it has some serious limitations that mainly derive from the unconstrained nature of Web 2.0 applications. Users are prone to make mistakes and they often suggest invalid metadata (tag spamming). The lack of (hierarchical) structure of information results in tag ambiguity (a tag may have many senses), tag synonymy (two different tags may have the same meaning) and granularity variation (users do not use the same description level, when they refer to a concept).

There is a growing number of research efforts that attempt to overcome the aforementioned limitations and exploit the dynamics of social tagging systems to facilitate different types of multimedia applications. In [2], the authors claim that the intrinsic shortcomings of collaborative tagging can be tackled by employing content-based image retrieval technique. The user is facilitated in image database browsing and retrieval by exploiting both the tag and visual features in a supplementary way. In [25] a number of clustering techniques were employed in order to couple tagging information with content-based features. The clustering was tag-oriented and occurred in two steps. In the first step the resources were assigned to clusters, depending on the similarity of their accompanying tags. In the second step, visual features were employed, in an effort to increase the purity of already created clusters. The second step of the process could be regarded as a "misleading tags tracking phase". Another work that combines user data with feature-based approaches is presented in [24], that is used to rank the results of a video retrieval system. The authors use this knowledge, along with a multimedia ontology to build a learning personalized environment.

There are also works that address the problem of identifying photos from social tagging systems that depict a certain object, location or event [35], [57]. In [35] the authors make use of community contributed collections and demonstrate a location-tag-vision-based approach for retrieving images of geography-related landmarks. They use clustering for detecting representative tags for landmarks, based on their location and time information. Subsequently, they combine this information with vision-assisted process for presenting the user with a representative set of images. In [57] the authors are concerned with images that are found in community photo collections and depict objects (such as touristic sights). The presented approach is based on geo-tagged photos and the task is to mine images containing objects in a fully unsupervised manner. The retrieved photos are clustered according to different modalities including visual content and text labels.

In all cases the authors are trying to benefit from the Collective Intelligence that emerges from the content contributed to STSs and improve the efficiency of certain tasks. However the correlations between the tag and visual information space that are established when the users suggest tags for the uploaded visual content, are mostly treated as complementary sources of information that both contribute to the semantic description of the resources. In contrast to the above this chapter investigates whether the aforementioned correlations can be used to facilitate the learning process of multimedia analysis models. For this reason in the following Section we provide a short introduction to some of the techniques used for multimedia analysis.

3 Multimedia Analysis and Management

3.1 The Need for Semantics

The efficient management of multimedia data poses many technological challenges in terms of indexing, querying and retrieving, that require a deep understanding of the information at a semantic level. Driven by this need and given that machines' perception is limited to numbers and strings, there have been many research efforts that try to map semantic concepts or events to low level features, an issue addressed as bridging the "semantic gap".

The very first attempts for image retrieval were based on keyword search [50] applied either on the associated annotations (assuming that annotations existed) or on the images' file names. However, these approaches apart from requiring textual annotations of the multimedia data, they are barely as descriptive as the multimedia content itself. To overcome these limitations, the use of the image visual characteristics has been proposed. In this case, the visual content is utilized by extracting a set of visual features from each image or image region. By comparing the visual features an algorithm can decide whether the two images/regions represent the same semantic concept. Then, image retrieval is performed by comparing the visual features of an example image/region that is associated with a semantic concept by the user, to the visual features of all images in a given collection [20] (known as Query By Image Content systems).

Subsequently, more sophisticated methods were proposed that aimed at simulating the functionality of human visual system by allowing the machines to mimic the procedure followed by a human when identifying semantics in visual content. In this direction pattern classification has been brought to the core of most image analysis techniques in order to render a kind of meaning on visual patterns. A typical pattern classification problem can be consider to include a series of sub-problems the most important of which are: a) determining the optimal feature space, b) removing the noisy data that can be misleading, c) avoid overfitting on training data, d) use the most appropriate distribution for the model, e) make good use of any prior knowledge that may help you in making the correct choices, f) perform meaningful segmentation when the related task requires to do so, h) exploit the analysis context, etc. All the above are crucial for initiating a learning process that aims at using the available training samples to estimate the parameters of a model representing a semantic concept. In the following section we discuss and provide related references for some of the aforementioned sub-problems, giving special emphasis on the mechanisms of learning.

3.2 Visual Features Extraction and Regions Identification

Many problems derive from the fact that it is very difficult to describe visual content effectively in a form that can be handled by machines. In general, feature extraction is a domain dependant problem and it is unlikely that a good feature extractor for a specific domain will work as good for another domain. The extraction of features for efficient image representation has attracted a lot of interest in the scientific community of image analysis. Motivated by the principles of human perception, most researchers have tried to describe images/regions using color, shape and texture characteristics. Some of the most widely adopted techniques for representing images/regions include the descriptors proposed by the MPEG-7 standard [1] that capture different aspects of color, texture and shape. Other approaches rely on the corners and edges that can be found inside an image in order to describe the image using a set of interest points. The Scale-Invariant Feature Transform (SIFT) proposed in [44] and its modifications (i.e., color SIFT, opponent SIFT, etc [59]) are considered some of the most representative algorithms of this category. Particularly important is also considered the vector quantization approach initially proposed in [64] where in analogy to text, images/regions are represented as bags-of-visual words that have been learned through an extensive training process using representative data.

Additionally, many problems derive from the fact that images tend to include more than one objects in their content, which decreases the descriptiveness of the feature space and raises the need for segmentation. The segmentation of images into regions and the use of a separate set of features for each region was introduced to address the aforementioned issue. Segmentation techniques seek to detect groups of pixels sharing similar visual characteristics and identify in this way meaningful objects (similar to the ones identified by human visual system). In the field of segmentation one of the most commonly used methods is Normalized Cuts [62] which is a graph partitioning algorithm using a global criterion, the normalized cut, for segmenting the graph. Other approaches include [52] that segments color images by applying a variance of K-means on intensity, position and texture features, as well as [12] that is based on the Expectation-Maximization (EM) algorithm. Both segmentation and feature extraction are two very important techniques for identifying patterns in visual content. However, in order to bridge the semantic gap these patterns will have to be classified into meaningful concepts. This is where the role of learning takes place since it is used to estimate the parameters of a model that will be sub-sequently used to classify new, unseen images or regions.

3.3 Learning Mechanisms

Humans can classify images through models that are built using examples for every single semantic concept. Based on this assumption, researchers have been trying to simulate human visual system by using machine learning algorithms to classify the visual content. A set of training samples plays the role of the examples that a person uses to learn a concept. Based on the prior knowledge that we have on the training samples during the learning process, we can distinguish between the following basic categories; unsupervised, strongly supervised, semi-supervised and weakly supervised learning.

3.3.1 Un-supervised Learning

Unsupervised learning is a class of problems in which one seeks to determine how the data are organized. It is distinguished from supervised learning in that the learner is given only unlabelled examples. One of the most known forms of unsupervised learning is clustering. The clustering output can be hard (a partition of the data into groups) or fuzzy (where each data point has a variable degree of membership in each output cluster) [6]. Clustering algorithms can be divided in two major categories, hierarchical [32] and partitional [47]. Hierarchical methods produce a nested series of partitions whereas partitional methods produce only one partition. Many clustering algorithms require the specification of the number of clusters to produce in the input data set, prior to the execution of the algorithm. Barring knowledge of the proper value beforehand, the appropriate value must be automatically determined, a problem for which a number of techniques have been developed [13], [23]. Another important step in any clustering scheme is the selection of a distance measure, which determines how the similarity of two elements is calculated. The distance measure influences the shape of the clusters, as some elements may be close to one another according to one distance and farther away according to another.

3.3.2 Strongly-Supervised Learning

In strongly-supervised learning there is prior knowledge about the labels of the training samples and there is one-to-one relation between a sample and its label (e.g., each region of the image depicted in Fig. 2 is associated with a label). The aim of strongly-supervised learning is to generate a global model that maps input objects to the desired outputs and generalize from the presented data to unseen situations in a "reasonable" way. Some of the most widely used types of classifiers that typically rely on strongly annotated samples are the Neural Network (Multilayer perceptron) [19], Support Vector Machines [51], naive Bayes [17], decision tree [8] and radial basis function classifiers [46]. A known issue in supervised learning is overfitting (i.e. the model describes random error or noise instead of the underlying relationship) which is more probable to occur when the training samples are rare and the dimensionality of the feature space is high. In order to avoid overfitting, it is necessary to use additional techniques (e.g. cross-validation, regularization, early stopping, Bayesian priors on parameters or model comparison), that can indicate when further training is not resulting in better generalization.

3.3.3 Semi-supervised Learning

Semi-supervised learning algorithms try to exploit unlabeled data, which are usually of low cost and can be obtained in high quantities, in conjunction with some supervision information. In this case, only a small portion of the data is labeled and the algorithm aims at propagating the labels to the unlabeled data. The earliest idea about using unlabeled data when learning a classification model is self-learning. In self-learning, the classification model is initially trained using only the labeled data and at each step a part of the unlabeled data is labeled according to the output of the current model. Then, a new classification model is trained using both the labeled as well as the data that were labeled as positive from the previous step.

Another category of semi-supervised learning algorithms is based on the cluster assumption, according to which the points that are in the same cluster belong to

Fig. 2. An image depicting the object sea that is manually annotated at region level

the same class. So there should be regions with high density of points (which formulate the clusters) and low-density regions where the decision boundary lies in. Most of the recent semi-supervised classification approaches aim at creating new specialized learning algorithms that are able to combine labeled and unlabeled data. More specifically, in order to have the ability to choose a learning algorithm with the required attributes most state-of-the-art methods combine semi-supervised learning with boosting techniques. Boosting is part of the ensemble learning family algorithms and aims at building an ensemble by training each new model instance to emphasize the training instances that previous models mis-classified [5]. The most popular algorithm that utilizes the boosting method is Adaboost (short for Adaptive Boosting) presented in [22]. Algorithms that adopted the SemiBoost approach, which employs the boosting method in order to improve any existing supervised learning algorithm with unlabeled data, include [48], [37], [9].

3.3.4 Weakly-Supervised Learning

By weakly-supervised we refer to the process of learning using weakly labeled data (i.e., samples labeled as containing the semantic concept of interest, but without indication of which segments/parts of the sample are observations of that concept, (as shown in Fig. 3). In this case, the basic idea is to introduce a set of latent variables that encode hidden states of the world, where each state induces a joint distribution on the space of semantic labels and image visual features. New images are annotated by maximizing the joint density of semantic labels, given the visual features of the new image [11]. The most indicative weakly-supervised learning algorithms are the ones that are based on aspect models like probabilistic Latent Semantic Analysis (pLSA) [63], [21] and Latent Dirichlet Allocation (LDA) [39], [58]. These models are typically applied on weakly annotated datasets to estimate the joint distribution of semantic labels and visual features.

3.4 Annotation Cost for Learning

Object detection schemes always employ some form of supervision as it is practically impossible to detect and recognize an object without using any semantic information during training. However, semantic labels may be provided at different levels of granularity (global or region level) and preciseness (one-to-one or many-to-many relation between objects and labels), imposing different requirements on the effort required to generate them. Indeed, there is a clear distinction between the strong and accurate annotations that are usually generated manually and constitute a laborious and time consuming task, and the weak and noisy annotations that are usually generated by web users for their personal interest and can be obtained in large quantities from the Web or collaborative tagging environments like flickr[5]. Due to the fact that the annotation cost is a critical factor when designing an object detection scheme with the intention to scale in many different objects and domains, in the

[5] www.flickr.com

Fig. 3. An image depicting the object sea that is manually annotated at global level

following we distinguish the object detection methods based on the characteristics of the dataset that they employ and the effort required for its annotation. Our goal is to highlight the tradeoff between the annotation cost for preparing the necessary training samples and the quality of the resulting models.

In the first category we classify the methods that use manually annotated images at region level as the one depicted in Fig. 2. These methods rely on strongly supervised learning and are usually developed to recognize certain types of objects with very high accuracy. In [70] and [66] manual annotations of faces are used in order to train the classifiers. In [43] a method for the recognition of buildings is proposed and in [36] an implicit shape model for the detection of cars is presented. In [71] manual annotations at region level are used to train a probabilistic model integrating both visual features and spatial context. Annotating images at region level is probably the task with the highest annotation cost.

Image annotations at global level, even manual ones, are easier to obtain than region level annotations. This fact has motivated many researchers in developing algorithms that rely on weakly-supervised and semi-supervised learning, and are able to exploit global annotations for performing object detection. The Corel database is probably the most widely used set of images annotated manually at global level (as shown in Fig. 3) and has been used in numerous works. Jia Li and James Z. Wang [40] used the Corel dataset to train models for each concept separately, while in [69] it was used to evaluate the performance of an algorithm that considers the recognition of visual concepts to be part of the segmentation process. The Corel dataset has been also used by Duygulu et. al. that presented a methodology for mapping words to image regions using an algorithm based on EM [18], as well as in [10] where a label propagation algorithm that incorporates time, location and visual similarity for event and scene detection has been proposed. The widespread use of Corel and other datasets of similar type can be mainly attributed to the fact that the global annotations associated with the images was noise free and accurate. This allowed the researchers to derive some probabilistic relations between objects and labels and use these relations to perform object detection on new images. However, labels accuracy comes with the cost of manual annotation which is something that limits the scaling potentials of the schemes relying on such labels. This was the reason that researchers turned their interest on the Web and started to investigate whether it could be used to obtain globally annotated images.

Using the Web as a source many approaches have been proposed that obtain globally annotated images through search engines, using the name of the object as argument (see Fig. 4 for some example images obtained using the query word "sea"). Keiji Yanai uses visual content from the Web as training images for a generic classification system [73] and in [21] the authors learn object categories from Google's image search. However, since search engines in their current form rely primarily on the image filename or the surrounding text to decide whether to return an image or not, the quality of the obtained annotations is very low. Thus, although these type of global annotations can be obtained at practically no cost, the high level of noise renders particularly difficult the extraction of reliable probabilistic relations between objects and labels.

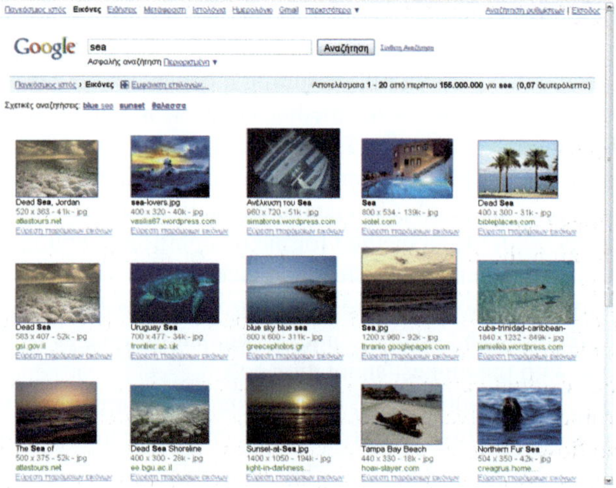

Fig. 4. Images depicting the object sea, obtained automatically from the Google Image Search engine using "sea" as the query word

For this reason, the most recent research efforts are focusing on the content that is being massively contributed by Web users in the context of Web 2.0 applications. In [57] object and event detection is performed by clustering images downloaded from flickr based on textual, visual and spatial information and verified through Wikipedia[6] content. Similarly a framework that probabilistically models geographical information for event and activity detection using geo-tagged images from flickr

[6] www.wikipedia.com

is presented in [33]. Although the tag annotations that accompany the images contributed by social users are less noisy from the ones obtained via the search engines, they are still considered to be rather noisy for directly extracting the necessary probabilistic relations between objects and labels, see Fig. 5 for an example image obtained from Flickr along with the associated tags.

Fig. 5. Image depicting the object sea, obtained automatically from Flickr along with the associated tags

In Table 1 we summarize the pros and cons for each of the aforementioned types of annotation. As a general conclusion we can say that manual image annotation (either at region or global level) is a time consuming task and as such it is particularly difficult to be performed on the desired volumes of content that are needed for building robust and scalable classifiers. On the other hand, the STSs and the Web provides cost free annotations that are very noisy to be used directly for extracting the necessary probabilistic relations between objects and labels. The Collective Intelligence that emerges from the tagged images aggregated in STSs would have to be exploited towards removing the existing obstacles. In this direction we present a method that transforms global image tags into region level annotations, in a form suitable to be used by a strongly-supervised learning algorithm for object detection.

4 Leveraging Social Media for Training Object Detectors

As already described, machine learning algorithms fail in two main categories in terms of the annotation granularity, the algorithms that are designed to learn from strongly annotated samples (i.e., samples in which the exact location of an object within an image is known) and the algorithms that learn from weakly annotated samples (i.e., samples in which it is known that an object is depicted in the image, but its location is unknown). In the first case, the goal is to learn a mapping from visual features f_i to semantic labels c_i given a training set made of pairs (f_i, c_i). On the

Table 1. Pros & Cons for the different types of annotation

Annotation Type	Automated Annotation	Scaling Capability	Training Efficiency	Learning Mechanism	Related Techniques
Region-level (manual) (Fig. 2)	Poor	Poor	Excellent	strongly-supervised	Viola & Jones [70], Sung & Poggio [66], Li et al. [43], Leibe et al. [36], Wand et al. [71]
Global-level (manual) (Fig. 3)	Fair	Fair	Good	weakly-supervised	Li & Wang [40], Vascooncelos et al. [69], Duygulu et al. [18], Cao et al. [10]
Global-level (automatically via Search Engines) (Fig. 4)	Excellent	Excellent	Poor	weakly-supervised	Yanai [73], Fergus et al. [21]
Global-level (automatically via Social Networks) (Fig. 5)	Excellent	Excellent	Fair	weakly-supervised	Quack et al. [57], Dhiraj & Lue [33]

other hand, in the case of weakly annotated training samples, the goal is to estimate the joint probability distribution between the visual features f_i and the semantic labels c_i given a training set made of pairs between sets $\{(f_1,\dots,f_n),(c_1,\dots,c_m)\}$.

While model parameters can be estimated more efficiently from strongly annotated samples, such samples are very expensive to obtain. On the contrary, weakly annotated samples can be found in large quantities especially from sources related to social networks. Motivated by this fact, our work aims at combining the advantages of both strongly supervised (learn model parameters more efficiently) and weakly supervised (learn from samples obtained at low cost) methods, by allowing the strongly supervised methods to learn object detection models from training samples that are found in collaborative tagging environments.

4.1 Problem Formulation

The problem can be formulated as follows. Drawing from a large pool of weakly annotated images, our goal is to benefit from the knowledge that can be extracted from social tagging systems, in order to automatically transform some of the weakly annotated images into strongly annotated ones. In order to do this, we consider that if the set of weakly annotated images is properly selected from the repository of a collaborative tagging environment, the most populated tag-"term" and the most populated visual-"term" will be two different representations/expressions (i.e., textual and visual) of the same object. We define tag-"terms" to be sets of tags that are provided by social users to describe an image and are grouped based on their semantic affinity (e.g., synonyms, derivatives, etc). Respectively, we define visual-"terms" to be sets of image regions that are identified by an automatic segmentation algorithm and are grouped based on visual similarity. The most populated tag-"term" (i.e., the most frequently appearing tag, counting also its synonyms, derivatives, etc) is used to provide the semantic label of the object that the developed classifier is trained to identify, while the most populated visual-"term" (i.e., the cluster of image regions containing the most instances) is used to provide the set of strongly annotated samples for training the classifier. It is clear that the process of leveraging weakly annotated images to become the strongly annotated training samples of a supervised learning scheme, is primarily achieved through the semantic clustering of image regions to objects (i.e., each cluster consists of regions that depict only one object). Using the notation of Table 2 semantic clustering can be formulated as follows. Given a large set of images $I_q \in S^c$ with annotation information of the type $\{(f_d(r_1^{I_q}),\ldots, f_d(r_m^{I_q})), (c_1,\ldots, c_t)\}$, semantic clustering would produce pairs (w_i, c_i) where each w_i is a set of regions extracted from all images in S^c that depict only c_i. Semantic clustering can only be made feasible in the ideal case where the image analysis techniques employed by our framework works perfect. However, this is highly unlikely due to the following reasons. In case of over or under segmentation we will have more or fewer regions from the actual objects in image, making perfect semantic

Table 2. Legend of Introduced Notations

Symbol	Definition
S	The complete social dataset
N	The number of images in S
S^c	An image group, subset of S that emphasizes on object c
n	The number of images in S^c
I_q	An image from S
$R_{I_q} = \{r_i^{I_q}, i = 1,\ldots,m\}$	Segments identified in image I_q
$f_d(r_i^{I_q}) = \{f_i, i = 1,\ldots,z\}$	Visual features extracted from a region $r_i^{I_q}$
$C = \{c_i, i = 1,\ldots,t\}$	Set of objects that appear in the images of group S^c
$W = \{w_i, i = 1,\ldots,o\}$	Set of clusters created by the region-based clustering algorithm
p_{c_i}	Probability that tag-based image selection draws from S an image depicting c_i
TC_i	Number of regions depicting object c_i in S^c

clustering impossible. Similarly, the inadequacy of visual descriptors to perfectly discriminate between different semantic objects is likely to lead the clustering algorithm in creating a different number of clusters than the number of actual semantic objects, or even mix regions depicting different objects into the same cluster. Thus, instead of requiring that each w_i is mapped with a c_i, we only search for a single pair (w_k, c_z) where the majority of regions in cluster w_k depicts c_z. Given that both w_i (i.e., visual-"term') and c_i (i.e., tag-"term") are sets (of images regions and user contributed tags, respectively), we can apply the $Pop(\cdot)$ function on them, that calculates the population of a set (i.e., number of members). Eventually, the problem addressed in our approach is what should be the characteristics of S^c so as the pair (w_k, c_z) determined using $k = \arg\max_i(Pop(w_i))$ and $z = \arg\max_i(Pop(c_i))$ satisfies our objective i.e., that the majority of regions included in w_k depicts c_z. Our approach in using user contributed content to create S^c is motivated by the fact that due to the common background that most users share, the majority of them tend to contribute similar tags when faced with similar type of visual content [49]. This is the point where our approach benefits from the Collective Intelligence that emerges from an STS, in the sense that it would be over-ambitious to rely on such an assumption if tags were to be contributed by just one or a few users. However, since the tags in an STS originate from a significantly large amount of users, it is statically safe to conclude that the majority of tag assignments will conform to the aforementioned rule. Then, given this assumption it is expected that as the pool of the weakly annotated images grows, the most frequently appearing "term" in both tag and visual information space will converge into the same object.

4.2 Framework Description

The framework we propose for leveraging the weakly annotated data in order to train object detection models, is depicted in Fig. 6. The analysis components that can be identified in our framework are, tag-based image selection, image segmentation, extraction of visual features from image regions, region-based clustering using their visual features and learning of object detection models using strongly annotated samples.

More specifically, given an object c that we wish to train a detector for, our method starts from a large collection of user tagged images and performs the following actions. Images are selected based on their tag information in order to formulate image group(s) that correspond to thematic entities. Given the tendency of social tagging systems to formulate knowledge patterns that reflect the way content is perceived by the web users [49], [27], tag-based processing is expected to identify these patterns and create image group(s) each one emphasizing on a certain object. By emphasizing we refer to the case where the majority of the images within a group depict different instances of a certain object and that the linguistic description of that object can be obtained from the most frequently appearing tag (see Section 4.3.1 for more details). Subsequently, region-based clustering is performed on all images belonging to the image group that emphasizes on object c, that have been pre-segmented by an automatic segmentation algorithm. During

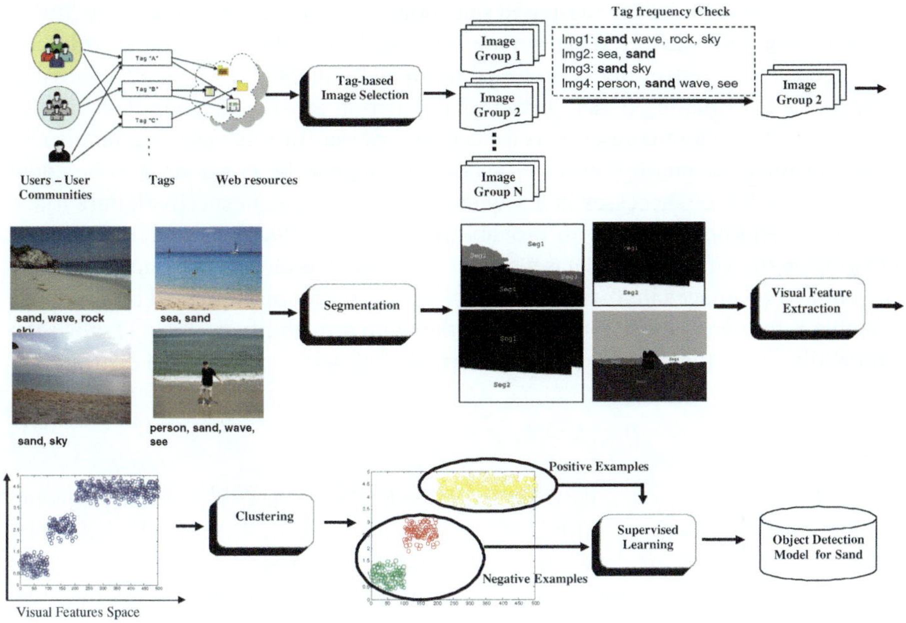

Fig. 6. Actions performed by our framework in order to train a model for detecting the object sand

region-based clustering the image regions are represented by their visual features and each of the generated clusters contains visually similar regions. Since the majority of the images within the selected group depicts instances of the desired object c, we anticipate that the majority of regions representing the object of interest will be gathered in the most populated cluster, pushing all irrelevant regions to the other clusters. Eventually, we use as positive samples the visual features extracted from the regions belonging to the most populated cluster to train (in a strongly supervised manner) a model detecting the object c. Although noisy tags and inaccurate segmentation are likely to prevent the most populated cluster from gathering all regions depicting object c, the fact that the collection of user tagged images can be arbitrary large (due to its "social" origin) can compensate for the loss in accuracy.

We can view the process of using image tag information to create an image group S^c that emphasizes on object c, as the process of selecting images from a large pool of weakly annotated images using as argument a query tag t_q. t_q is the linguistic description of the object c. The selection criteria can be keyword-based search (in the trivial case), pre-annotated groups, or more sophisticated approaches (see Section 4.3.1). Although misleading and ambiguous tags always hinders this process, the expectation is that as the number of selected images grows large and the tag-proportions for each image crystallizes [27], there will be a connection between what is depicted in the majority of the selected images and what is described by the majority of the contributed tags.

Let us assume that using tag-based selection we construct an image group $S^c \subset S$ emphasizing on object c. What we are interested in is the frequency distribution of objects $c_i \in C$ appearing in S^c based on their frequency rank. We can view the process of constructing S^c as the act of populating an image group with images selected from a large dataset S using certain criteria. In this case, the number of times an image depicting object c_i appears in S^c, can be considered to be equal with the number of successes in a sequence of n independent success/failure trials, each one yielding success with probability p_{c_i}. Given that S is sufficiently large, drawing an image from this dataset can be considered as an independent trial. Thus, the number of images in S^c that depict object $c_i \in C$ can be expressed by a random variable K following the binomial distribution with probability p_{c_i}. Eq. (1) shows the probability mass function of a random variable following the binomial distribution.

$$Pr_{c_i}(K=k) = \binom{n}{k} p_{c_i}^k (1-p_{c_i})^{n-k} \tag{1}$$

Given the above, we can use the expected value $E(K)$ of a random variable following the binomial distribution to estimate the expected number of images in S^c that depict object $c_i \in C$, if they are drawn from the initial dataset S with probability p_{c_i}. This is actually the value of k maximizing the corresponding probability mass function, which is:

$$E_{c_i}(K) = np_{c_i} \tag{2}$$

If we consider α to be the average number of times an object appears in an image, then the number of appearances (*#appearances*) of an object in S^c is:

$$TC_i = \alpha np_{c_i} \tag{3}$$

Moreover, we accept that there will be an object c_1 that is drawn (i.e., appears in the selected image) with probability p_{c_1} higher than p_{c_2}, which is the probability that c_2 is drawn, and so forth for the remaining $c_i \in C$. This assumption is experimentally verified in Section 4.4.2 where the frequency distribution of objects for different image groups are measured in a manually annotated dataset. Finally, using eq. (3) we can estimate the expected number of appearances (*#appearances*) of an object in S^c, $\forall c_i \in C$. Fig. 7(a) shows the *#appearances* $\forall c_i \in C$ against their frequency rank, given some examples values of p_{c_i} with $p_{c_1} > p_{c_2} > \ldots$. It is clear from eq. (3) that if we consider the probabilities p_{c_i} to be fixed, the expected difference, in absolute terms, on the *#appearances* between the first and the second most highly ranked objects c_1 and c_2, increases as a linear function of n, see Fig. 7(b) for some examples. Additionally, apart from increasing the expected absolute difference on the *#appearances* between the two most frequently appearing objects, the high values of n also minimize the probability of the case where c_2 although drawn with probability smaller that c_1 appears more times in the generated image group. In Fig. 8 we draw the probability mass function of two random variables that correspond to objects c_1, c_2 of Fig. 7(b) (i.e., $p_{c_1} = 0.8$, $p_{c_1} = 0.6$) for three different values of n (i.e., $n = 50$, $n = 100$ and $n = 200$). The probability of experiencing the case where c_2, although drawn with smaller probability, appears more times than c_1

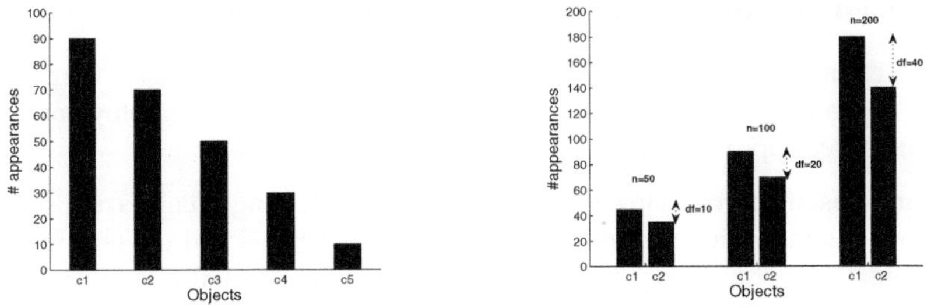

(a) (b)

Fig. 7. a) Distribution of *#appearances* $\forall c_i \in C$ based on their frequency rank, for n=100 and p_{c_1}=0.9, $p_{c_2} = 0.7$, $p_{c_3} = 0.5$, $p_{c_4} = 0.3$, $p_{c_5} = 0.1$. b) Difference of *#appearances* between c_1, c_2, using fixed values for $p_{c_1} = 0.8$ and $p_{c_2} = 0.6$ and different values for *n*.

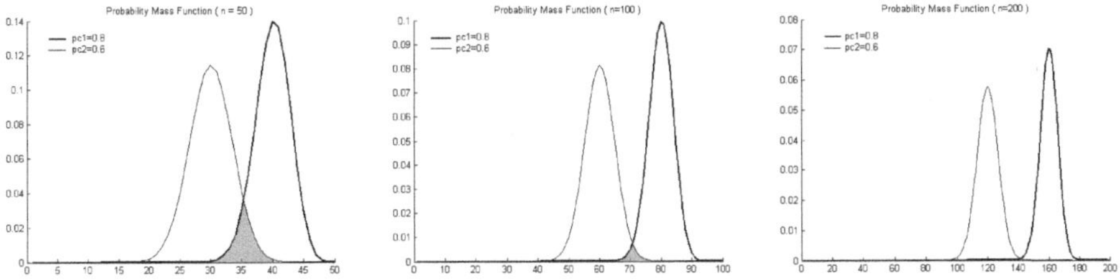

(a) (b) (b)

Fig. 8. a) Probability mass function for $n = 50$ trials and $p_{c_1} = 0.8$, $p_{c_2} = 0.6$. b) Probability mass function for $n = 100$ trials and $p_{c_1} = 0.8$, $p_{c_2} = 0.6$. c) Probability mass function for $n = 200$ trials and $p_{c_1} = 0.8$, $p_{c_2} = 0.6$.

in S^c, is proportional to the surface where the two curves overlap. It is clear from Fig. 8 that as n increases, the variance of the two random variables decrease, forcing the surface of the overlapping region to also decrease (e.g., for $n = 200$ the surface of the overlapping region is almost zero). Based on these observations, we reach the theoretical expectation that there is higher probability in w_k being a set of regions the majority of which depict c_z as n increases.

4.3 Implementing the Framework

In this section we provide details for the analysis components that are used by the proposed framework. Due to the fact that a necessary pre-requisite for our framework to work efficiently is operating on a large number of images, a discussion about the complexity of each analysis component is also included.

4.3.1 Tag-Based Image Selection

In this section we specify the approaches that are used to select images from a large dataset of arbitrary content, based on their tag information. We employ one of the following three approaches based on the associated annotations:

Keyword-based search: This approach is used for selecting images from strongly annotated datasets. These datasets are usually hand-labeled and the tags provided by the annotators can be considered to be mostly accurate and free of ambiguity. Thus, in order to create S^c we need only to select the images that are tagged with the linguistic expression of the object c.

Flickr groups: are virtual places hosted in collaborative tagging environments that allow social users to share content on a certain topic. Although managing flickr groups still involves some type of human annotation (i.e., a human assigns an image to a specific group) it can be considered weaker than the previous case since this type of annotation does not provide a full description of the objects depicted in the image. In this case, S^c is created by taking the images contained in a flickr group titled with the name of the object c. From here on we will refer to those images as roughly-annotated images.

SEMSOC: stands for SEmantic, SOcial and Content-based clustering and is applied in our framework on weakly annotated images (i.e., images that have been tagged by humans in the context of a collaborative tagging environment, but no rigid annotations have been provided) in order to create semantically consistent groups of images. SEMSOC was introduced by Giannakidou et. al. in [25], [26] and is an un-supervised model for the efficient and scalable mining of multimedia social-related data that jointly considers social and semantic features. The reason for adopting this approach in our framework is to overcome the limitations that characterize collaborative tagging systems such as tag spamming, tag ambiguity, tag synonymy and granularity variation (i.e., different description level). The outcome of applying SEMSOC on a large set of images S, is a number of image groups $S^{c_i} \subset S, \quad i = 1, \dots, m$, where m is the number of created groups. This number is determined empirically, as described in [25]. Then in order to obtain the image group S^c that contains the images depicting the desired object c, we select the SEMSOC-generated group S^{c_i} where its most frequent tag relates with c. Fig. 9 shows four examples of image clusters generated by SEMSOC, along with the corresponding most frequent tag.

4.3.2 Segmentation

Segmentation is applied on all images in S^c with the aim to extract the spatial masks of visually meaningful regions. In our work we have used a K-means with connectivity constraint algorithm as described in [53]. The output of this algorithm is a set of segments $R_{I_q} = \{r_i^{I_q}, i = 1, \dots, m\}$, which roughly correspond to meaningful objects, $c_i \in C$. The time efficiency of the segmentation process depends mainly on the size of the image. The segmentation of low-resolution images is performed

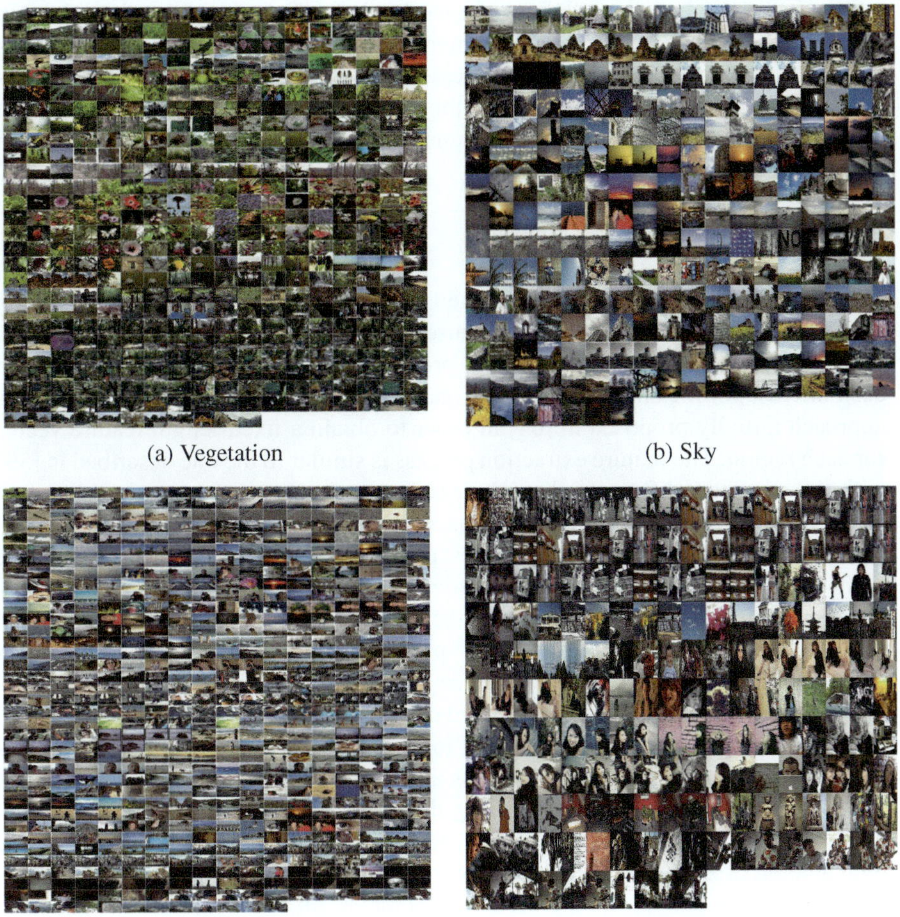

(a) Vegetation (b) Sky

(c) Sea (d) Person

Fig. 9. Examples of image groups generated using SEMSOC (in caption the corresponding most frequent tag). It is clear that the majority of images in each group include instances of the object that is linguistically described by the most frequent tag. The image is best view in color and with magnification.

considerably fast but the time efficiency of the algorithm degrades quickly as the image size increases. Therefore in order to cope with high-resolution images the authors of [53] make the reasonable assumption that the regions falling below the 0.75% of the total image area are insignificant. Based on this assumption the segmentation algorithm is applied on a reduced version of the image (i.e., by downscaling its original version). This improves the time efficiency of the algorithm but at the expense of the quality of the segmentation result. To alleviate this, the pixels belonging to blocks on edges between regions are reclassified using the Bayes classifier. Applying the segmentation algorithm on reduced images with reclassification using the Bayes classifier, delivers the same segmentation quality with segmentation time significantly reduced.

4.3.3 Visual Descriptors

In order to visually describe the segmented regions we have employed the following: a) the Harris-Laplace detector and a dense sampling approach for determining the interest points, b) the SIFT descriptor as proposed by Lowe [45] in order to describe each interest point using a 128-dimensional feature vector and c) the bag-of-words approach initially proposed in [64] in order to obtain a fixed-length feature vector for each region. The feature extraction process is similar to the one described in [59] with the important difference that in our case descriptors are extracted to represent each of the pre-segmented image regions, rather than the whole image.

More specifically, for detecting interest points we have applied the Harris-Laplace point detector on intensity channel, which has shown good performance for object recognition [74]. In addition, we have also applied a dense-sampling approach where interest points are taken every 6th pixel in the image. For each interest point (identified both using the Harris-Laplace and dense sampling approach) the 128-dimensional SIFT descriptor is computed using the version described by Lowe [45]. SIFT descriptors have been been found to be particularly robust against variations in scale, rotation, changes in brightness and contrast, etc. A Visual Word Vocabulary (Codebook) was created by using the K-Means algorithm to cluster in 300 clusters, approximately 1 million SIFT descriptors that were sub-sampled from a total amount of 28 million SIFT descriptors, extracted from 5 thousand training images. The Codebook allows the SIFT descriptors of all interest points contained in an image region to be vector quantized against the set of Visual Words and create a histogram. Thus, $\forall r_i^{I_q} \in R_{I_q}$ and $\forall I_q \in S^c$ a 300-dimensional feature vector $f(r_i^{I_q})$ is extracted, that contains information about the presence or absence of the Visual Words included in the Codebook. All feature vectors were normalized so as the sum of all dimensions to be equal with 1.

4.3.4 Clustering

For performing feature-based region clustering we applied the affinity propagation clustering algorithm on all extracted feature vectors $f(r_i^{I_q})$, $\forall r_i^{I_q} \in R_{Iq}$ and $\forall I_q \in S^c$. This is an algorithm that takes as input the measures of similarity between pairs of data points and exchanges messages between data points, until a high-quality set of centers and corresponding clusters is found. Affinity propagation was proposed by Frey and Dueck [23] and was selected for our work due to the following reasons: a) The requirements of our framework imply that in order to learn a robust object detection model, clustering will need to be performed on a considerably large number of regions, making computational efficiency an important issue. The common approach followed by most clustering algorithms is to determine a set of centers such that the sum of squared errors between data points and their nearest centers is minimized. This is done by starting with an initial set of randomly selected centers and iteratively refining this set so as to decrease the sum of squared errors. However, such approaches are sensitive to the initial selection of centers and work well only when the number of clusters is small and the random initialization is close to a good solution. This is the reason why these algorithms need to re-run many times with different initializations in order to find a good solution. In contrast to this, affinity propagation simultaneously considers all data points as potential centers. By viewing each data point as a node in a network, affinity propagation recursively transmits real-valued messages along edges of the network until a good set of centers and corresponding clusters emerges. In this way, it removes the need to re-run the algorithm with different initializations which is very beneficiary in terms of computational efficiency. b) The fact that the number of objects depicted in the images of an image group can not be known in advance, poses the requirement for the clustering procedure to automatically determine the appropriate number of clusters based on the analyzed data. Affinity propagation, rather than requiring that the number of clusters is pre-specified, takes as input a real number for each data point. This number is called preference and has the meaning that data points with larger preferences are more likely to be chosen as centers. In this way the number of identified centers (number of clusters) is influenced by the values of the input preferences but also emerges from the message-passing procedure. If a priori, all data points are equally suitable as centers, as in our case, the preferences should be set to a common value. This value can be varied to produce different numbers of clusters and taken for example to be the median of the input similarities (resulting in a moderate number of clusters) or their minimum (resulting in a small number of clusters). Given that it is better for our framework to handle noisy rather than inadequate (in terms of indicative examples) training sets, we opt for the minimum value in our experiments.

4.3.5 Learning Model Parameters

Support Vector Machines (SVMs) [61] were chosen for generating the object detection models, due to their ability in smoothly generalizing and coping efficiently with high-dimensionality pattern recognition problems. All feature vectors assigned to the most populated of the created clusters are used as positive examples for training a binary classifier. Negative examples are chosen arbitrary from the remaining dataset. Tuning arguments include the selection of Gaussian radial basis kernel and the use of cross validation for selecting the kernel parameters. Considering that the size of available samples can grow arbitrary big, training a model could become a particularly costly procedure. The SVM^{light} implementation of SVMs was used to address the problem of large scale tasks. The algorithmic and computational improvements that were incorporated by the SVM^{light} implementation as well as the complexity issues are analyzed in [31].

4.4 Experimental Study

The goal of our experimental study is twofold. On the one hand, we wanted to get an experimental insight on the cluster-to-object assignment error introduced by the visual analysis algorithms and check whether our expectation on the most populated cluster holds. On the other hand, we aimed at comparing the quality of object models generated by the proposed framework, against the models trained by manually provided strong annotations.

4.4.1 Datasets

To carry out our experiments we have relied on three different types of datasets. The first type includes the strongly annotated datasets constructed by asking people to provide region detail annotations of images pre-segmented with the automatic segmentation algorithm of Section 4.3.2. For this case we have used a collection of 536 images from the *Seaside* domain annotated in our lab, denoted as S^B. The second type refers to roughly-annotated datasets like the ones formed by flickr groups. In order to create a dataset of this type, for each object of interest, we have downloaded 500 member images from a flickr group that is titled with a name related to the name of the object, we refer to this dataset as S^G. The third type refers to the weakly annotated datasets like the ones found in collaborative tagging environments. For this case, we have crawled 3000 S^{F3K} and 10000 S^{F10K} images from flickr using the wget[7] utility and the flickr API facilities, in order to investigate the impact of the dataset size on the robustness of the generated models. Depending on the annotation type we use the selection approaches presented in Section 4.3.1 to construct the necessary image groups S^c. Table 3 summarizes the information of the datasets used in our experimental study.

4.4.2 Tag-Based Image Selection

As a result of our assumption on the tagging habits of social users, we expect the absolute difference between the number of appearances (*#appearances*) of the first (c_1) and second (c_2) most highly ranked objects within an image group S^c, to increase as the volume of the initial dataset S increases. This is evident in the case of keyword-based search since, due to the fact that the annotations are strong, the probability that the selected image depicts the intended object is equal to 1, much greater than the probability of depicting the second most appearing object. Similarly, in the case of flickr groups, since a user has decided to assign an image to a

[7] wget: http://www.gnu.org/software/wget

Table 3. Datasets Information

Symbol	Annotation Type	No. of Images	objects	Selection approach
S^B	strongly annotated	536	sky, sea, vegetation, person, sand, rock, boat	keyword based
S^G	roughly-annotated	4000 (500 for each object)	sky, sea, vegetation, person, car, grass, tree, building	flickr groups
S^{F3K}	weakly annotated	3000	cityscape, seaside, mountain, roadside, landscape, sport-side	SEMSOC
S^{F10K}	weakly annotated	10000	jaguar, turkey, apple, bush, sea, city, vegetation, roadside, rock, tennis	SEMSOC

group titled with the name of the object, the probability of this image to depict the intended object should be close to 1. On the contrary, for the case of SEMSOC that operates on ambiguous and misleading tags, this claim is not evident. For this reason and in order to verify our claim experimentally, we plot the distribution of objects' *#appearances* in four image groups created to emphasize on objects *sky*, *sea*, *vegetation*, *person*, respectively. These image groups were generated from both S^{F3K} and S^{F10K} using SEMSOC. Each of the bar diagrams depicted in Fig. 10, describes the distribution of objects' *#appearances* inside an image group S^c, as evaluated by humans. This annotation effort was carried out in our lab and its goal was to provide weak but noise-free annotations in the form of labels for the content of the images included in both S^{F3K} and S^{F10K}. It is clear that as we move from S^{F3K} to S^{F10K} the difference, in absolute terms, between the number of images depicting c_1 and c_2, increases in all four cases, advocating our claim about the impact of the dataset size on the distribution of objects' *#appearances*, when using SEMSOC.

4.4.3 Clustering Assessment

The purpose of this experiment is to provide an insight on the validity of our approach in always selecting the most populated cluster for training a model recognizing an object described by the most frequently appearing tag. In order to do so we evaluate the content of each of the formulated clusters using the strongly annotated dataset S^B. More specifically, $\forall c_i$ depicted in S^B we obtain $S^{c_i} \subset S^B$ and apply

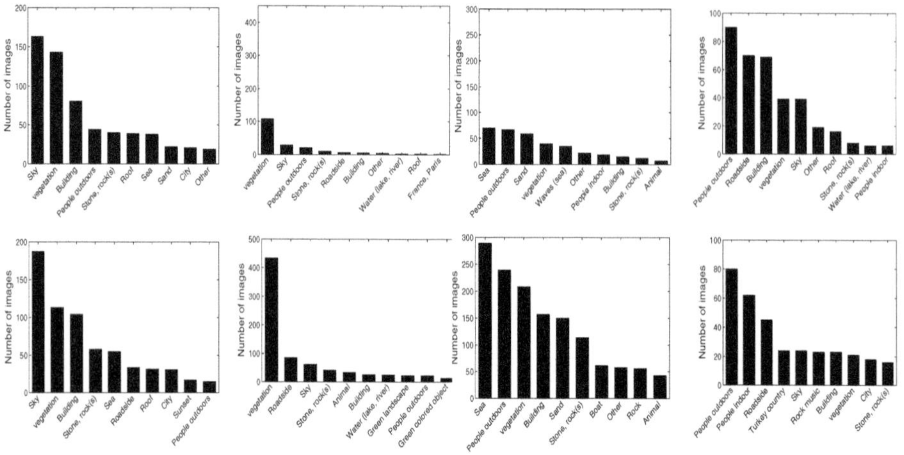

(a) Sky (b) Vegetation (c) Sea (d) Person

Fig. 10. Distribution of objects' #appearance in an image group S^c, generated from S^{F3K} (upper line) and S^{F10K} (lower line) using SEMSOC

clustering on the extracted regions. In Fig. 11 we visualize regions distributions among the generated clusters by projecting their feature vectors in three dimensions using PCA (Principal Component Analysis). The regions depicting the object of interest c_i are marked in squares, while the other regions are marked in dots. Color code is used to indicate a cluster's rank according to their population (i.e., red: 1st, black: 2nd, blue: 3rd, magenta: 4rth, green: 5th, cyan: 6th). Thus, in the ideal case all squares should be painted red and all dots should be colored differently. Squares being painted in colors other than red, indicate false negatives and dots painted in red indicate false positives. We can see that our claim is validated in 4 (i.e., *sand, vegetation, rock, boat*) out of 7 examined cases. In the cases of objects *sea*, *sky* and *person*, the error introduced from visual analysis, prevents clustering from assigning the regions of interest into the same cluster.

4.4.4 Comparing Object Detection Models

In order to compare the efficiency of the models generated using training samples with different annotation type (i.e., strongly, roughly, weakly), we need a set of objects that are common in all three types of datasets. For this reason after examining the contents of S^B, reviewing the availability of groups in flickr and applying SEMSOC on S^{F3K} and S^{F10K}, we ended up with 4 object categories C^{bench}={**sky, sea, vegetation, person**}. These objects exhibited significant presence in all different datasets and served as benchmarks for comparing the quality of the different models. The factor limiting the number of benchmarking objects is on the one hand the need to have strongly annotated images for these objects and from the other hand the un-supervised nature of SEMSOC that restricts the eligible objects to the ones

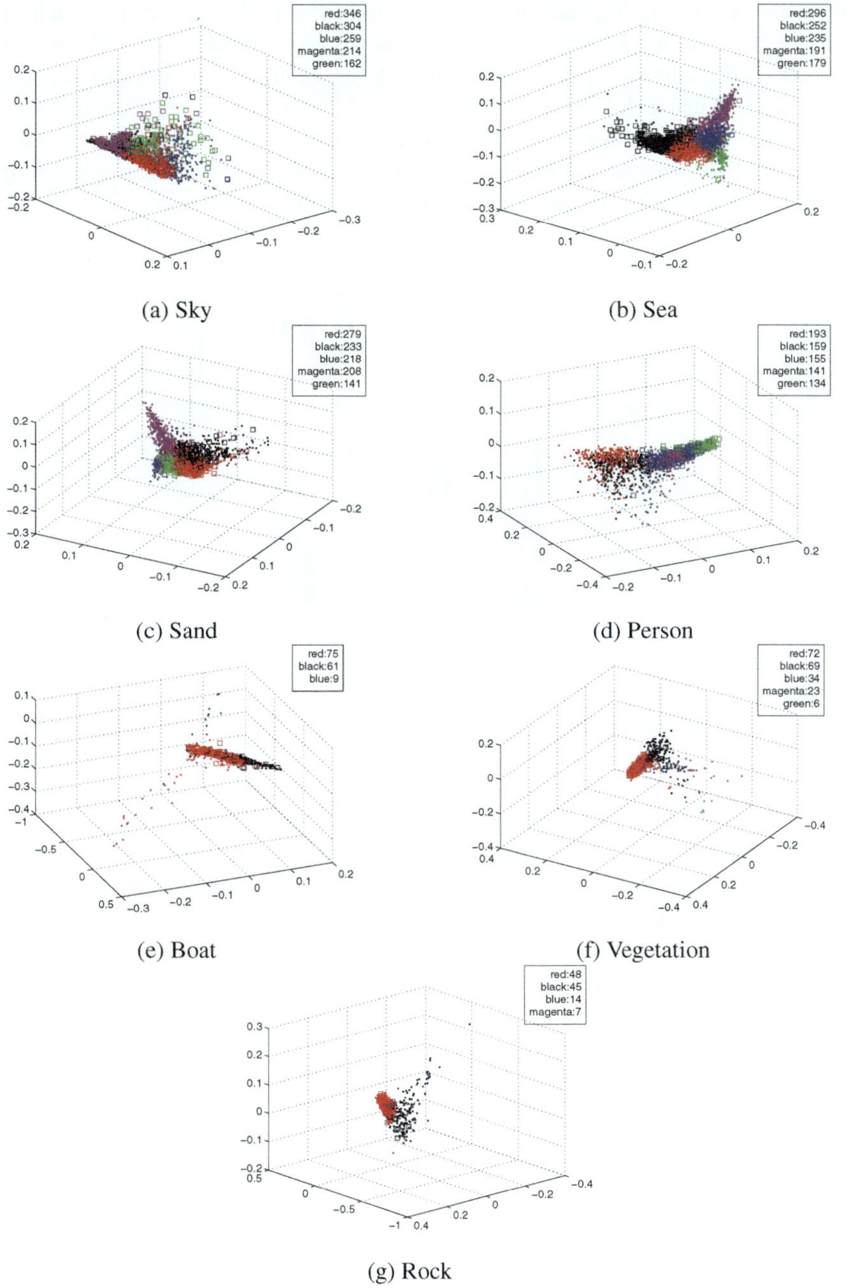

(a) Sky (b) Sea

(c) Sand (d) Person

(e) Boat (f) Vegetation

(g) Rock

Fig. 11. Regions distribution amongst clusters. The regions depicting the object of interest are marked in squares, while the other regions are marked in dots. Squares being painted in colors other than red, indicate false negatives and dots painted in red indicate false positives. This Figure is best viewed in color with magnification.

identified by clustering the images in the tag information space. For each object $c_i \in C^{bench}$, one model was trained using the strong annotations of S^B, one model was trained using the roughly-annotated images contained in S^G, and two models were trained using the weak annotations of S^{F3K} and S^{F10K}, respectively. In order to evaluate the performance of these models we test them using a subset (i.e., 268 images) of the strongly annotated dataset $S^B_{test} \subset S^B$, not used during training. F-Measure was used for measuring the efficiency of the models.

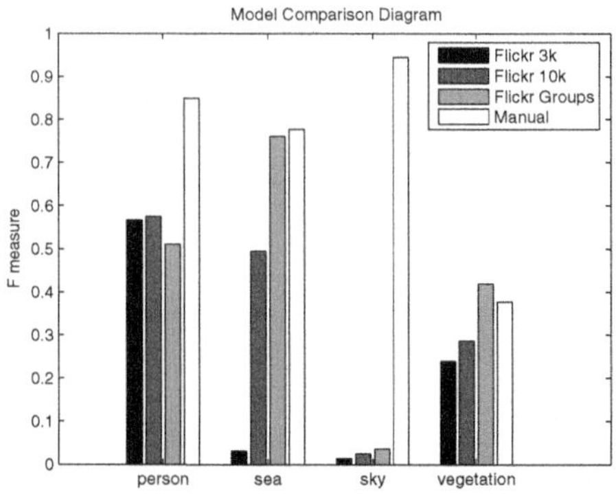

Fig. 12. Performance comparison between four object recognition models that are learned using samples of different annotation quality (i.e., strongly, roughly and weakly)

By looking at the bar diagram of Fig. 12, we derive the following conclusions: a) Model parameters are estimated more efficiently when trained with strongly annotated samples, since in 3 out of 4 cases they outperform the other models and sometimes by a significant amount (e.g., sky, person). b) Flickr groups can serve as a less costly alternative for learning the model parameters, since using the roughly-annotated samples we get comparable and sometimes even better (e.g., vegetation) performance than manually trained models, while requiring considerable less effort to collect the training samples. c) The models learned from weakly annotated samples are usually inferior from the other cases, especially in cases where the proposed approach for leveraging the data has failed in selecting the appropriate cluster (e.g., *sea* and *sky*). However, the efficiency of the models trained using weakly annotated samples is likely to be improved if the size of the dataset is increased.

From the bar diagram of Fig. 12 it is clear that when using the S^{F10K} the incorporation of more indicative examples into the training set improves the generalization ability of the generated models in all four cases. However, in the case of object *sea* we note also a drastic improvement of the model's efficiency. This is attributed to the fact that the increment of the dataset size alleviates the error introduced by visual analysis algorithms and allows the proposed method to select the appropriate cluster for training the model. In order to visually inspect the content of the generated clusters we have implemented a viewer that is able to read the clustering output and simultaneously display all regions included in the same cluster. Using this viewer to inspect the content of the formulated clusters, we realize that the selected cluster is not the one containing the regions depicting sea when using the S^{F3K}, whereas the correct cluster is selected when using the S^{F10K}. Fig. 13 and Fig. 14 show indicative images for some of the generated clusters for the object *sea* obtained using the S^{F3K} and S^{F10K} dataset respectively. The clusters' rank (#) refers to their population. We can see that when using the S^{F3K} dataset the regions depicting *sea* are split in two clusters (ranked #4 and #5), while the most populated cluster #1 consists of regions primarily depicting people. On the other hand, in the case of the S^{F10K} dataset, where the correct cluster is selected (see Fig. 14), it seems that the larger size of the utilized dataset compensates for the error introduced by the visual analysis algorithms.

5 Related Methods

The presented method can be considered to relate with various works in the literature in different aspects. From the perspective of exploring the trade-offs between analysis efficiency and the characteristics of the dataset, we find similarities with [34], [16]. In [34] the authors explore the trade-offs in acquiring training data for image classification models through automated web search as opposed to human annotation. The authors set out to determine when and why search-based models manage to perform satisfactory and design a system for predicting the performance trade-off between annotation- and search-based models. Essentially what the authors are trying to do is to learn a model that operates on prediction features (i.e., cross-domain similarity, model generalization, concept frequency, within-training-set model quality) and provide quantitative measures on when the cheaply obtained data is of sufficient quality for training robust object detectors. In [16] the authors investigate both theoretically and empirically when effective learning is possible from ambiguously labeled images. They formulate the learning problem as partially-supervised multiclass classification and provide intuitive assumptions under which they expect learning to succeed. This is done by using convex formulation and showing how to extend a general multiclass loss function to handle ambiguity.

There are also some works [72], [65], [68] that rely on the same principle assumption with our method. In [72] the authors are based on social data to introduce the concept of flickr distance. Flickr distance is a measure of the semantic relation between two concepts using their visual characteristics. The authors rely on the assumption that images about the same concept share similar appearance features and use images obtained from flickr to represent a concept. Subsequently, the distance between two concepts is measured using the Jensen-Shannon (JS) divergence between the constructed models. Although different in purpose from our approach the authors present some very interesting results demonstrating that the collaborative tagging environments like flickr can serve as a particular valuable source for mining the necessary information for implementing various computer vision tasks. In [65] the authors make the assumption that semantically related images usually include one or several common regions (objects) with similar visual features. Based on this

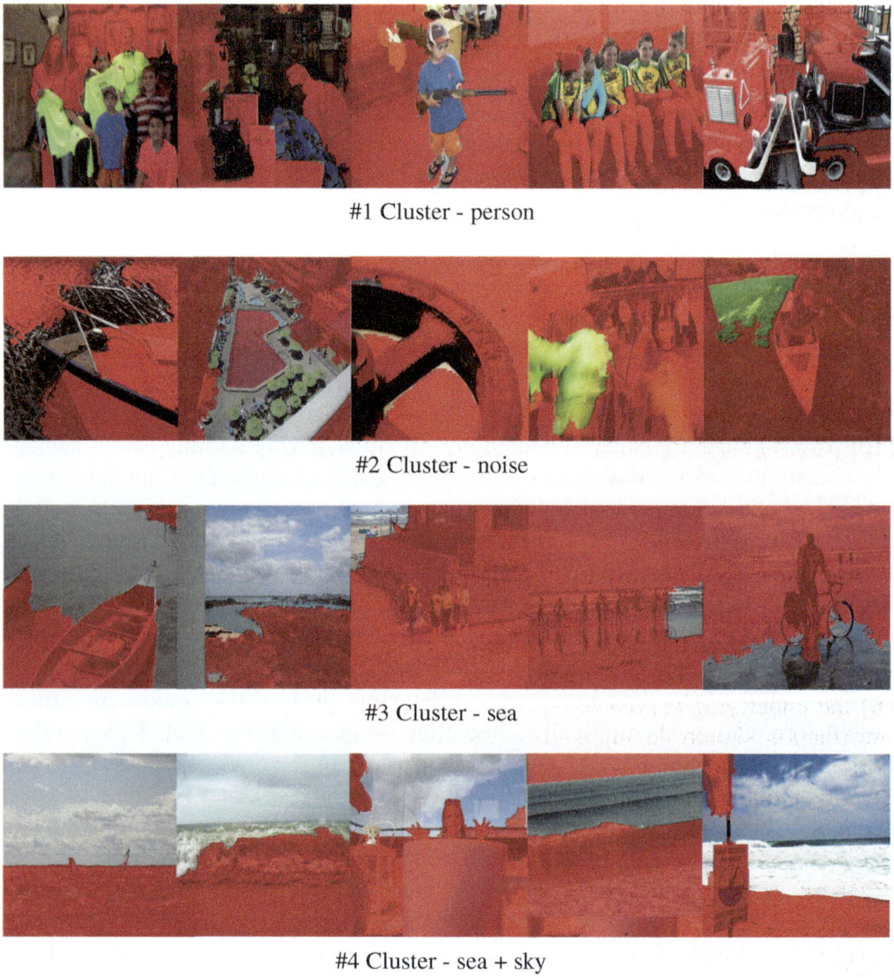

Fig. 13. Indicative regions from the clusters generated by applying our approach for the object *sea* generated by the S^{F3K} dataset. The regions that are not covered in red are the ones that have been assigned to the corresponding cluster.

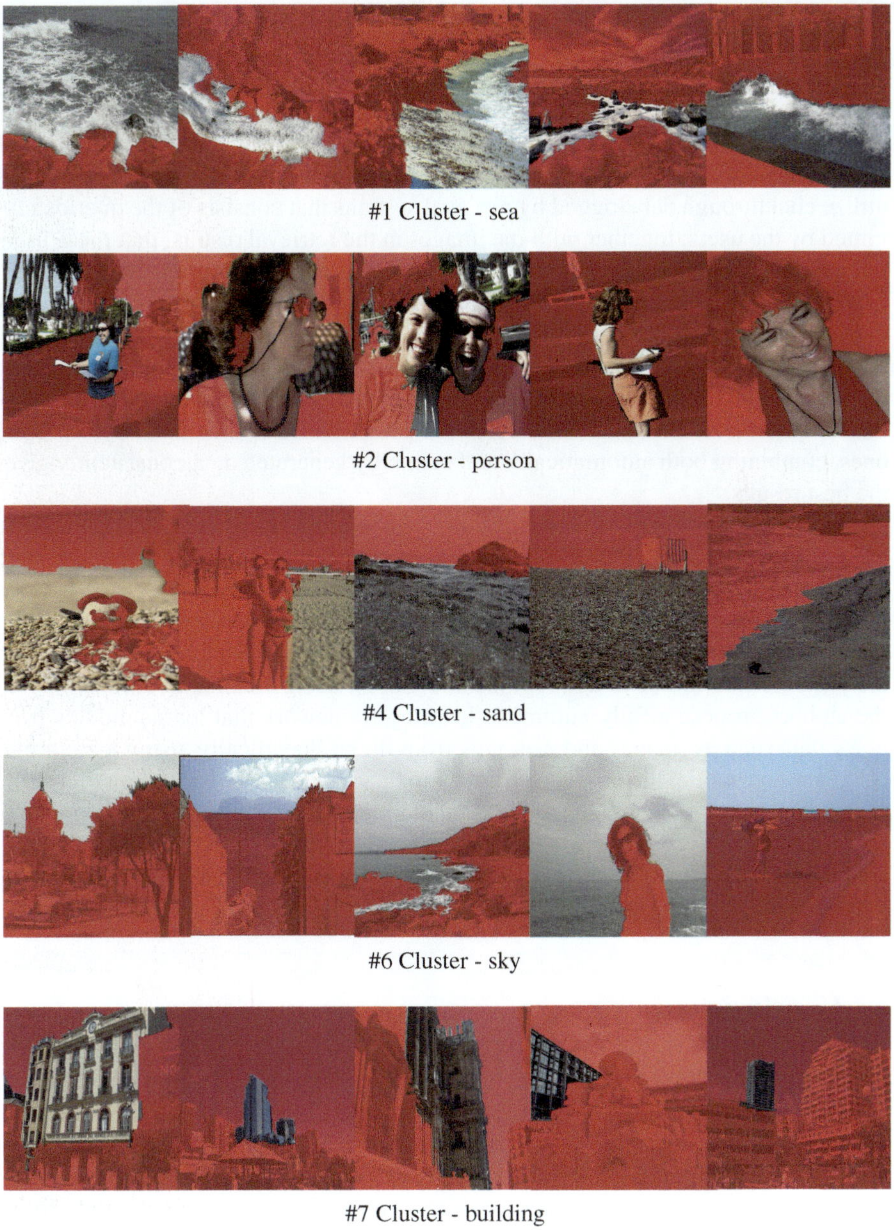

Fig. 14. Indicative regions from the clusters generated by applying our approach for the object *sea* generated by the S^{F10K} dataset. The regions that are not covered in red are the ones that have been assigned to the corresponding cluster.

assumption they build classifiers using as positive examples the regions clustered in a cluster that is decided to be representative of the concept. They use multiple region-clusters per concept and eventually they construct an ensemble of classifiers. They are not concerned with object detection but rather with concept detection modeled as a mixture/constellation of different object detectors. In the same lines the work presented in [68] investigate non-expensive ways to generate annotated training samples for building concept classifiers using supervised learning. The authors utilize clickthrough data logged by retrieval systems that consists of the queries submitted by the users, together with the images in the retrieval results, that these users selected to click on in response to their queries. Although the training data collected in this way can be potentially noisy, the authors rely on the fact that clickthrough data exhibit noise reduction properties, given that they encode the collective knowledge of multiple users. The method is evaluated using global concept detectors and the conclusion that can drawn from the experimental study is that although the automatically generated data cannot surpass the performance of the manually produced ones, combining both automatically and manually generated data consistently gives the best results.

Finally our work bares also similarities with works like [3] and [42] that operate on segmented images with associated text and perform annotation using the joint distribution of image regions and words. In [3] the problem of object recognition is viewed as a process of translating image regions to words, much as one might translate from one language to another. The authors develop a number of models for the joint distribution of image regions and words, using weak annotations. In [42] the authors propose a fully automatic learning framework that learns models from noisy data such as images and user tags from flickr. Specifically, using a hierarchical generative model the proposed framework learns the joint distribution of a scene class, objects, regions, image patches, annotation tags as well as all the latent variables. Based on this distribution the authors support the task of image classification, annotation and semantic segmentation by integrating out of the joint distribution the corresponding variables.

6 Conclusions

Although the quality of object detection models trained using the described method is still inferior from the one achieved using manually trained data, we have shown that under certain circumstances Social Media can be effectively used to facilitate effortless learning. Particularly encouraging was the experimental observation concerning the size of the dataset that showed a consistent improvement on all different types of objects. Given that the size of publicly available content is constantly increasing in the context of social networks, we can claim that by using a larger collection of Social Media we will eventually achieve performance similar to the one obtained using manually trained models. As a general conclusion we can say that social networks can provide more semantically enhanced media than search engines, in pretty much the same effort. Although the noise present in the tags hinders the direct use of these media for training machine learning algorithms, the Collective Intelligence that emerges from the massive participation of users in social networks can be used to remove the need for dedicated human supervision in machine learning.

Another important issue is the computational cost of the proposed framework, especially when the size of the social dataset is large. On a core 2 duo processor running on $3.33GHz$ with $3.25GB$ of RAM, image segmentation takes place in a few seconds and the time needed for extracting the SIFT features and creating the bag-of-words representation of each region is of the same order. Similarly, the clustering of regions and the calculation of the necessary support vectors are also executed within a few seconds, on average. Thus, the time needed for analyzing a single image for the presence of a certain concept is less than a minute, enabling the proposed framework to be used in real life applications.

Acknowledgement. This work was sponsored by the European Commission as part of the Information Society Technologies (IST) programme under EC grant number IST-FP6-026978 - X-Media and the European Community's Seventh Framework Programme FP7/2007-2013 under grant agreement n215453 - WeKnowIt and the contract FP7-248984 GLOCAL.

References

1. MPEG-7 Visual Experimentation Model (XM). Version 10.0, ISO/IEC/JTC1/SC29/WG11, Doc. N4062 (2001)
2. Aurnhammer, M., Hanappe, P., Steels, L.: Augmenting navigation for collaborative tagging with emergent semantics. In: International Semantic Web Conference (2006)
3. Barnard, K., Duygulu, P., Forsyth, D.A., de Freitas, N., Blei, D.M., Jordan, M.I.: Matching words and pictures. Journal of Machine Learning Research 3, 1107–1135 (2003)
4. Begelman, G.: Automated tag clustering: Improving search and exploration in the tag space. In: Proc. of the Collaborative Web Tagging Workshop at WWW 2006 (2006)
5. Bennett, K.P., Demiriz, A., Maclin, R.: Exploiting unlabeled data in ensemble methods. In: KDD 2002: Proceedings of the eighth ACM SIGKDD international conference on Knowledge discovery and data mining, pp. 289–296. ACM, New York (2002), `http://doi.acm.org/10.1145/775047.775090`
6. Bezdek, J.C.: Pattern Recognition with Fuzzy Objective Function Algorithms. Kluwer Academic Publishers, Norwell (1981)
7. Biederman, I.: Recognition-by-components: A theory of human image understanding. Psychological Review 94, 115–147 (1987)
8. Breiman, L., Friedman, J., Olshen, R., Stone, C.: Classification and Regression Trees. Wadsworth and Brooks, Monterey, CA (1984)
9. d'Alché-Buc, F., Grandvalet, Y., Ambroise, C.: Semi-supervised marginboost. In: NIPS, pp. 553–560 (2001)
10. Cao, L., Luo, J., Huang, T.S.: Annotating photo collections by label propagation according to multiple similarity cues. In: MM 2008: Proceeding of the 16th ACM international conference on Multimedia, pp. 121–130. ACM, New York (2008), `http://doi.acm.org/10.1145/1459359.1459376`

11. Carneiro, G., Chan, A.B., Moreno, P.J., Vasconcelos, N.: Supervised learning of semantic classes for image annotation and retrieval. IEEE Trans. Pattern Anal. Mach. Intell. 29(3), 394–410 (2007)
12. Carson, C., Belongie, S., Greenspan, H., Malik, J.: Blobworld: Image segmentation using expectation-maximization and its application to image querying. IEEE Transactions on Pattern Analysis and Machine Intelligence 24, 1026–1038 (1999)
13. Comaniciu, D., Meer, P.: Mean shift: a robust approach toward feature space analysis. IEEE Transactions on Pattern Analysis and Machine Intelligence 24(5), 603–619 (2002), doi:10.1109/34.1000236
14. Conrady, R.: Travel technology in the era of Web 2.0. Trends and Issues in Global Tourism 2007. Springer, Heidelberg (2007)
15. Cour, T., Sapp, B., Jordan, C., Taskar, B.: Learning from ambiguously labeled images. In: IEEE Computer Society Conference on Computer Vision and Pattern Recognition, pp. 919–926 (2009), `http://doi.ieeecomputersociety.org/10.1109/CVPRW.2009.5206667`
16. Cour, T., Sapp, B., Jordan, C., Taskar, B.: Learning from ambiguously labeled images. In: IEEE Computer Society Conference on Computer Vision and Pattern Recognition (CVPR 2009) (2009)
17. Domingos, P., Pazzani, M.J.: On the optimality of the simple bayesian classifier under zero-one loss. Machine Learning 29(2-3), 103–130 (1997), `citeseer.ist.psu.edu/domingos97optimality.html`
18. Duygulu, P., Barnard, K., de Freitas, J.F.G., Forsyth, D.A.: Object recognition as machine translation: Learning a lexicon for a fixed image vocabulary. In: Heyden, A., Sparr, G., Nielsen, M., Johansen, P. (eds.) ECCV 2002. LNCS, vol. 2353, pp. 97–112. Springer, Heidelberg (2002)
19. Egmont-Petersen, M., de Ridder, D., Handels, H.: Image processing with neural networks–a review. Pattern Recognition 35(10), 2279–2301 (2002), doi:10.1016/S0031-3203(01)00178-9
20. Faloutsos, C., Barber, R., Flickner, M., Hafner, J., Niblack, W., Petkovic, D., Equitz, W.: Efficient and effective querying by image content. J. Intell. Inf. Syst. 3(3-4), 231–262 (1994), `http://dx.doi.org/10.1007/BF00962238`
21. Fergus, R., Li, F.F., Perona, P., Zisserman, A.: Learning object categories from google's image search. In: ICCV, pp. 1816–1823 (2005)
22. Freund, Y., Schapire, R.E.: A decision-theoretic generalization of on-line learning and an application to boosting. J. Comput. Syst. Sci. 55(1), 119–139 (1997), `http://dx.doi.org/10.1006/jcss.1997.1504`
23. Frey, B.J., Dueck, D.: Clustering by passing messages between data points. Science 315, 972–976 (2007), `www.psi.toronto.edu/affinitypropagation`
24. Ghosh, H., Poornachander, P., Mallik, A., Chaudhury, S.: Learning ontology for personalized video retrieval. In: MS 2007: Workshop on multimedia information retrieval on The many faces of multimedia semantics, pp. 39–46. ACM, New York (2007), `http://doi.acm.org/10.1145/1290067.1290075`
25. Giannakidou, E., Kompatsiaris, I., Vakali, A.: Semsoc: Semantic, social and content-based clustering in multimedia collaborative tagging systems. In: ICSC, pp. 128–135 (2008)
26. Giannakidou, E., Koutsonikola, V.A., Vakali, A., Kompatsiaris, Y.: Co-clustering tags and social data sources. In: WAIM, pp. 317–324 (2008)
27. Golder, S.A., Huberman, B.A.: The structure of collaborative tagging systems. CoRR abs/cs/0508082 (2005)

28. Grahl, M., Hotho, A., Stumme, G.: Conceptual clustering of social bookmarking sites. In: 7th International Conference on Knowledge Management (I-KNOW 2007), Know-Center, Graz, Austria, pp. 356–364 (2007)
29. Gruber, T.: Ontology of folksonomy: A mash-up of apples and oranges (2005), http://tomgruber.org/writing/ontology-of-folksonomy.htm
30. Jaschke, R., Hotho, A., Schmitz, C., Ganter, B., Stumme, G.: Trias–an algorithm for mining iceberg tri-lattices. In: ICDM 2006: Proceedings of the Sixth International Conference on Data Mining, pp. 907–911. IEEE Computer Society, Washington (2006), http://dx.doi.org/10.1109/ICDM.2006.162
31. Joachims, T.: Making large-scale support vector machine learning practical, pp. 169–184 (1999)
32. Johnson, S.: Hierarchical clustering schemes. Psychometrika 32(3), 241–254 (1967)
33. Joshi, D., Luo, J.: Inferring generic activities and events from image content and bags of geo-tags. In: CIVR 2008: Proceedings of the 2008 International Conference on Content-based Image and Video Retrieval, pp. 37–46. ACM, New York (2008), http://doi.acm.org/10.1145/1386352.1386361
34. Kennedy, L.S., Chang, S.-F., Kozintsev, I.: To search or to label?: predicting the performance of search-based automatic image classifiers. In: Multimedia Information Retrieval, pp. 249–258 (2006)
35. Kennedy, L.S., Naaman, M., Ahern, S., Nair, R., Rattenbury, T.: How flickr helps us make sense of the world: context and content in community-contributed media collections. In: ACM Multimedia, pp. 631–640 (2007)
36. Leibe, B., Leonardis, A., Schiele, B.: An implicit shape model for combined object categorization and segmentation. In: Toward Category-Level Object Recognition, pp. 508–524 (2006)
37. Leistner, C., Grabner, H., Bischof, H.: Semi-supervised boosting using visual similarity learning. In: CVPR (2008)
38. Li, F.F., Fergus, R., Perona, P.: One-shot learning of object categories. IEEE Trans. Pattern Anal. Mach. Intell. 28(4), 594–611 (2006)
39. Li, F.F., Perona, P., Technology, C.I: A bayesian hierarchical model for learning natural scene categories. In: CVPR, vol. 2, pp. 524–531 (2005)
40. Li, J., Wang, J.Z.: Real-time computerized annotation of pictures. In: MULTIMEDIA 2006: Proceedings of the 14th Annual ACM International Conference on Multimedia, pp. 911–920. ACM, New York (2006), http://doi.acm.org/10.1145/1180639.1180841
41. Li, J., Wang, J.Z.: Real-time computerized annotation of pictures. IEEE Trans. Pattern Anal. Mach. Intell. 30(6), 985–1002 (2008), http://dx.doi.org/10.1109/TPAMI.2007.70847
42. Li, L.-J., Socher, R., Fei-Fei, L.: Towards total scene understanding: Classification, annotation and segmentation in an automatic framework. In: IEEE Conference on Computer Vision and Pattern Recognition (2009)
43. Li, Y., Shapiro, L.G.: Consistent line clusters for building recognition in cbir. In: ICPR, vol. (3), pp. 952–956 (2002)
44. Lowe, D.: Object recognition from local scale-invariant features. In: The Proceedings of the Seventh IEEE International Conference on Computer Vision, vol. 2, pp. 1150–1157 (1999), doi:10.1109/ICCV.1999.790410
45. Lowe, D.: Distinctive image features from scale-invariant keypoints. Int. J. Comput. Vision 60(2), 91–110 (2004), http://dx.doi.org/10.1023/B:VISI.0000029664.99615.94

46. Lukaszyk, S.: A new concept of probability metric and its applications in approximation of scattered data sets. Computational Mechanics 33, 299–304 (2004), http://www.ingentaconnect.com/content/klu/466/2004/00000033/00000004/art00007
47. MacQueen, J.B.: Some methods for classification and analysis of multivariate observations. In: Cam, L.M.L., Neyman, J. (eds.) Proc. of the fifth Berkeley Symposium on Mathematical Statistics and Probability, vol. 1, pp. 281–297. University of California Press, Berkeley (1967)
48. Mallapragada, P.K., Jin, R., Jain, A.K., Liu, Y.: Semiboost: Boosting for semi-supervised learning. IEEE Transactions on Pattern Analysis and Machine Intelligence 31(11), 2000–2014 (2008), doi:10.1109/TPAMI.2008.235
49. Marlow, C., Naaman, M., Boyd, D., Davis, M.: Ht06, tagging paper, taxonomy, flickr, academic article, to read. In: Hypertext, pp. 31–40 (2006)
50. Meadow, C.T.: Text Information Retrieval Systems. Academic Press, Inc., Orlando (1992)
51. Meyer, D., Leisch, F., Hornik, K.: The support vector machine under test. Neurocomputing 55(1-2), 169–186 (2003)doi:10.1016/S0925-2312(03)00431-4, http://www.sciencedirect.com/science/article/B6V10-49CRCBP-1/2/346ddc665b1b67be089a7d5d46edca07
52. Mezaris, V., Kompatsiaris, I., Strintzis, M.G.: Still image segmentation tools for object-based multimedia applications. IJPRAI 18(4), 701–725 (2004)
53. Mezaris, V., Kompatsiaris, I., Strintzis, M.G.: Still image segmentation tools for object-based multimedia applications. IJPRAI 18(4), 701–725 (2004)
54. Mika, P.: Ontologies are us: A unified model of social networks and semantics. Web Semant. 5(1), 5–15 (2007), http://dx.doi.org/10.1016/j.websem.2006.11.002
55. O'Really, T.: What is Web 2.0: Design Patterns and Business Models for the Next Generation of Software. O'Reilly Media Inc., Sebastopol (2005)
56. Palen, L., Hiltz, S.R., Liu, S.B.: Online forums supporting grassroots participation in emergency preparedness and response. Commun. ACM 50(3), 54–58 (2007), http://doi.acm.org/10.1145/1226736.1226766
57. Quack, T., Leibe, B., Gool, L.J.V.: World-scale mining of objects and events from community photo collections. In: CIVR, pp. 47–56 (2008)
58. Russell, B.C., Freeman, W.T., Efros, A.A., Sivic, J., Zisserman, A.: Using multiple segmentations to discover objects and their extent in image collections. In: CVPR, vol. (2), pp. 1605–1614 (2006)
59. van de Sande, K., Gevers, T., Snoek, C.: Evaluating color descriptors for object and scene recognition. IEEE Transactions on Pattern Analysis and Machine Intelligence 99(1) (doi:5555), http://doi.ieeecomputersociety.org/10.1109/TPAMI.2009.154
60. Schmitz, P.: Inducing ontology from flickr tags. In: Proc. of the Collaborative Web Tagging Workshop (WWW 2006) (2006), http://www.rawsugar.com/www2006/22.pdf
61. Scholkopf, B., Smola, A., Williamson, R., Bartlett, P.: New support vector algorithms. Neural Networks 22, 1083–1121 (2000)
62. Shi, J., Malik, J.: Normalized cuts and image segmentation. In: IEEE Computer Society Conference on Computer Vision and Pattern Recognition, vol. 0, p. 731 (1997), http://doi.ieeecomputersociety.org/10.1109/CVPR.1997.609407
63. Sivic, J., Russell, B.C., Efros, A.A., Zisserman, A., Freeman, W.T.: Discovering objects and their localization in images. In: ICCV, pp. 370–377 (2005)

64. Sivic, J., Zisserman, A.: Video google: A text retrieval approach to object matching in videos. In: ICCV 2003: Proceedings of the Ninth IEEE International Conference on Computer Vision, p. 1470. IEEE Computer Society, Washington (2003)
65. Sun, Y., Shimada, S., Taniguchi, Y., Kojima, A.: A novel region-based approach to visual concept modeling using web images. In: ACM Multimedia, 635–638 (2008)
66. Sung, K.K., Poggio, T.: Example-based learning for view-based human face detection. IEEE Trans. Pattern Anal. Mach. Intell. 20(1), 39–51 (1998)
67. Torralba, A.B., Murphy, K.P., Freeman, W.T.: Contextual models for object detection using boosted random fields. In: NIPS (2004)
68. Tsikrika, T., Diou, C., de Vries, A.P., Delopoulos, A.: Image annotation using click-through data. In: 8th ACM International Conference on Image and Video Retrieval, Santorini, Greece (2009)
69. Vasconcelos, M., Vasconcelos, N., Carneiro, G.: Weakly supervised top-down image segmentation. In: CVPR, vol. (1), pp. 1001–1006 (2006)
70. Viola, P.A., Jones, M.J.: Rapid object detection using a boosted cascade of simple features. In: CVPR, vol. (1), pp. 511–518 (2001)
71. Wang, Z., Feng, D.D., Chi, Z., Xia, T.: Annotating image regions using spatial context. In: International Symposium on Multimedia, vol. 0, pp. 55–61 (2006), `http://doi.ieeecomputersociety.org/10.1109/ISM.2006.32`
72. Wu, L., Hua, X.-S., Yu, N., Ma, W.-Y., Li, S.: Flickr distance. In: ACM Multimedia, 31–40 (2008)
73. Yanai, K.: Generic image classification using visual knowledge on the web. In: ACM Multimedia, 167–176 (2003)
74. Zhang, J., Marszalek, M., Lazebnik, S., Schmid, C.: Local features and kernels for classification of texture and object categories: A comprehensive study. Int. J. Comput. Vision 73(2), 213–238 (2007), `http://dx.doi.org/10.1007/s11263-006-9794-4`

Chapter 10
From Extensional Data to Intensional Data: AXML for XML

Viet Binh Phan, Eric Pardede, and J. Wenny Rahayu

Department of Computer Science and Computer Engineering
La Trobe University, Melbourne VIC 3086, Australia
{vbphan@students,e.pardede@,w.rahayu@}latrobe.edu.au

Abstract. As a data representation language, eXtensible Markup Language (XML) needs to satisfy new requirements from the rapid development of Peer-to-Peer architectures and wide use of Web Services. The solution of this problem is the proposal of Intensional XML Data, such as Active XML (AXML). AXML consists of normal XML data and embedded Web Service calls. This extension of XML promises many advantages such as: (1) reusable data, (2) dynamic and fresh data, (3) user-oriented and intensive data, (4) time and bandwidth saving, (5) Web Service invocations simplifications, (6) sharing computing tasks as well as (7) support for distributed computing.

This chapter focuses on the introduction of intensional XML data, a new extension for XML that is able to integrate XML data and Web Services. Moreover, a current solution for intensional XML that is known as Active XML (AXML) will be introduced and discussed. Advantages of the new XML extension and its comparison to traditional XML will be presented through examples.

1 Introduction

Traditional database systems are well suited to managing and organizing structured data, and they are also used as one of foundations for building and publishing websites over centralized network architectures and internet. The strength of these database systems is derived from strict requirements of structured data as well as their own data formats. However, there is a considerably high cost implication for data exchange and integration between these database systems, applications and websites built on top of these database systems. Moreover, there is a high demand for storing non-structured data which is not well-suited to be stored inside these database systems. These problems were solved by the emergence of *eXtensible Markup Language* (XML).

XML [1] is a text-based language. It is flexible, scalable and it can capture semi-structured data. These features have caused it to become very popular for data representation and exchange over the internet. XML is applied as a means to communicate between applications over networks, regardless of different platforms.

A. Vakali & L.C. Jain (Eds.): New Directions in Web Data Management 1, SCI 331, pp. 273–310.
springerlink.com

XML is also widely used as a repository for non-structured data. It can be said that XML successfully met the expectations of database communities in the early phase of the internet, along with centralized architectures.

In the early development stage of the internet, client-server architecture is a typical model for web applications. Nevertheless, the centralized network models do not support various properties needed in current web applications, such as exchangeability, heterogeneity, scalability of data and autonomous systems. Therefore, peer-to-peer architectures are proposed as an alternative solution. Many considered this to be the third developmental stage of the internet. In addition, the development of peer-to-peer architecture is also supported and facilitated by the emergence of web services technologies that help to remove the antagonism between diversified systems and platforms.

In decentralized architectures, each peer is information consumer as well as provider. Information exchanged between peers is including extensional XML data as well as XML data that are dynamically generated, constructed and provided by web services. The issues of how to manage dynamic XML data has become a research interest by database community. INRIA has proposed ideas that apply "intensional" XML data to organize and integrate regular XML data and dynamic XML part. The ideas for intensional data is derived from current well-known implementations such as embedded JavaScript codes inside HTML files and active database, triggers as well as stored procedures in Relational Database systems.

In the context of this chapter, intensional XML data are defined as intensive XML data which are representatives for actual extensional XML data. Intensional data are information about web services as well as their concrete parameters which will return corresponding and normal XML data after being activated. In INRIA proposal, intensional XML data are embedded inside XML documents, those XML documents then be called Active XML (or AXML for short) [2, 3]. This new XML extension with special characteristics offers advantages to web technology, particularly in peer-to-peer architectures. Moreover, it promises to facilitate and provide a declarative frame work to integrate data and service.

This chapter will provide basic knowledge about XML as well as AXML and its applicability. After the introduction, we discuss XML in general in section 2. In section 3, we talk about Intensional XML Data, which is the bridge between XML and the solution of AXML. In section 4, details on AXML are provided. In section 5, we briefly discuss alternative solutions to AXML and we conclude this chapter in section 5.

2 eXtensible Markup Language (XML)

When internet is rapidly widespread, the demands to communicate and exchange information between applications and data repositories are imperative. With the rapid increase in the popularity of the internet, the need to facilitate the exchange of communication between applications and data repositories is imperative. However, often data formats of diversified database systems are not compatible. Database communities must spend time and money to find a solution for converting information between database systems for integration purposes. It is also necessary to specify concrete applications to convert data between database systems because each application is only suitable for a specific database system, and when a new database system is proposed, we have to create new programs to convert data from the new to the old database system. Therefore, the *World Wide Web Consortium (W3C)* proposed a new data representation that is simple, exchangeable, flexible, and can be easily transformed to and from old database systems. This new kind of data format is known as *eXtensible Markup Language (XML)*. It is derived from SGML, and was proposed and has been, developed as well as standardized by the W3C since 1996 [1].

XML can be applied in two main domains: document publishing and information exchange. Initially, XML was proposed to confront the challenges of publishing a very large scale of data and exchanging diversified information over the web and traditional database systems, such as Relational, Object-Oriented and Object-Relational databases.

2.1 Why XML?

XML is a markup language like SGML and HTML. It has tags, which are flexibly used to define structure and contain information. However, the advantage of XML compared with SGML and HTML is that it is not as complex as SGML, and is not only used to publish documents in web browsers like HTML. To make it more clear, we will look at the simple example below.

```
<p>
<b>Mr.</b><br>
<b>Viet Binh Phan</b><br>
<b>PhD Candidate</b>
</p>
```

In HTML, tags are predefined and fixed. Each is used for a specified purpose. In the example, tag `<p>` is used for a paragraph, `<br>` specifies the start of a new line, `<b>` is used for bold font. HTML tags are not designed to carry and manage data. For humans, it is easy to see that information between the first pair of `<b>` and `</b>` is a person's name. For machines to extract a name from this HTML document, it is possible to find the second pair of tags `<b>`. Unfortunately, in practice, this is not simple. For example, there are many tags `<b>` and `</b>` in HTML documents but these tags will contain various information, not only people's names. Therefore, it can be said that we cannot propose an algorithm to extract expected data from HTML documents. However, with XML, this can be solved simply by covering people's name between tags `<name>` and `</name>`, for instance. This technique allows a computer to recognise and process information mechanically and effectively.

```
<person>
  <title >Mr.</title>
  <name>
    <first_name>Viet Binh</first_name>
    <last_name>Phan</last_name>
</name>
  <status>PhD Candidate</status>
</person>
```

It can be seen again that tags in XML are arbitrarily defined by users for their purposes. XML only facilitates to describe tags and specify the related relationship between them [4]. For example, in the XML document above, there is a root element, namely <person>. This element contains three child elements <title>, <name> and <status>. Element <name> has two children </first_name> and <last_name>. These children are descendants of <person>.

From these two examples of HTML and XML, it is clear that HTML will not be replaced by XML; these two markup languages complement each other. XML will perform the task of describing data and the data will be formatted and displayed by HTML over the web.

2.2 Basic Concepts of XML

XML is a ***meta-language*** to define other languages such as SVL, MathML and XSLT [5-7]. The most important characteristic of XML is applying tags to ***markup***, defining structure and containing data. XML allows users deploying text to express data and construct the data under a tree-liked structure (see Fig.1).

The ***extensible*** feature of XML means that we can define our own set of tags, organise data according to our preference and design types, relationships and hierarchies of data to serve our purposes. We can also decide orders, positions and occurrence of tags and its own enclosed data.

As ***semi-structured data,*** XML is different to a traditional data model where the data structure must be defined beforehand with fixed data types and lengths. In addition, all data must comply with constrains specified by their schema. Data structures in XML are flexible and may not need any pre-defined schema. The number of data structures is only dependent on our purposes. Therefore, XML can represent both kinds of data models: structured and semi-structured data. Another strength of XML is that it provides flexible data representation that assists the exchange of information between systems and different database platforms. Besides, XML schema is changeable and XML data is portable because of its natural characteristics inherited from the semi-structure data model.

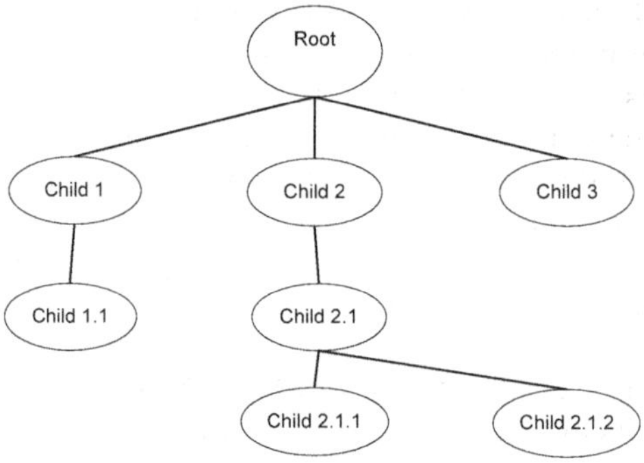

Fig. 1. Tree-like structure of XML document

The XML framework is centred around XML documents that are plain text and formed by tags. A tag is an XML name delimited by two angle brackets. ***Tags*** are used to markup data. Each opening tag is accompanied with a corresponding closing tag. An XML ***element*** is a combination of a pair of opening and closing tags, enclosing any content such as its attributes data. Data in XML documents can be represented as a hierarchical tree-like structure that is constituted by nodes. An XML ***node*** can be an element, attribute, processing instruction or text.

We can also see that each element in XML documents has a name identified by its bounded tags that depict the content. It can be said that elements describe themselves. Therefore, XML is also called a self-describing language.

2.2.1 XML Documents Structure

In XML 1.0, XML data is considered to be data objects. However, the term "XML document" is used because it was based on an earlier standard to compose narrative text for publications [4, 8], regardless of being deployed for information exchange or presented to a human audience. XML documents comprise of processing instructions, comments, elements, attributes, parsed entities and un-parsed entities. The structure of XML documents are explained as follows.

- XML declaration is always placed at the beginning of a document. It starts by **`<?xml>`** and is closed by **`<?>`**. It includes ***version***, ***encoding*** and ***standalone*** strictly in that order. However, ***version*** is the only compulsory component. If there is no declaration, the XML document will be considered as version 1.0. An XML declaration will look like this *`<?xml version="1.0" encoding="UTF-8"?>`*
- A processing instruction is application-specific information to instruct XML processors. It is used to indicate that a piece of information is not data in the document, but it is needed to be passed to the application. A processing instruction must be bounded by **`<?xml>`** and **`<?>`**. The first part of a processing instruction is the name of the application and the next will be the application instruction.
- Elements are text encircled by an opening and a closing tag. Elements are basic data units in XML documents. The first element is called the root node. Other elements must be contained inside the first element.
- Namespace must be immediately followed by its related element.
- An attribute is a pair of a name and a value related to an element. Attributes must be placed inside start-tags and close-tags of the elements. Attributes must have their own value which is enclosed by two single-quotes or two double-quotes.
- Comments begin with **`<!-->`** and end with **`<-->`**. An XML processor will ignore parsing information in a comment since it is only used by human readers.

2.2.2 XML Namespace

As previously mentioned, tags in XML are defined by their authors. This offers flexibility for creating XML documents, but issues arise when we need to integrate XML documents from various sources, for example, if we combine two documents together.

```
<!— Document 1 -->
<person>
    <title >Mr.</title >
    ................................
</person>

<!— Document 2 -->
<book>
    <title> Data Quality </title>
    ................................ .
</book>
```

In this instance, a name conflict will occur between *title* in *<person>* and *title* in *<book>*. These tag names are designed to express diverse meanings and contents. To solve this problem, W3C introduces the concept of XML namespace to avoid name conflicts. The idea is that an element name is constructed from two components. The first is a global name, namely URI namespace and the second is local name that is the ordinary element's name. Namespace is declared as attributes of associated elements by ***xmlns*** keyword with the syntax *xmlns:prefix="URI"*, shown as follows.

```
<rootElement>
  <p:person xmlns:p="http://www.latrobe.edu.au/personName">
    <p:title >Mr.</p:title >
  ................................
  </p:person>

  <b:book xmlns:b="http://www.latrobe.edu.au/bookTitle" >
    <b:title>Data Quality</b:title>
    ................................ .
  </b:book>
</rootElement>
```

In common practice, XML namespaces are defined as attributes inside root elements. For illustration, the same XML document above can be rewritten as follows.

```
<rootElement    xmlns:p="http://www.latrobe.edu.au/personName"
    xmlns:b="http://www.latrobe.edu.au/bookTitle">
  <p:person>
    <p:title >Mr.</p:title >
    ................................
  </p:person>

  <b:book>
    <b:title> Data Quality </b:title>
    ................................
  </b:book>
</rootElement>
```

2.2.3 XML Model

A data model is a set of building blocks to construct the data structure and a set of regulations to stipulate the data relationship. In XML, basic units are elements and data is organized as a tree-like structure. Each element has a parent-child and sibling-sibling relationship to other elements. Elements that have both text and other elements are called to have mixed content such as in the following Fig.2.

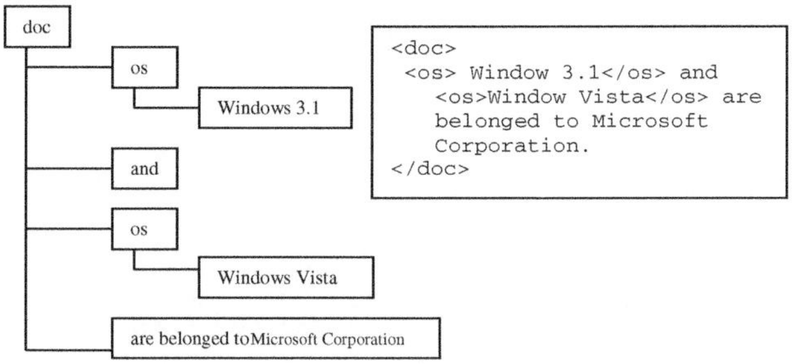

Fig. 2. The conceptual XML document structure

2.2.4 Well-Formed and Valid XML Documents

Although XML is self-describing flexible language, it is still controlled by some basic requirements. Based on how well XML documents conform to the approved requirements, these XML documents are classified into two possible levels: well-formed and valid XML documents. A well-formed document is one which meets all basic rules to declare its entities. The rules are listed below.

- Every XML document must have one and only one root element.
- Each open-tag must be accompanied by and enclosed with a corresponding close-tag.
- Every element has to be appropriately nested. Intersection of two arbitrary elements must be empty or be one of these elements.
- Elements' name must obey naming convention. For example, the first character must be a letter, but not numbers or punctuation characters. The name must not contain any space or reserved characters such as <. The name is not allowed to start with phrase xml (uppercase, lowercase or mixed).

Documents which do not satisfy these rules are not XML documents [9]. Only well-formed documents can be parsed and interpreted by XML parsers. If the structure of a well-formed XML document obeys a DTD or XML schema, the document is called a valid XML document.

2.2.5 XML Classifications

In respect to contained data, XML documents can be divided into two classes: data-centric and document-centric. Data-centric documents convey structured data in which the order of data is not important and mixed content is not allowed, such as information on product orders. It can be said that data-centric documents describe and organize information similar to traditional database systems.

Document-centric documents contain unstructured data such as book contents, for instance. In document-centric documents, the order of data and elements are very important and must be obeyed. Moreover, these documents can contain mixed contents.

The following two XML documents, ***orders.xml*** and ***books.xml,*** will provide an example of typical data-centric and document-centric classes.

```
<!--orders.xml is data centric XML -->
<orders>
  <order>
    <orderNumber></orderNumber>
    <orderDate></orderDate>
    <customer customerId= "1000">
      <customerName>John</customerName>
      <customerAddress>John</customerAddress>
    </customer>
    <item itemId= "1234">
      <itemName>Table</itemName>
      <price>200</price>
      <quantity>3</quantity>
    </item>
    <item itemId= "4321">
      <itemName>Chair</itemName>
      <price>100</price>
      <quantity>12</quantity>
    </item>
  </order>
</orders>

<!--books.xml is data centric XML -->
<books>
  <book isbn="12345678">
<title>Open and Novel issues in XML Database Applica-
    tion</title>
<editor>Eric Pardede</editor>
  <chapter chapterNo="1" chapTitle="Closing the gap between
    XML and Relational Database Technology: State-of-the-
    Practice, State-of-the-Art and Future Directions">
    <author>Mary ann Malloy</author>
    <author>IrenaMlynkova</author>
    <abstract>AS XML technologies have become a standard
    for data representation .........</abstract>
    <content> ........................................ </content>
  </chapter>
  </book>
</books>
```

2.3 XPath and XQuery

In the previous sections, we have discussed basic concepts and the logical and physical structure of XML. Next, we will investigate how to exploit XML data with common XML languages using XPath and XQuery.

XPath is one of the most typical XML query languages. It was recommended and released on 16 November 1999 by W3C. Primitively, this language applies XPath expressions for locating and exacting fragments of XML document. The notion of an XPath expression is quite similar to the concepts of path in DOS. The abbreviation dot (**.**) is used to specify a self-node and double dots (**..**) for a parent-node of a context node.

XPath can be used to specify path to node(s), and extract and return one or many nodes that satisfies users' pre-conditions. There are two types of XPath expressions: relative and absolute expression. Absolute expressions specify the path to locate XML fragments from root nodes and always starts with a slash (/), whereas, relative expressions will start from the location of the current element.

For example, `/element1/element2/…/elementN` is an absolute XPath expression, and `element1/element2/…/elementN` is a relative one. In XPath expressions, every element following an `elementK` must be descendants of the `elementK`. To illustrate XPath, we will consider some examples of XML documents `orders.xml` as follows:

```
<?xml version="1.0"?>
<Orders>
  <Order OrderNo="1047">
    <OrderDate>2002-03-26</OrderDate>
    <Customer>John Costello</Customer>
    <Item>
    <Product ProductID="1" UnitPrice="70">Chair </Product>
      <Quantity>6</Quantity>
    </Item>
    <Item>
    <Product ProductID="2" UnitPrice="250">Desk </Product>
      <Quantity>1</Quantity>
    </Item>
  </Order>
</Orders>
```

XPath expression `/Orders` selects all child node of the root node `Orders`. The expression can also be written in unabbreviated `/child::Orders`. Similarly, these two XPath expressions `/child::Orders/child::Order/child::Item` and `/Orders/ Order/Item` are equivalent. These expressions will locate all child nodes of `item`.

To extract a descendant of a node, we can declare keyword descendant or use double slash (//). To illustrate, `child::Orders/child::Order/child :Item/ child::Quantity` is the same as `Orders//Quantity`.

To extract attributes, the key word `attribute` or abbreviated as `@` is needed to be declared before the attribute name. Expression `/Orders/ Order/Item/Product /attribute::ProductId` or `/Orders/Order/Item/ Product/@ProductId` will return the value of attribute `ProductId`.

The key word `parent` or abbreviated as double dots (`..`) can be used to go backward and return to the parent node of the context node. The expression `../Quantity` or `parent::Quantity` will return node `Item`, which is the parent node of `Quantity`.

An asterisk abbreviator (`*`) is applied to specify nodes with arbitrary names. For example, `Order/*/*` will return all grandchild nodes of `Order` element.

In the previous example, every specified node in the XPath expressions will be returned. However, we can restrict the nodes returned by attaching conditions to the XPath expression. These conditions must be placed inside an open and a closed square bracket.

General XPath expressions can be formalized as a non-empty string **</or//><step/>… <step/><step>**. Abbreviations in this expression are explained as follows:

- Everything inside a pair of angle brackets may exist or may not exist.
- In each XPath expression, double slash only appears once.
- </or//> can be / or //
- <Step> can be
  - `Parent` or `child descendant`
  - `ancestor` or `following-sibling`
  - `preceding-sibling`
  - `following` or `preceding` or `attribute`
  - `namespace`
  - `self` or `descendant-or-self` or `ancestor-or-self` then `::element-name`
  - *
  - `element-name`
  - `element-name[Condition]` *Where* `[Condition]` *can be a position test expression or a contained element expression test or an attribute test expression or a Boolean test expression or a combination of these tests*/or//

XPath also provides functions to assist in extracting information and filtering nodes. These functions are divided into four groups, namely set functions, string functions, boolean functions and number functions. They can be applied in `condition` to restrict the nodes returned. More information about XPath can be found at *http://www.w3.org/TR/xpath*.

XQuery is another XML query language which is an extension of XPath and recommended by W3C in January 2007. XQuery is considered to be the most popular and important query language for XML-like SQL in traditional relational database systems. Both XPath and XQuery are applied to navigate, locate and extract XML fragments. These XML languages can be applied to extract data for web services, transform to HTML or XHTML by XSLT, search web documents and generate summary reports [7].

XQuery and XPath have the same data model that is described as a "node-labelled, tree-constructor representation" [10]. Each XQuery expression can be a combination of XPath expressions, element constructors, FLWOR expressions, list expressions, data-type expressions and conditional and qualified expressions. It is also noted that XQuery supports all XPath operators and functions.

Path expressions are one of the most frequently used expressions in XQuery. These expressions are usually used with `collection()`and `doc()` functions. The two functions are employed to open an XML collection or an XML file. The XQuery path expression `doc("oders.xml")/Orders/Order` will return all items in `orders.xml`.

Predicate is also applied in XQuery to restrict the data returned. For example, to extract items that have `Quantity>100`, we will use the expression `doc("oders.xml")/Orders/Order/Item[Quantity>100]`.

FLWOR expression is considered to be the spirit of XQuery. The abbreviated word "FLWOR" stands for `For`, `Let`, `Where`, `Order by` and `Return`. This expression is very similar to the `Select ... From` clause in traditional database systems. The `for` clause in XML is like the `select` clause in SQL. Path expression with the `doc()` function is similar to `From` in SQL. The `where` and `order by` clauses appear in both languages and have the same meaning. The `let` clause is used to assign a value to a variable or to declare a variable. The `return` clause is applied to give satisfying results back to users. For example, to list all items in `orders.xml`, we can use the XQuery expression below

```
for $x in doc("orders.xml")/Orders/Order
Order by $x
Return $x/Item
```

An XQuery expression does not only return fragments of XML documents, it can also construct a new valid XML element by inserting literal start and end tags in proper positions inside the XQuery. An XQuery expression is also able to define results ordered in HTML form

```
<!--XQuery Expression-->
<List_Of_Customers>
  {for $x in doc("orders.xml")/Orders/Order
   return $x/Customer}
</List_Of_Customers>


<!--XQuery Result-->
<List_Of_Customers>
  <Customer>John Costello</Customer>
<List_Of_Customers>

<!--XQuery Expression for HTML result format-->
<html>
  <h1>Ordered Items</h1>
  <body>
    <ul>
      {for $x in doc("orders.xml")/Orders/Order
       return <li>{$x/Item}</li>}
    </ul>
  </body>
</html>
```

```
<!--XQuery Results in HTML format-->
<html>
  <h1>Ordered Items</h1>
  <body>
    <ul>
      <li>
        <Item>
          <Product ProductID="1" UnitPrice="70">Chair</Product>
          <Quantity>6</Quantity>
        </Item>
        <Item>
          <Product ProductID="2" UnitPrice="250">Desk</Product>
           <Quantity>100</Quantity>
        </Item>
      </li>
    </ul>
  </body>
</html>
```

Conditional clauses in XQuery are defined by the `if-then-else` statement. Other functions and operators can also be used in conditional clauses to filter data as shown in the example below.

```
<Large_Quantity>
  {for $x in doc("orders.xml")/Orders/Order
   return
     if ($x/Item/Quantity >50)
         then data($x/Item)
         else ()}
</Large_Quantity>
```

In this expression, the `return` statement is governed by an `if` statement. Only `Item` that has `Quantity` greater than 50 will be extracted. Note that parentheses must be used to cover `if` and `else` expressions.

List expressions in XQuery can be constant lists `(1, 3, 5, 7)`, integer ranges `1 to 10`, or an XPath expression. Moreover, the list can be constructed from `union, intersect, except` operators, functions such as `remove, index-of, count`, etc. The expression below shows an example of using list expressions.

```
<!--XQuery Expression using list-->
<html>
  <body>
  <h1>List Example</h1>
    <ul>
    {for $x in (1, 2, 3, doc("orders.xml")/Orders/
      Order/Item, 4, 5)
     return <li>{$x}</li>}
    </ul>
  </body>
</html>
```

```
<!--XQuery list result-->
<html>
  <body>
  <h1>List Example</h1>
    <ul>
      <li>1</li>
      <li>2</li>
      <li>3</li>
      <li>
        <Item>
          <Product ProductID="1" UnitPrice="70">Chair</Product>
          <Quantity>6</Quantity>
        </Item>
      </li>
      <li>
        <Item>
          <Product ProductID="2" UnitPrice="250">Desk</Product>
          <Quantity>100</Quantity>
        </Item>
      </li>
      <li>4</li>
      <li>5</li>
    </ul>
  </body>
</html>
```

Quantified expressions including `some-in-satisfies` and `every-in-satisfies` are used in XQuery as a combination with the `list` expression. As an example, the following XQuery expression will list all the customers who have a first name "John" as well as their order date.

```
<Customer_John>
  {for $x in doc("orders.xml")/Orders/Order
 where some $y in $x/Customer satisfies contain($y, "John")
   return
    $x/Customer
    $x/OrederDate}
</Customer_John >
```

In the first section of this chapter, we briefly described XML foundations such as XML document structure, XML namespace, XML data models and well-formed documents. XQuery and XPath, two typical XML languages are also introduced in the latter part of this section. As discussed at the beginning of this chapter, XML must extend its capabilities in order to be applied in the third developmental stage of the internet. In the next section, we will investigate the reasons and the motivation for the development of the new XML extension.

3 Intensional XML Data

The history of internet development can be divided into three distinct periods: (1) the primitive stage, (2) the appearance of the World Wide Web and (3) the innovation of Peer-to-Peer networking technology [11]. In the first stage, the internet was mostly used by researchers and IT professionals. In the second stage, WWW technologies impacted a large number of internet applications in the real world. XML and XML-based techniques were invented, developed and widely deployed in this period. It is clear that the interests of researchers and IT technologies were predominantly centred around client/server architectures.

However, the rapid development of peer-to-peer networking technology as well as the development of the capability to transfer data over the internet increased the need to share information in the commerce community, for research and for personal uses [12, 13]. Besides, there are many different platforms, diversified programs used to produce information so these can cause many problems to data exchange over the internet.

Therefore, web services were proposed as a new paradigm to integrate web-based applications, and were used as an effective tool to remove the antagonism of diversified systems and platforms. Web services have become one of the most important information providers over the internet. These sources of information are heterogeneous, independent, autonomous and very scalable.

Hence, it can be seen that the third period of internet development raises issues regarding the integration and management of large scale data from autonomous and heterogeneous resources. Problems such as how to reduce bandwidth usage, to apply a single query to retrieve information across multiple and diversified sources of databases etc. and the cost of sharing files motivated the proposal for a new XML extension. This extension must be able to manage regular XML data as well as dynamic parts of XML data that is so-called intensional data.

Intensional data can be seen as meta-data that provides information used to retrieve explicit data from web services. The intensional data facilitates, instructs and provides the means for retrieving explicit XML data from multiple sources and web services. Based on ideas about intensional data, in Active XML proposed by the INRIA group [2, 3], intensional data are information about web services, their parameters and other properties to activate those web services and materialize returned results. Using Active XML instead of regular one, many applications can be built more effectively. The following scenarios will illustrate how intensional data can be used in the context of a local newspaper, a Real Estate Agent and a hospital.

Scenario 1: A Local Newspaper Homepage Content [14]

The homepage of a newspaper comprises: (1) explicit data such as the newspaper's name, the current date and time, etc. and (2) implicit data such as the weather forecast and local events. In this scenario, it is easy to see that information about weather and local events should not be given extensionally because this information only makes sense at a specific time (see [14]). This information is usually retrieved by web service calls.

Scenario 2: Collaborative Management for a Real Estate Agency [15]

A real estate agency wants to design a database to manage, facilitate and provide collaborative and mobile workspaces for their staff as well as information to customers [15]. Each staff member can access their related documents, create new information and insert these into documents from their portable devices such as mobile phones and laptops via internet.

In this scenario, the most suitable solution is using intensional data on peer-to-peer architecture because it reduces overload on the central server and each peer can have their own initiative and independency. [15] proposes and organizes the following approach. The database is divided into 3 separate documents including ***properties, requests and status*** (see Fig. 3). These documents contain comprehensive information about properties for sale managed by the real estate. This information will be updated on the central repository. Staff can also examine this information on the central database. All information will be exchanged and organized between peers and the central repository by intensional information via web services (see [15]).

AXML Document name	Content	Enclosed Services to Access Document
Properties.xml	<properties> <property propID="344" type="studio"> <status> ...</status> <assign>...</assign> .... <maybeClients>...</maybeClients> </property> </properties>	**getProperties**(assignedTo, type, location. price) **getPicture**(proID) **addClient**(propID, clientDescr, by)
Requests.xml	<requests> <type> Villa</type> <price> ....</price> .... </requests>	**getRequest**(reqDescr) **putRequest**(reqDescr , han-dleBy)
allStatus.xml	<allStatus> <propStatus propID="344" up-dated="3/2/03" status="open"/> .... <allStatus>	**getStatus**(propID) **updStatus**(propID)

Fig. 3. AXML Documents and Web services for a Real Estate Agency

Scenario 3: Electronic Patient Record Management [16]

Electronic Patient Record (EPR) is a document under the control of a number of peers such as hospitals, patients, insurance companies, and the Department of Health. These peers exist on remote mobile devices, computers and servers. These peers can be information providers as well as consumers. Patients can add or remove their information. Doctor and Nursers can access this information with some restrictions. Moreover, all data can be controlled and monitored by hospitals, insurance companies and Department of Health. (see [16], Fig.4). In this scenario, each peer works independently of other peers. The patients' information is distributed by these peers. That information can also be arbitrarily updated at anytime by peers such as patient's monitoring device, hospital. Therefore, the scenario is suitable to applying intensional data. Patients' information will be contributed by peers via web services. Using dynamic XML data assists and guarantees that data retrieved from each peer is up-to-date and correct at anytime.

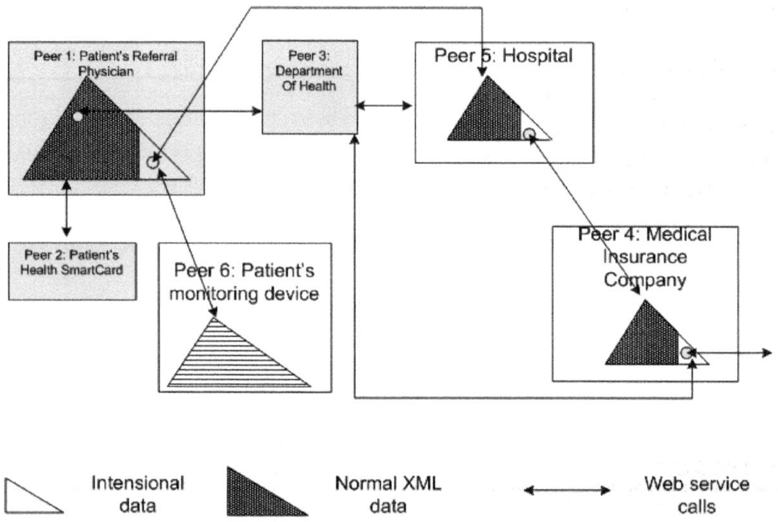

Fig. 4. Models and Relation between peers in EPR

Scenario 4: Organizing Personal Auction Page [17]

A person wants to use a simple page and available web services provided by *www.auction.com* to organize and manage information of specified categories of auction as well as current offers in those categories. This simple page will be used for their personal observation on their interested auction categories. It can be seen that categories for auction are stable but items offered in those categories can be changed every day. Therefore, the database for the personal page should be AXML because active XML guarantees fresh data at any time.

Intensional data is used and accompanied by support of web services. Intensional data can be applied in centralized architectures but in that case, the power of intensional data may not be fully exploited because only the central servers take the responsibility to provide web services as well as information.

Intensional data has a more powerful impact in peer-to-peer models. It will apply and take advantages of the power of peers and web services in contributing and consuming data. From the scenarios presented above, some advantages of applying intensional data can be seen as follows.

- Intensional data assists data re-use. For example, in scenario 1, the editors do not need to regularly update information about weather or currency exchange rates. They only need to subscribe to web services such as Weather Forecast web sites and credible banks and offer this corresponding service to their readers. This allows organizations and individuals to share and re-use data.

- The dynamic data enables end-users to receive both extensional information and intensional data. This allows the users to retrieve corresponding data only when they need that information. Furthermore, it is known that data in real life are changeable such as information about weather, exchange rates and headline news. Therefore, using static data such as static XML may lead to outdated information whereas using an intensional data approach ensures dynamic and fresh data.
- Intensional data can carry more information than extensional data. It is because intensional information is concise so it provides end users with high and intensive contents. This kind of information enables users to obtain their needed extensional data independently and directly from the original sources. Popular example is the use of intensional data in search engines such as Google and Yahoo.
- In most cases, full extensional information contains a large amount of complex data that may distract users. Moreover, in many cases, lots of data found and retrieved are not really useful and even never used. For those reasons, data should be summarized and sent to end users. Common example are the returned results from search engines such as in Google and Yahoo. Another example is the organization of information in online newspapers. Fully listed news is not necessary as this will make it difficult for the readers to easily find the news topics in which they are interested. Therefore, news in the newspaper is given intensionally by being classified in groups. Each group contains a headline (intensional data) of concrete news.
- Another advantage from using intensional data is that it can assist to share computing tasks between information providers and consumers because server sides (information providers) can send intensional data to client (information requesters) instead of computing, processing and materializing results, then requesters will be responsible to process and materialize their needed data. It may help to reduce overload at server sides or information providers.
- Applying an intensional data approach reduces communication costs. To illustrate, we will consider the example below.

Using an example from scenario 1, with a traditional database, when a reader requests information for local cultural events in Cabramatta, Sydney – Australia, the newspaper server will send the request to ***TimeOut.com*** server. The ***TimeOut.com*** server has to look for local cultural events at the http://www.timeoutsydney.com.au/ server and so on until the requested information has been found. Then, extensional data for the request will be gradually sent back (see Figure 5 A). However, using intensional data, the result will be returned intensionally (see Figure5 B). The extensional data retrieved do not need to be conveyed over intermediate servers and the readers will be given extensional data directly from the system *An*. Therefore, it can be said that intensional data assists in reducing communication costs [17].

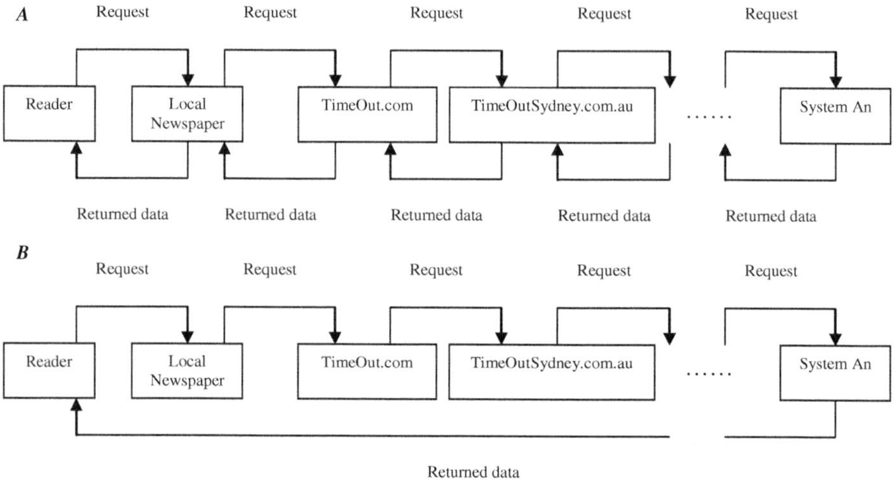

Fig. 5. Communication Cost Saving through intensional XML

- Intensional data assists to simplify web service invocation. Instead of programming complex software to call web services, we only need to create simple intensional documents that contain embedded web services calls. Programmers do not need to control piping. It is simple to create the intensional documents that consist of web services where the parameters of these web service are other web services [2]. In addition, end users can generalize the use of web services because they reemploy these services frequently.

Briefly, intensional XML data can offer advantages including: applicable to share computing tasks between peers, integrate and manage heterogeneous, autonomous, large scale and dynamic XML data, data re-use, user-oriented and intensive data, save time and bandwidth, support distributed computing and simplifying web services uses.

From the scenarios and corresponding advantages discussed, it is easy to see that XML should be extended through the inclusion of new features such as supporting intensional data to satisfy the needs of the development of web technologies, peer-to-peer architecture and web services in the third developmental period of the internet. In the next section, we will introduce a solution to support, manage, and integrate both regular and dynamic XML data known as active XML.

4 Active XML Solution

In this section, we discuss Active XML for the management of static and dynamic XML data.

4.1 AXML Basic Concepts

As discussed in previous sections, the development of the internet, network architectures and web services lead to the proposal for a new class of XML called Active XML (AXML) [2], a declarative framework to connect and exploit advances in web services in peer-to-peer architecture. AXML framework provides the means for data and services integration, facilitates for peer-to-peer architectures which are considered as effective approaches to solve heterogeneity, interoperability, autonomy, scalability and independence of sources [17]. In addition, it can be said that AXML arose from ideas to employ concepts of stored procedures and triggers in Relational Database Systems to XML.

The core of AXML is AXML documents, which are a particular class of XML documents. They contain both regular XML data and extraordinary and abstract data known as embedded web service calls or programs. These web services and programs are intensional data, and can be considered as remote procedures like triggers in Relational Database systems.

These embedded service calls will enrich the original document by returning results when services are activated. One of the most important reason for the existence of intensional documents is to enable web services invocation from anywhere on the web. Therefore, it is not essential to exchange and send whole documents with fully extensional data.

When an embedded service is invoked, a process known as *Materialization* will be started. This process will evaluate program(s) corresponding with the service call. Then, returning results will be added or will replace the intensional data. It is noted that the returned results can be both extensional and other intensional information. Figure 6 is a description for a fragment of AXML tree before and after invoking a service call.

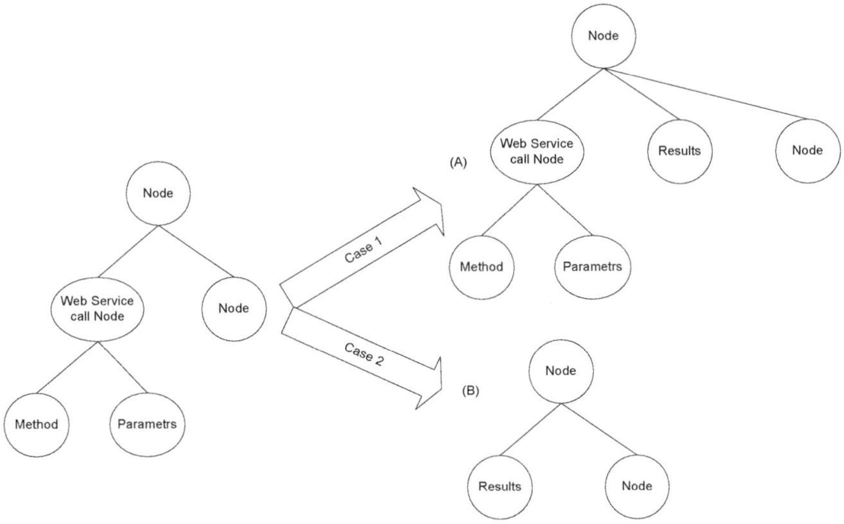

Fig. 6. AXML Tree Before and After Web Service Activation

The following example is an AXML document [14] containing two dynamic parts which are the elements <weather> and <exhibits>. These elements are two web service calls "forecast@weather.com" and "getEvents@TimeOut.com" together with their corresponding parameters.

```
<?xml version="1.0" encoding="UTF-8" ?>
<newspaper xmlns="http://lemonde.fr"
  xmlns:rss="http://purl.org/rss"
  xmlns:axml="http://activexml.net">
  <title>Le Monde</title>
  <date>2-Apr-2003</date>
  <edition>Paris</edition>
  <weather>
  % service call
    <axml:call service=forecast@weather.com >
      <city>Paris</city>
      <unit>Celsius</unit>
    </axml:call>
  </weather>
  <exhibits>
  % service call
    <axml:call service=getEvents@TimeOut.com>
      exhibits
    </axml:call>
  </exhibits>
  <stories>
    <rss:item id="cx_ah_0_218">
      <rss:title>Google goes Blog-Crazy</rss:title>
      <rss:pubDate>
        Feb 18, 2003 10:36:03 GMT
      </rss:pubDate>
      <rss:description>
        Google just acquired <b>Pyra labs</b>,
        the company that makes <b>Blogger</b>.
      </rss:description>
    </rss:item>
...
  </stories>
</newspaper>
```

After activating the web service forecast@weather.com [14], the web service call in the element <weather> will be replaced by its corresponding returned results. The piece of AXML document will be:

```
...
<weather>
  <temp>16</temp>
</weather>
...
```

Service call elements are the heart of AXML documents. They are special elements with functions to manage intensional information. Therefore, elements consist of particular attributes and elements. Generally, service call elements are denoted by the <sc> tag. These elements contain children elements and attributes to define the web service to invoke, the invocation methods and the parameters and methods for handling returned results.

Service calls stored inside AXML documents can be classified into three groups:

- Web service calls to any web services on the web, for example, the invocations to web services provided by Amazon.com.
- Web service calls to a peer-to-peer network.
- Continuous service calls that can deliver a stream of data including requests, parameters and returning results. End of streams will be notified by special and adopted signals like End-Of-Stream [18]. These embedded web services can be available web services that are provided by third parties. They can also be declared by XML Query Languages inside AXML documents [3, 17]. For example, the AXML document below contains service calls defined by X-OQL [18].

```
<!-- Original AXML Document -->
<example xmlns:axml="http://futurs.inria.fr/gemo/axml/">
<axml:sc axml:id="simple">
  <axml:return>
    <axml:append/>
  </axml:return>
    <axml:ws-soap endpoint="http://localhost:6969
     /MyPeer/services/GenericQueryService">
      <q:executeGenericQuery xmlns:q="http:// fu
       turs.inria.fr/gemo/axml/service/Query">
      <q:declaration>
        for $x in doc('/db/bookstore3.xml')/*/*
        return $x
      </q:declaration>
      </q:executeGenericQuery>
    </axml:ws-soap>
  </axml:sc>
</example>

<!—Document after service call invocation -->

<example xmlns:axml="http://futurs.inria.fr/gemo/axml/">
  <axml:sc axml:id="simple"
   activated="2008-04-08T11:30:01.656+02:00">
    <axml:activation status="TERMINATED"/>
    <axml:return>
      <axml:append/>
    </axml:return>
    <axml:ws-soap endpoint="http://localhost:
     6969/MyPeer/services/GenericQueryService">
      <q:executeGenericQuery xmlns:q="http://futurs.
       inria.fr/gemo/axml/service/Query">
        <q:declaration>
          for $x in doc('/db/bookstore3.xml')
          /*/* return $x
        </q:declaration>
      </q:executeGenericQuery>
    </axml:ws-soap>
  </axml:sc>
```

```
    <book axml:origin="simple" axml:timestamp="2008-04-
     08T12:40:42.765+02:00"category="COOKING3">
      <title lang="en">Everyday Italian2</title>
      <author>Giada De Laurentiis3</author>
  </book>
    <book axml:origin="simple" axml:timestamp="2008-04-
     08T12:40:42.765+02:00"category="CHILDREN3">
      <title lang="en">Harry Potter2</title>
      <author>J K. Rowling2</author>
    </book>
  </example>
```

Based on the AXML initiative proposed by INRIA, currently three AXML systems have been built. The first is an experimental AXML system built by an Active XML group from INRIA. The second is an AXML system built on top of an Object Relational Database system. The final one is an implementation of the AXML peer-for -2ME-peer platform. We will introduce each of these systems in the following section.

4.2 AXML Projects

This is the first project to propose the idea of deploying intensional data in XML database systems. Researchers from INRIA used a local newspaper scenario as an experimentation set-up for their AXML framework. The home page of a local newspaper will expose extensional information such as the name of the newspaper and the current date. The weather forecast and a list of exhibitions are displayed by intensional information by invoking web services provided by *timeOut.com* and *weatherForecast.com*. This intensional information will be materialised when it is needed by calling the corresponding web services.

Architectures. The experimental AXML system is built by relying on current technologies including XML, Web, Web service, SOAP and WSDL [14]. To build the AXML system, four software programs are deployed: Tomcat 5.5, eXist, Axis2 and AXML service calls execution engine [18]. There are three cases to combine these components, and the AXML system is configured with a peer-to-peer network (See Fig.8).

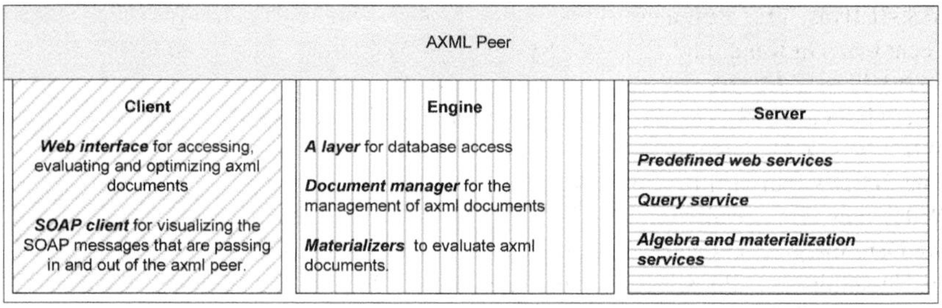

Fig. 7. AXML Peer Architecture

Each AXML peer is a repository of AXML documents. In the system, each AXML peer performs three functions: web server, client and engine (See Fig.7). To play the role of server side, each peer provides services including algebra (SendOperator, ReceiveOperator and NewNodeOperator), a query service (GenericQueryService) to execute queries, and a materialization service that provides an entry point to activate AXML documents. Playing the role of client side, each peer is equipped with a SOAP client, and web interfaces to assess, optimize and access AXML documents. As an engine, AXML peer includes a ***database access layer***, a ***document manager*** for organizing AXML documents, and ***materializers*** that provide information on how to evaluate AXML documents [18].

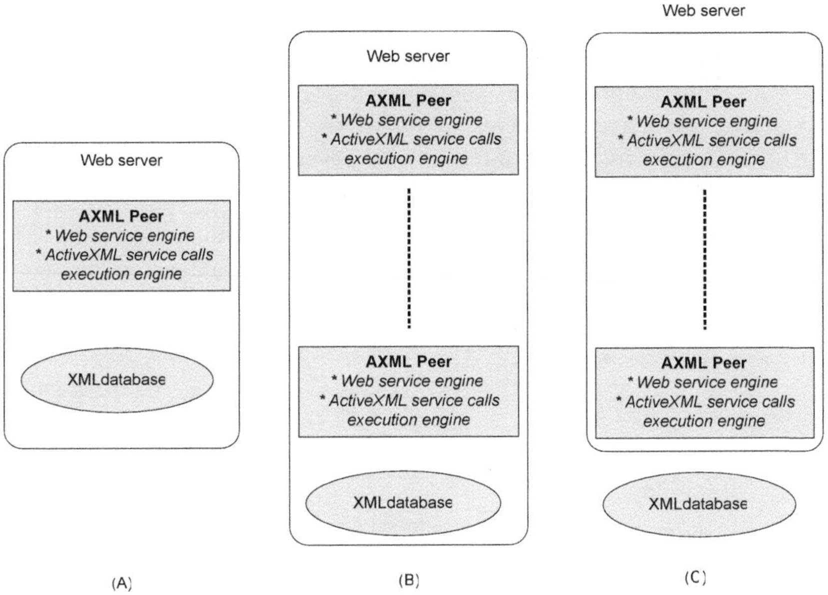

Fig. 8. Components Configurations

Important Operators. In this AXML system (see [18]), the AXML peer is equipped with essential web services :

- RECEIVE: to get and manage returning results from a service call.
- SEND: to pass data to a specified address.
- NEW NODE: to create and install the new AXML data peer's repository
- MATERIALIZE: to contain various operations such as evaluate (to materialize the document in a depth first manner), evaluateNode (to materialize the document starting at a specified node), activate (to activate a specified service call), etc.
- GenericQueryService: to implement a stream processing engine. It receives a query declaration and parameter streams and answers XQueries over the database

- DummyStreamService: to test the streamed back results of a specified query.
- OptimaxService: to act as a distributed query optimizer. Given an active document, the web service transforms it into a distributed plan. The documents that have service calls to GenericQueryService are candidates for optimization with this web service.

AXML peers collaboration (see Fig.9 & Fig.10). In this AXML system, AXML peers will be able to collaborate with other peers, as well as:

- Exchange a message, a peer can send request messages to other peers and get returning results that are conveyed by SOAP messages.
- Materialize some service calls before responding to produce more results.
- Optimize requested queries. Incoming SOAP messages will be optimized to bring results from the whole peer-to-peer network quicker.
- Evaluate active documents.
- Serve as an integrator of results from several web services.

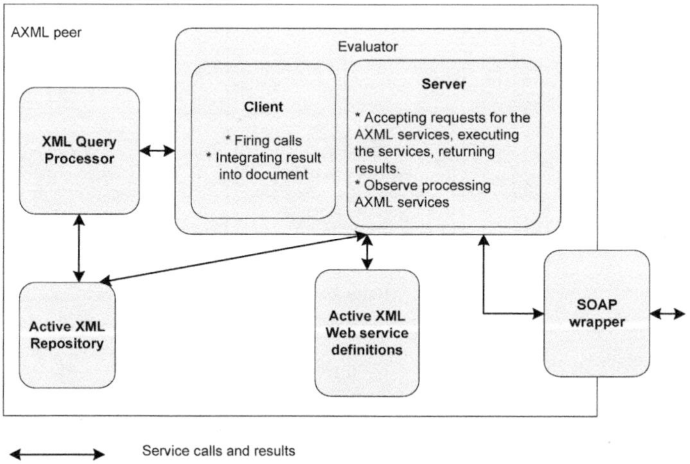

Fig. 9. AXML architecture

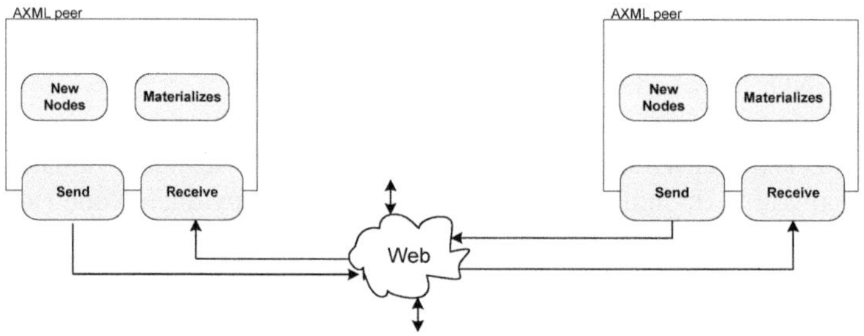

Fig. 10. Peer Interactions

Service calls element. The syntax of service calls complies with normal regulations to declare XML elements. However, the service call elements contain some special attributes. The service call is tagged as `<sc>` with axml prefix. Their attributes include services to call, methods of service invocation and methods to manage returned results. Parameters of service calls are usually stored as children of the service calls.

Services calls consist of five attributes: (i) Service URL, which is the end point URL of the service; (ii) serviceNameSpace, which is used for the body of the SOAP message; (iii) methodName, which is the name of operation to invoke the service; (iv) signature, which is the URL of WSDL file for the service; and (v) useWSDLD, which is a boolean value that stipulates whether or not to perform type validation.

Methods of service invocation are specified by six attributes: (i) id, which is a unique value of the service call; (ii) name, which assigns the service call with a name; (iii) frequency, which stipulates the interval to invoke the service; (iv) callable, which is a boolean value which specifies whether or not to allow another AXML peer to activate the service; (v) lastCalled, which is the storing time of the last service activation; and (vi) followedBy, which indicates the next service call that will be activated after completing activation of the current service.

Methods to manage returned results consist of two attributes: (i) mode, which is the stipulation of storing returned results, and (ii) doNesting, which is the history of previous results inserted.

Parameters must be presented for every service call. If a service does not have any parameter, this parameter element will be declared with an empty content. Parameters can be concrete values or an XPath expression. If there are many parameters for a service call, these parameters must be in order as specified.

Management of activation of service calls. AXML systems embed service calls and manage activation of service calls. Therefore, it is easy to see that controlling service calls activation is one of the most important issues in AXML systems. Methods to activate service calls are specified by a ***frequency*** attribute. The attribute can be (i) ***once***, where the service call is being activated only one time when start-up an AXML peer, (ii) ***lazy***, where the service call is being activated by user demands, (iii) on "***DATE***", where the service is activated exactly on a specified date and time and (iv) every "***TIME INTERVAL***", where the service is activated after a certain time interval in seconds [19].

The attribute ***callable*** has two values, "true" and "false". If this attribute is "false", the corresponding service will never be activated.

The history of the last service call activated is also kept by the attribute ***lastCalled*** and set by the AXML system. This attribute is usually used for frequently activated services.

For the chain activation, the attribute ***followedBy*** is applied to indicate that the service will be invoked after completing the current service.

It is noted that an invocation of a service call can retrieve results containing other service calls that may need to be activated and so forth. Therefore, this will result in recursive and infinite calls, so systems need to restrict these activations by limited intervals (see [2]).

Continuous service calls. In an auction scenario, to have information about offers, the service getOffers("Toy") will be automatically activated daily. These kinds of service calls are called 'continuous service' calls. The process of a continuous service call generally consists of three steps [19].

1. A requester peer registers a subscription to a service provider containing desired continuous service calls,
2. The service provider will automatically evaluate the service after TIME INTERVAL and send the results to the requester peer,
3. The requester peer receives and merges the results as stipulation specified by ***mode*** attribute.

Continuous service calls assist client sides not to repeatedly activate the same service calls. However, the service can result in overhead for the provider sides because of periodically implementing the service call, generating data, and sending data back to the requesters.

Managing and Passing Parameters. When a peer invokes a service, it will use SOAP protocol as an envelop to send requests. The invocation may enclose parameter(s). Normally, the parameters in AXML documents will be organized as children of a service call (see Fig.6). Parameters of service calls can be explicitly stated or given intensionally by using XPath. Two AXML documents (see [17]) listed below are examples for extensional and intensional parameters.

```
<!-- Stating parameters explicitly -->
  <axml id=user"25"> Welcome to mypeer com!
    <category name="Toys">
      <sc>auction.com/getOffers("Toy")</sc>
    </category>
    <category name="Glassware">
      <sc>eBay.com/getAuctions("Glassware")</sc>
    </category>
  </axml>

<!-- Stating parameters using XPath -->
  <axml id=user"25"> Welcome to mypeer com!
    <category name="Toys">
      <sc>auction.com/getOffers([../@name/text()])</sc>
    </category>
    <category name="Glassware">
      <sc>eBay.com/getAuctions("Glassware")</sc>
    </category>
  </axml>
```

It is noticed that when there is more than one instance matched with the given XPath expression, the service call will be repeatedly activated for every matched parameter. Particularly, parameters may not only be a simple string like the previous examples, but can also be an AXML document. It also contains other web service calls. As an example of an auction, before allowing user "25" to bid for an item, it may be necessary to check the user's balance. Hence, an additional service call will be invoked by activating a service call bank.com/getBudget(user"25"). The illustration is shown in the example below.

```
<axml id=user"25"> Welcome to mypeer com!
<category name="Toys">
<sc>
auction.com/getOffers([../@name/
text()], <sc>bank.com/getBudget(user"25")</sc>)
</sc>
<category>
<category name="Glassware">
<sc>eBay.com/getAuctions("Glassware")</sc>
<category>
</axml>
```

Managing returned results. Results that are returned after activating a service call can be managed in several different ways. In [17-19], received results can replace or be appended as a sibling of a corresponding service call. The method of management of results is specified by the attribute mode, with two values including replace and merge. The ***doNesting*** attribute is employed to keep track of previously inserted results. To manage the life-span of these results, [17] specifies that a time-stamp can be added to the results as a special attribute, namely ***expiresOn***.

[2, 18] summarize that decisions must be made in relation to *performance, capabilities, securities* and *functionality* when undertaking materialization of intensional data: To control the materialization process, AXML project developers only employ XML schema.

Storing AXML documents: In the AXML project, all AXML documents are stored by using a file-systems directory which is defined by an internal program [14]. This approach does not support access control, data compression and the indexing of data. These AXML documents are only processed by the operational system [20].

This section is a brief introduction to Active XML systems, including scenario, architectures and other aspects of the systems such as the structure of the service call element, service call activations and management, result management as well as storage aspects. It is noted that in the first Active XML project, file-systems are used as a repository, which offers advantages and disadvantages on the implementation of the AXML system.

4.3 ARAXA Project

The ARAXA project proposed another approach to store and query AXML documents by employing the Object Relational Database System (ORDB). Intensional data will be accommodated to objects using user-defined types and methods. To manage continuous services, an additional agent will be created, the agent will observe the system clock to activate continuous service calls at appropriate times automatically.

Mapping methods proposed by [21] are applied to map AXML documents to ORDB. However, these mapping methods do not support the reconstruction of original AXML documents because attributes and elements are stored equally. Hence, [20] applies two additional solutions for the mapping methods [21]. Firstly, attributes will be distinguished from elements by adding the symbol @ to the attribute name and its parent Dewey code will be recorded. Secondly, a new relation will be added to mapping schema and a new column doc id is also added to Edge relation (see [20], Fig.11).

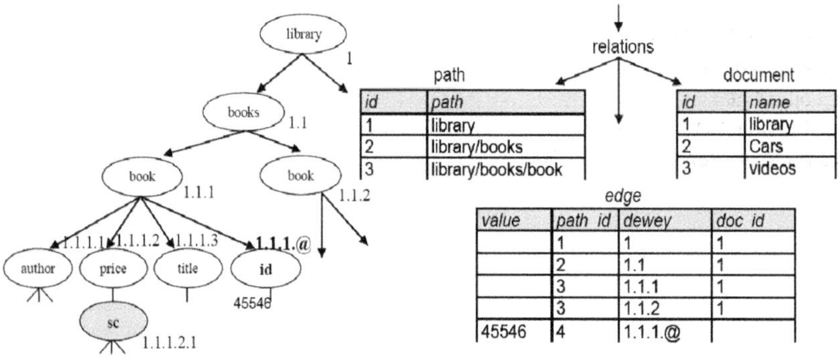

Fig. 11. Tatarinov's Mapping Method Modification [20]

Architectures. Each AXML peer in the ARAXA project consists of two main modules which are *Control Module* and *Integration Module* (see Fig.12) The Integration Module is a client application. It is operated independently with a DBMS. In this module, the *XML Relational Mapper* is responsible for mapping and storing AXML documents. In the mapping process, service calls will be identified and stored in the Service Call Catalog of the Control Module. The *Query/Result Processor* is used for translating XQuery and Xpath to SQL as well as to reconstruct the returned results to XML data.

The Control Module is composed of four modules: (i) a *Service Call Catalog* to keep service calls which is updated after each service invocation; (ii) a *Service Manager* to manage and execute service calls; (iii) a *Results Manager* to manage and materialize obtained results; and (iv) a *Monitor Agent* to observe the system clock to execute continuous service calls. The full architecture of the system is shown in Figure 12.

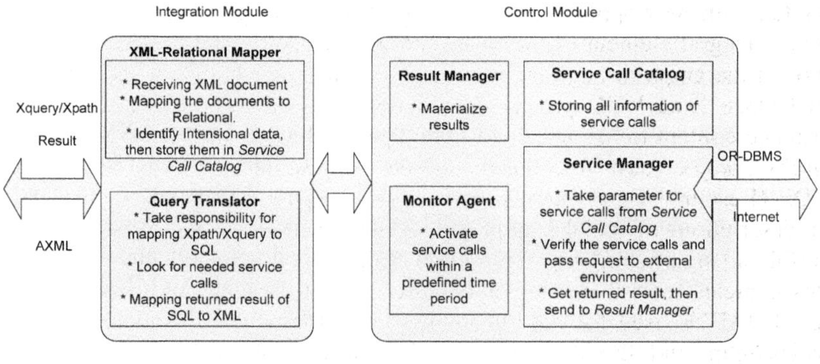

Fig. 12. ARAXA Architecture

Management of service calls. As mentioned, service calls are divided into two groups based on modes to execute: 1) service calls with lazy mode and 2) continuous service. The former needs to be observed during the query evaluation process because the obtained results will be used to answer given queries. In the latter, service calls must be monitored at all times to activate them regularly. ARAXA researchers use two relations to store service calls [20].

```
Service call
   (id, path id, dewey, doc id, serviceURL, methodName, service-
    NameSpace, useWSDLDefinition, signature, callable, frequency,
    lastcalled, followed, mode, doNesting)
Parameter
     (id, service id, path id, type, name)
```

A new method to activate web services is created and applied, namely *execute_service( ),* which is operated independently from the database system. Then, the services will be invoked by a query. The class diagram and the following example (see Fig. 13) depict the association of ARAXA infrastructures with the DBMS [20].

Materialization. The materialization process in ARAXA includes seven steps [20]: (i) looking for appropriate services to answer a given query; (ii) translating the given query into SQL; (iii) finding dependency of the service calls; (iv) arranging these service calls; (v) storing obtained results in corresponding relational tables by mapping; (vi) evaluating the given query, and (vii) reconstructing results to XML and sending to users.

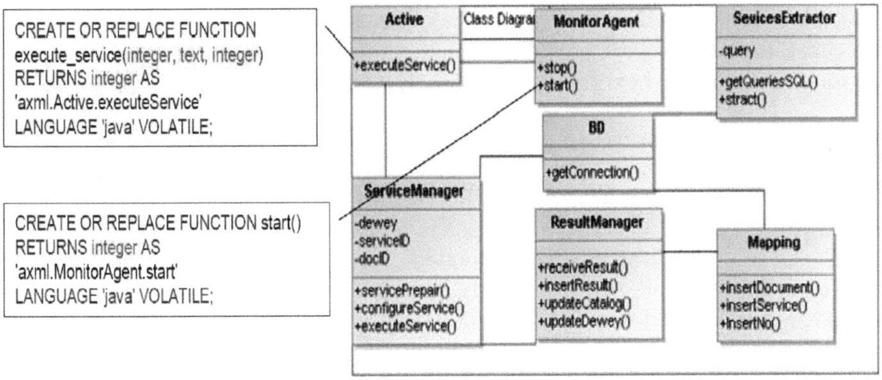

Fig. 13. Infrastructure Class Diagram [20]

It is noticed that there are some different ways to invoke service calls during the materialization process (see [20, 22]). Therefore, optimization strategies are essential in order to make the performance effective. ARAXA applies techniques proposed by [22, 23] to optimize the materialization process and avoid invoking unnecessary service calls.

For example, in relational table book(author, price, ISBN), price is dynamic data provided by a service call with the parameter ISBN. When evaluating a query related to price information, dynamic data price must be materialized by invoking a price service. Figure 14 demonstrates the translation process in ARAXA.

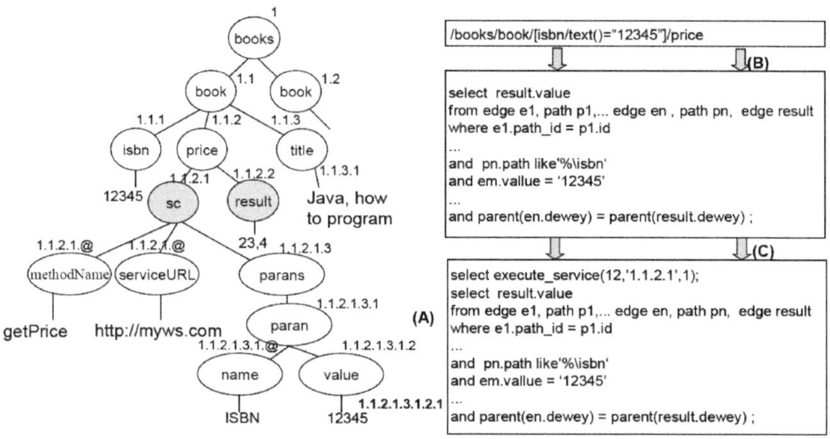

Fig. 14. ARAXA Translation Mechanism [20]

This section introduces the object-oriented approach to implement the AXML system. The section mostly focuses on three issues, namely architecture, management of service calls and materialization that is different from the INRIA project. To deploy Active XML on mobile devices that own limited resources and computing capability (in comparison with normal computers), we will investigate proposals for AXML on the next project on a J2ME platform.

4.4 AXML for J2ME Platform

In the AXML project proposed by INRIA and the ARAXA project, AXML peers are developed for computers. Each peer is built on complex components such as Tomcat, Axis2, eXist or an object-oriented database system. Therefore, it can be said that these Active XML systems are computing and resource consumers. However, Internet and web services are not only used and exploited by computers but also by hand-held devices that have limited resources and computing capability, particularly mobile phones. Therefore, the idea of employing AXML to mobile phones on J2ME platform is raised. The implementation of AXML on mobile phones is different to that for computers. This is because mobile phones have considerably weak processors, limited resources such as memory and storage space, and a lack of applications support for processing XML on mobiles. These specific features require particular designs to apply AXML to mobile phones based on the J2ME platform.

Architecture. AXML mobile peers consist of two main parts. The first is the AXML peer including the AXML repository for managing AXML documents, the client for materializing service calls and the server for receiving and responding to requests from the outside world via proxy. The second is the proxy, which is responsible for controlling communication between mobile devices and the outside environment.

Management of service calls. In this architecture (see Fig. 15),*Web service reference* contain information of <sc> elements that will be provided to *Service executor*. *Static analysis* is a module that takes responsibility for choosing service call(s) to activate, arranging these service calls in order and provide parameters for them. Then, the information of service call(s) will be transformed to XML-RPC message and sent to AXML proxy.

The activation of web service calls is controlled by *Static Analysis* module and *Inspector* module. The *Static Analysis* module will be in charge for triggering lazy modes whereas *Inspector* is applied for immediate mode (when service call is expired and data need to be refresh).

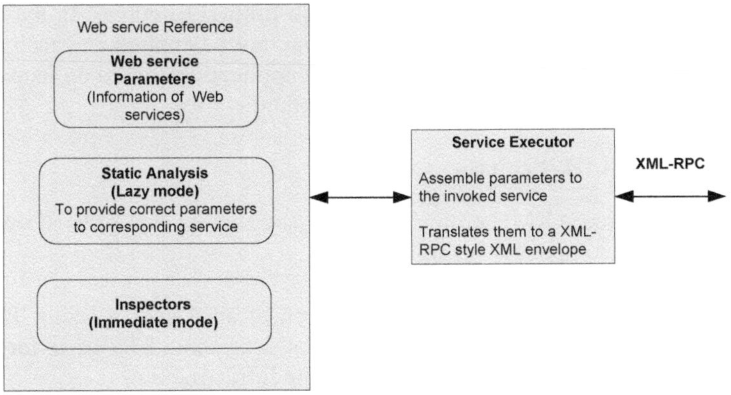

Fig. 15. Service Manager Architecture

Storage. The limitation of resources and applications is overcome by storing the AXML document inside the Phone Specific Record Store with a J2ME platform. The AXML repository components are *Persistent memory manager*, *File Retriever*, and *DOM cache* [24] (see Fig.16).

Persistent Memory Manager uses a mechanism to store and retrieve data that is modelled upon the Records Management System in RDB. This system is comprised of Stores, with each Store being divided into Records. Every AXML document is assigned a unique initial location URL.

File Retriever takes responsibility for searching AXML files in Record Stores or at the URL location. The *DOM cache* tries to retain information about DOM trees, which have already been generated, until such memories are revoked by the *Garbage Collector*.

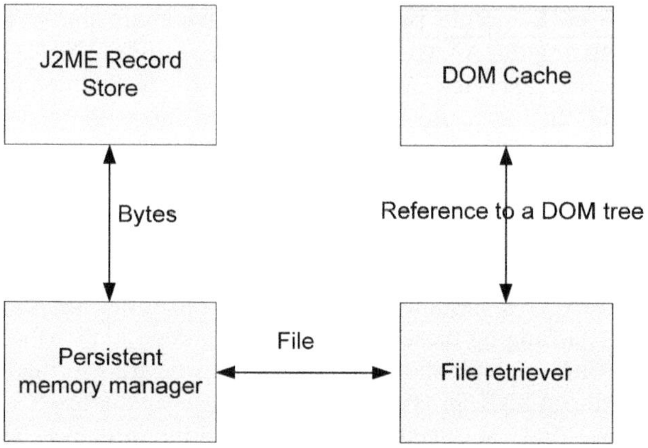

Fig. 16. AXML Repository in Mobile Phones

The AXML on the J2ME platform presents proposals and gradually clarifies issues as to how to apply AXML to systems with limited resources. In these cases, standard software or supportive applications may not be applied and each particular implementation may have to design its own applications based on its particular platforms and its ACML documents storage method.

4.5 Summary AXML Projects

Based on these three AXML projects, we can see that each AXML system must have some typical components as follows:

- AXML repository. This is the foundation of an AXML system. The most important issue to answer is which platform should be used for storing AXML documents to adapt with available infrastructure.
- AXML document manager applications. Based on the platform used for storage, there must be additional applications deployed to manage dynamic data.
- AXML engine or materialisation modules. These components process dynamic data (service calls) and generate appropriate results.
- Modules to manage continuous services. Being able to facilitate continuous service calls is one of the most important issues in managing AXML documents. Functionalities for continuous service have not been supported in previous systems.

The heart of AXML is the <sc> element. This element contains special attributes and child elements with specific regulations to organize, manage and activate web services. A brief summary of <sc> element is depicted on Figure 17.

	Type	Explanations	Required?
Web service information			
@Service URL	*URI*	*Service URL*	Y
@Service Namespace	*URI*	*For the body element of the SOAP message*	Y
@Method Name	*NCName*	*Method Name*	Y
Signature	*URI*	*To do type validation*	N
useWSDLDefinition	*boolean*	*For type validation*	N
Service call information			
Id	*ID*	*To identify service call*	Generated
Name	*String*	*Specify name of service*	N
Frequency	*String*	*How to activate service*	N
Callable	*Boolean*	*Allow to activate service*	N
LastCall	*unsignedLong*	*To observe last activation*	Generated
FollowedBy	*String*	*Next service being activated*	N
Service call result handling			
Mode	*String*	*To specify how to store results*	N
doNesting	*Boolean*	*To observe previous inserted results*	N
Service call parameters			
Value	*String*	*To specify parameters being value*	
XPath	*String*	*To specify parameters being XPath*	

Fig. 17. A summary of Sc Element [19]

Proposals and ideas from AXML projects introduce new approaches to data and service integration. However, these systems are quite complex for end-users to set up and employ. Moreover, these systems need to be improved in many aspects.

Reasons for materializing intensional information before or after sending to clients are specified but not completely solved [2, 14, 17]. There is no mechanism and ability to check when intensional data must be materialized before sending to receivers. It is also difficult when end-users employ hand-held devices or obtain all extensional data to work off-line. These problems have become a major concern and need to be solved in the future.

In addition, standards for declaring service calls and managing parameters, activation of service calls, as well as managing and organizing returning results have not been clarified. For example, in some cases, if service calls are replaced by corresponding results, the reasons for using intensional data will be not useful because it will lose its dynamic character. Always merging results with the service call in [17] can cause very large AXML documents as well as repeating and redundant data or overuse expensive XML update operation. Moreover, if the results contain other service call(s), the management of these services will be complex.

The issues of AXML storage and which database systems should be employed to manage AXML documents are not explicitly explained. Some researchers propose to use native XML database systems [2, 18], some apply RDB or OODB [14, 20]. Using native XML database systems can be appropriate for typical properties of AXML documents such as heterogeneity, inter-operability and large scales, but there are also some problems such as update, triggers and stored procedures facilities that do not have standards and are not well supported. Under traditional database systems, heterogeneity, inter-operability and large scales are the major concerns. Moreover, traditional database systems must transform native data to their special data types which is quite expensive. Furthermore, the effects of deploying AXML on different database systems have not yet been specified because these experimental AXML systems are not widely deployed.

The way to apply the effects of AXML deployment in general networks and on the internet has not been discussed. In addition, a comparison of AXML systems and non-AXML-based systems has not been proved by scientific experiments and statistics.

Storing AXML documents is one of the most important issues in AXML projects. It will determine how to build the AXML systems and the management of data and service calls. Most current proposals are based on existing database systems such as file-system storage and XML native databases [2, 14], RDB [25], ORDB [20] and specific Record Stored platform [24]. Each of these storage approaches has typical advantages and problems. For example, file-system storage does not support indexing and data compression. Native XML database storage approaches need additional applications to take reasonability for activating the service calls. RDB and ORDB storage must deal with scalability and the expensiveness of mapping from an XML document to their data structures and format. These approaches also need additional programs to manage continuous service calls. Moreover, an evaluation of performance (speed, usage of disk space, and ability of optimization) between AXML-based systems and other systems is needed to carefully analyse the advantages of Active XML systems.

This section introduces basic ideas for the implementation of AXML as well as principal elements of service calls in AXML. In the next section, we will discuss these implementations and propose some ideas for managing intensional data in native XML database systems.

	AXM Project	AR-AXAA	AXML for Mobile
Employs Database system	File-system storage	O-R DBMS	Record Stored that is provided by Mobile Information Device Profile
Evaluation Performances	✗	✓	✓
Stipulations for which dynamic part can be sent extensionally or intensionally	✓	✗	✗

Fig. 18. A Comparison of Three AXML Projects

5 Alternative Solutions to Intensional XML Data

From existing AXML proposals, we can see some alternative approaches to extend XML features to integrate data and services.

(1) **Trigger-like approach** stores web services and regular XML data together. Information on service call nodes is stored as a stored XML procedure or as an XML trigger. The node is also stored in the XML document as a simple and special node, namely *sc* node. When an arbitrary *sc* node is evaluated, the corresponding trigger (or procedure) will be fired. The trigger takes responsibility for the retrieval of extensional XML data by activating its service calls, and the returned data will be updated to the XML document.

Advantages. We may apply foundations of trigger and procedures from traditional database systems. The implementation may be simple over traditional database systems.

Disadvantages. We need a module to monitor the processing of each query. However, this observation is not supported by any database system. Triggers are only activated on update queries but not on retrieval queries. Moreover, triggers and stored procedures are still not implemented and support by W3C.

(2) **Plug-in Module** for XML database system(s) (see Figure 19) can be used and placed between users and XML database systems. For simplicity, the plug-in may have some module(s) including:

Web service call Manager Module (Module 1)

Input : address of a web service call.

Purpose : automatically get the WSDL of web service call, analyze this file to indicate required parameters and their types and analyse possible returned results

Output : general information about parameter and web service call response

Activator Module (Module 2)

Input : web service information

Purpose : create a SOAP message, send the SOAP to corresponding web service provider, and get response

Output : returned AXML results

Web Service Node Dectector Module (Module 3)

Input : XML node

Purpose : check the node. If that node is a web service node, this module will take information on the web service in that node and pass it to the second module. This module satisfies requirements such as: simplicity to avoid imposing overheads on queries that do not evaluate regular nodes.

Output : boolean (true if the second module activated and false if opposite)

Update Module (Module 4)

Input : XML node (from the third module)

Purpose : return result (from the second module), update document at the node (with some additional information: timestamp, etc.), and process returned web service fragments before updating the node.

Output : boolean (true if the document is updated and false if not)

Query Transformer Module (Module 5)

Input : XML query (from end-users)

Purpose : catch every query before implementing them, analyze the query, and then insert the third module inside the query at the possible place.

Output : the query with embedded second module

Advantages. It is simple, easy and mobile for end-users to deploy, install and apply.

Disadvantages. We may need to create a particular plug-in for each particular XML database system. It may be difficult to interfere with the inside code of XML database systems and it may have some side effects.

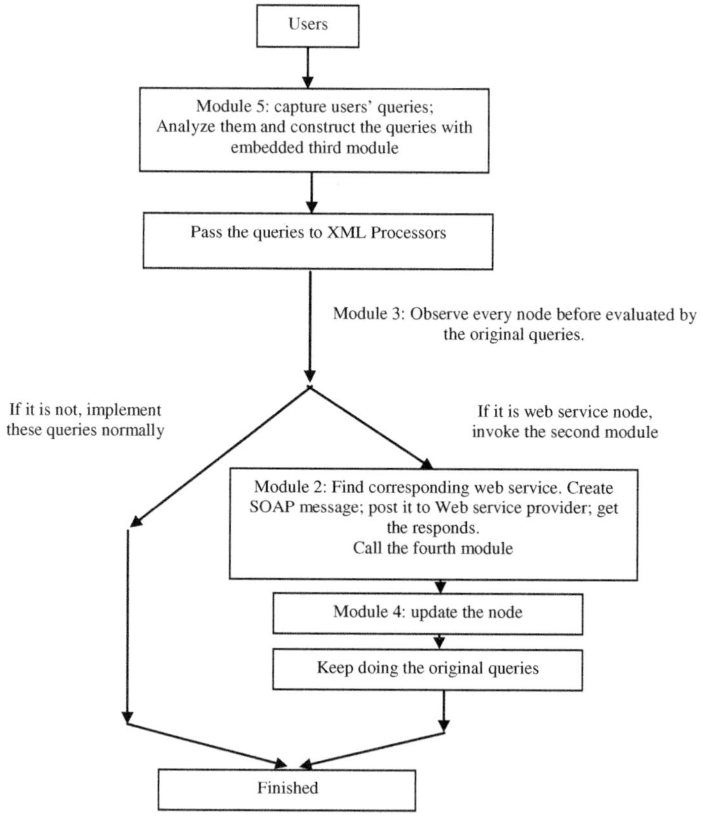

Fig. 19. Brief Description for AXML Database System Creation

6 Conclusion

The rapid development and widespread use of the Internet is partially influenced by XML and web services technology. However, it seems that XML still lacks the capability to integrate and manage data retrieved from diversified web services. XML does not fully satisfy the current needs for requirements from development of peer-to-peer architectures as well as data and service integration.

This chapter introduces the XML evolution to Active XML (AXML). This XML extension is not only used for exchanging and representing data, it combines new features to integrate regular data and intensional data from web services.

In the first section, a concise summarization of XML and related technologies such as XML namespace, XPath and XQuery is presented. Problems of XML in applying web services in web technologies are outlined. Then, AXML – a solution for data and web services integration is briefly introduced. In the last section a concise proposal to implement and improve the current AXML solution is given.

In the future, some issues in AXML still need to be improved. They include standards to store intensional data, database storage solution for storing the data, query optimizations, solutions for infinite and asynchronous activations, criterions for intensional data materializations, and particularly security issues because there can be codes embedded inside the data. Moreover, how to effectively organizes, manage and inform available web services from one peer to other peers is another area to be investigated.

In conclusion, the new XML extension, which can combine and manage web services, will facilitate and extend the capability of XML in data exchange as well as data and service integration. In addition, this new XML extension will be an effective motivation for the widespread use of web services in contemporary web technology.

References

1. W3C. Extensible Markup Language (XML) 1.0, 5th edn., (November 26, 2008), `http://www.w3.org/TR/REC-xml/` (cited 2009 20 February 2009)
2. Milo, T., et al.: Exchanging intensional XML data. ACM Trans. Database Syst. 30(1), 1–40 (2005)
3. Abitrboul, S., et al.: Active XML: peer-to-peer data and web services integration. In: Proceedings of the 28th International Conference on Very Large Data Bases, pp. 1087–1090. VLDB Endowment, Hong Kong (2002)
4. Hoque, R.: XML for Real Programmers, 3rd edn. Morgan Kaufmann, Sandiego (2000)
5. Goldfarb, C.F., Prescod, P.: The XML Handbook, 4th edn. Prentice Hall PTR, USA (2002)
6. Mercer, D.: XML: A Beginner's Guide. Obborne/McGraw-Hill, New York (2001)
7. Open and Novel Issues in XML Database Application: Future Directions and Advanced Technologies. In: Pardede, E. (ed.), Information Science Reference, Hershey - New York (2009)
8. Bradley, N.: The XML Companion, 3rd edn. Addison-Wesley, London (2002)

9. W3C. XQuery 1.0 and XPath 2.0 Data Model (XDM) (January 23, 2007), http://www.w3.org/TR/xpath-datamodel/#document-order (cited 2009 24 February 2009)
10. Mercer, D.: XML: A Beginner's guide. Osborne/McGraw-Hill, New York (2001)
11. Pras, A., et al.: Peer-to-Peer Technologies in Network and Service Management. J. Netw. Syst. Manage. 15(3), 285–288 (2007)
12. Wallach, D.S.: A Survey of Peer-to-Peer Security Issues. In: Okada, M., Babu, C. S., Scedrov, A., Tokuda, H. (eds.) ISSS 2002. LNCS, vol. 2609, pp. 42–57. Springer, Heidelberg (2003)
13. Tutschku, K., Tran-Gia, P.: Traffic Characteristics and Performance Evaluation of Peer-to-Peer Systems. In: Steinmetz, R., Wehrle, K. (eds.) Peer-to-Peer Systems and Applications, pp. 383–397. Springer, Heidelberg (2005)
14. Abiteboul, S., Benjelloun, O., Milo, T.: The Active XML project: an overview. The VLDB Journal 17(5), 1019–1040 (2008)
15. Abiteboul, S., et al.: Managing distributed workspaces with active XML. In: Proceedings of the 29th International Conference on Very Large Data Bases, vol. 29, pp. 1061–1064. VLDB Endowment, Berlin (2003)
16. Abiteboul, S., et al.: An electronic patient record on steroids: distributed, peer-to-peer, secure and privacy-conscious. In: Proceedings of the Thirtieth International Conference on Very Large Data Bases, vol. 30, pp. 1273–1276. VLDB Endowment, Toronto (2004)
17. Abiteboul, S., et al.: Active xml: A data-centric perspective on web services. In: Web Dynamics, pp. 275–300 (2004)
18. Ghitescu, A., Taroza, E.: ActiveXML documentation. INRIA Saclay Ile-de-France. p. 41 (2008)
19. TheActiveXMLteam, Active XML User's Guide. GEMO INRIA (2005)
20. Ferraz, C.A., Braganholo, V.P., Mattoso, M.: Storing AXML documents with ARAXA. In: SBBDSBC, SBC, pp. 255–269 (2007)
21. Tatarinov, I., et al.: Storing and querying ordered XML using a relational database system. In: Proceedings of the 2002 ACM SIGMOD International Conference on Management of Data, pp. 204–215. ACM, Madison (2002)
22. Ruberg, G., Mattoso, M.: XCraft: boosting the performance of active XML materialization. In: Proceedings of the 11th International Conference on Extending Database Technology: Advances in Database Technology, pp. 299–310. ACM, Nantes (2008)
23. Abiteboul, S., et al.: Lazy query evaluation for Active XML. In: Proceedings of the 2004 ACM SIGMOD International Conference on Management of Data, pp. 227–238. ACM, Paris (2004)
24. Cremarenco, C.: Implementation of the Active XML Peer for the J2ME platform. In: Faculty of Automatics Control and Computers, p. 88. University Bucharest (2003)
25. Vidal, V., Lemos, F., Porto, F.b.: Towards automatic generation of AXML web services for dynamic data integration. In: Proceedings of the 2008 EDBT Workshop on Database Technologies for Handling XML Information on the Web, pp. 43–50. ACM, Nantes (2008)

Chapter 11
Chapter Migrating Legacy Assets through SOA to Realize Network Enabled Capability

David Webster, Lu Liu, Duncan Russell, Colin Venters, Zongyang Luo, and Jie Xu

Distributed Systems and Services Group,
School of Computing, University of Leeds,
LS2 9JT, United Kingdom
D.E.Webster@leeds.ac.uk
http://www.comp.leeds.ac.uk/distsys/

Abstract. Network Enabled Capability (NEC) is the UK Ministry of Defence's aspiration to enhance the achievement of military effect through the networking of future and existing military capabilities. The NECTISE (NEC Through Innovative Systems Engineering) program responded to this need by investigating the question *'are you ready for NEC?'* on behalf of equipment and service providers. Research work on this project proposed Service Oriented Architectures (SOA) as an architectural approach to delivering dependable and sustainable military capability. Specifically the work looked at how loosely coupled services could be used to expose and reuse functions and databases and how to describe the quality of service for heterogeneous systems and networks. The System of Systems that NEC will be realized from will not be implemented from scratch, but rather will be migrated from legacy assets over time. These assets will provide both functionality and data/information services, for example a weather sensor.

The focus of this chapter is to layout an understanding of the challenges faced and lessons learned in realizing NEC when migrating legacy assets to an SOA (Service Oriented Architecture) based System of Systems over time in order to reuse their functionality and databases. This work was based around a Software Demonstrator to illustrate a situational awareness capability realized by dynamically discovering and aggregating sensor data. This focus is not specifically on sensors, however, the sensor example provides a good example of data integration to realize military capability.

An abstract decision process model for wrapping legacy components was proposed to guide how existing system components can be selected for integration into the system of systems that NEC will be realized from. This model can be used to assist in the integration process of system components when migrating to or between execution architectures. The process model provides decision support for trade-offs between the costs of reimplementation, system wrapping and those costs incurred as a consequence of System of Systems complexity and ongoing maintenance.

A. Vakali & L.C. Jain (Eds.): New Directions in Web Data Management 1, SCI 331, pp. 311–346.
springerlink.com

1 Introduction

Network Enabled Capability (NEC) is the UK Ministry of Defence's aspiration to enhance the achievement of military effect through the networking of future and existing military capabilities [39]. Military capability in the context of NEC has been described as *"the ability to achieve a specified 'wartime' objective"* [43]. It is important to state at this point that NEC is not a *system*, but rather is the integration of assets to fulfil a mission objective. The MOD defines NEC as *"the coherent integration of sensors, decision-makers, weapon systems and support capabilities to achieve the desired effect"* [39], therefore, the networking of communications is only part of the problem as human activities are required within this asset integration. Military capability is the whole integration of resources including people and equipment; this includes the systems that provide dependable inter-operation to support human activity. Examples of capability include: surveillance; capture and defend hilltop; and deliver and administer medical treatment in the field.

To realize NEC the Armed Forces need to be flexible, ready and rapidly deployable, with the application of controlled and precise force, to achieve realisable effects and should have the ability to support and co-operate with each other to deliver a real-time capability [35]. To be successful in achieving this goal, NEC requires system integration of independent components (both functional and data oriented) that can evolve, operate in a dependable manner, managing system and component changes, cost effectively and connecting industrial, defence and pan-defence environments. In NEC one of the challenges (amongst many) involves the through-life provision of military capability; acquisition, service and support.

The NECTISE (NEC Through Innovative Systems Engineering) program [37] responded to this need by investigating how loosely coupled services can be used to expose and reuse the functions and databases and describe the quality of service for heterogeneous systems and networks. The EPSRC and BAE Systems jointly funded NECTISE throughout its three year duration, which involved ten UK universities and aimed to address the question of how BAE Systems delivers systems to NEC to the UK MOD, taking account of the aims summarised in the 2005 Defence Industrial Strategy [38].

Systems developed by NEC oriented programs will be delivered over time and these will need to be updated rapidly as technology and ongoing capability needs develop. Systems that come out of NEC programs will possess varying timescales for their life-cycles, which, in order to provide integrated operation, need to be coordinated in some manner, be it implicit or explicit. It is generally acknowledged that some military systems will have a long service life; for example a ship may have a service life of forty years. During this time the operational environment will evolve and the operational concepts will evolve. As noted by [32] in the military context *"many currently fielded embedded information systems face readiness challenges imposed by evolving missions and extended service life spans"*. Likewise, assets will be used within multiple capabilities throughout their active service lives.

The focus of this chapter is to layout an understanding of the challenges faced and lessons learned in realizing NEC when migrating legacy assets to an SOA (Service Oriented Architecture) based System of Systems over time. This has come from an understanding that the System of Systems that NEC will be realized from will not be implemented from scratch for the NEC initiative, but rather migrated from from legacy assets over time. SOA technologies (for instance Web Services) allow data provided by legacy applications to be integrated in a manageable manner to support the realization of NEC. Whilst legacy systems have been discussed in research literature in the context of system lifecycles [15,14,53,10]; the drive for this chapter is that organizations operating legacy systems face ongoing maintenance risks of *'brittle systems'* [12,32] and that can be viewed as a false economy in the long run - this is particularly the case when the characteristics of NEC are taken into consideration.

This chapter further presents our experience in examining and testing the benefits and consequences of SOA applied to the NEC environment through conceptual understanding of existing practice and a proof-of-concept SOA Software Demonstrator to simulate a situational awareness capability where data is aggregated from multiple dynamically discovered sensors in a geographical region. It should be noted at this point that the focus of this work is not specifically on sensors, however, the sensor example provides a good example of data integration to realize military capability.

The structure of the remainder of this chapter is as follows. Section 2 recapitulates current research work from NECTISE conducted towards delivering NEC through the SOA architectural style and helps to justify the decisions for why SOA is a suitable architectural approach for realizing NEC. In Section 3 the incremental delivery within the NEC System of Systems is discussed. This leads to an examination of the problems faced when implementing NEC from existing assets and discusses techniques to reuse existing assets. A decision support model for wrapping legacy system components within an SOA environment is proposed. Section 4 discusses a scenario for NEC based on sensor data integration which was used for our developed Software Demonstrator system. In Section 5 we discuss the lessons learned our research work (partially based on the Software Demonstrator) when providing a sustainable and maintainable approach to NEC provision through wrapped legacy assets. Finally in Section 6 conclusions are drawn and the potential for future work is outlined.

2 Service Oriented Architectures in NEC

Within NECTISE, SOA has been proposed as an architectural style for the realization of military capability to aid the engineering process for scalable System of Systems [43,44,45].

Engineering at the capability and product system engineering levels is defined by the MOD in [18]. *Military Capability Engineering* is concerned with the management of system of systems coherence and is defined by a high level capability architecture, critical scenarios and mission threads. On the other hand, *Product Systems Engineering* involves mortal projects that deliver capability. The key linkage between the two is described that Product Systems' *"interoperability requirements should be derived from the system-of-system portfolio architecture to which they contribute."* NEC will be enabled by a networked system of systems and will be provided through multiple services from multiple providers. The engineering process for systems that contribute towards capability, therefore, will be affected by requirements and changes from the capability engineering level.

2.1 Service Oriented Architectures

'Service Oriented Architecture' (SOA) is an architectural style that has been proposed as a way for businesses to redefine their processes to take advantage of *'formerly isolated component activities'* [8] for building distributed systems that are cross-organisational. This marks the move away from building traditional *'stove-pipe'* systems with fixed requirements [5] to building distributed systems from reusable and discoverable components. This proposal reflects the Defence Industrial Strategy's statement that *"we are seeing a shift away from platform oriented programmes towards a capability-based approach"* [38].

The premise here from the above definitions is that a deployed software application component can be reused as a *'network-available service'* with published and discoverable interfaces. Part of this reusability comes from the ability to clearly define the software interface in a manner and the data that passes through it abstracted away from the underlying system technology to promote loose coupling.

Service Contracts - A service contract in the context of SOA is a technical service contract describing the service's interface, according to [21]. SOA contracts provide a description of technical constraints of using the service and any associated requirements. The motivation of providing a service contract is to ensure a consistent expression of service capabilities [20]. This can be viewed as the documentation of the benefits and obligations for each party in the service interaction, or as a promise from a service provider in order for a service client to be built around that promise [4].

Service contracts define the data required for service providers and clients to communicate and build upon what in the past has been known as an Application Programming Interface (API) [46]. Typically this information is described using an Interface Definition Language (IDL).

Service contracts can be described at different granularities. To give an example of this concept, describing a service with a fine granularity requires less logic processing of the data that passes through the interface than describing a service with a coarse granularity as this will map closer to internal business documents and handling logic. Describing the data granularity of services typically involves to how the input, output and fault messages of a service endpoint are described [21].

Loose Coupling - *"Loose Coupling"* is the degree to which software components depend on each other, according to [25] and similarly defined by [2] as:

> *"a feature of software systems that allows those systems to be linked without having knowledge of the technologies used by one another".*

The loose coupling that SOA provides (both through the standardised communication protocols and the logical separation of systems into services) delivers an architectural approach to limit the impact of changes to services' internal systems on other services that will each have their own internal life-cycle. SOAs provide a style of loose coupling by negating the need for providers and consumers to require access to each other's underlying implementation details and system components (for instance, databases) of these services.

With this in mind, one of the major attributes of SOA is in the interfaces for services. The interface can be described as a logical layer abstraction of the components that provide functionality within the service [2]. This can be further clarified as a provider who offers a service to a consumer/client through the use of an interface - stating obligations of the provider and the responsibilities that the consumer must agree to. A key to the SOA paradigm in software is that services should ideally share schemas and contracts, not classes as is the case with object-oriented languages [36]. As discussed, one benefit of achieving loose coupling is to ensure that when modifications are made to a software system, the effects on local systems are reduced. One tactic to achieve this is to use *'deferred binding'* by binding modules as late to run-time as possible [7].

2.2 Web Services

Web Services are a *de facto* set of industry standards for implementing an SOA and are used to define contracts that publicly describe a service's operations or capabilities. Web Services can be described as:

> *"self-contained, modular business applications that have open, Internet-oriented, standards-based interfaces"*; according to [49].

To give a more technical description, Web Services are composed of a group of specifications: WSDL, XML Schema, SOAP and WS-Profile. Web Services Definition Language (WSDL) is an XML-based machine-processable interface definition language for describing messages that are exchanged between a client and service provider [11] and is considered to be the most important of these standards according to [21]. XML Schema is a formal language used to describe and validate the structure of XML documents, for instance the WSDL document. The *'SOAP'* (initially standing for Simple Object Access Protocol) protocol specification defines an XML-based stateless protocol for exchanging structured and typed information across Web peers, typically in a Web Services environment [50]. WS-Policy is a specification used for describing behaviour characteristics of a service in a machine processable manner.

WSDL - WSDL is composed of abstract and concrete parts [51]. The abstract part of the WSDL description describes the messages that are exchanged. This can be broken down into types and service definitions (see **Figure 1**). The concrete part of the WSDL document defines the network protocol and message format of messages sent and received and defines an endpoint which associates a network address with a binding [11,51]. The advantage of separating the abstract from concrete part of the WSDL is that multiple service providers can provide similar services that implement the same abstract service definition.

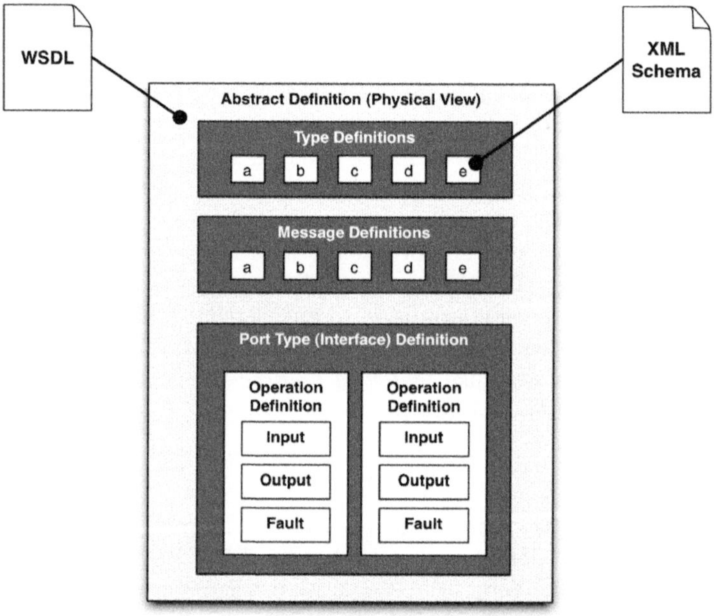

Fig. 1. WSDL abstract document part and technology mapping. Adapted from [21].

2.3 NEC Requirements and Realizing NEC through an SOA

In the delivery of military capability enabled by networks, dynamic integration based on SOA has the following characteristics [45,43]:

- *Service Integration* - Services are defined as composable functions, similar to component architecture [48], and can be combined to form higher levels of functionality and deliver capability.
- *Service Discovery* - Service providers offer services in a loosely coupled architecture to consumers for dynamic composition. The consumer requires discovery mechanisms to locate and bind before utilising services. For NEC, discovery is the means to identify service types for integration before forces are operational, and to enable dynamic binding in service integration during operations.
- *Service Reconfiguration* - Services can be adapted to meet consumer requirements at binding time. During service discovery, the consumer and provider may negotiate terms of service delivery involving QoS parameters. For example, using redundancy, such as replication of resources, may improve service dependability.
- *Service Evolution* - By abstracting the interface from the service implementation, a service can adapt to changes in its environment and the demands of the service consumer. Selecting appropriate resources at the time of service execution allows the resources to be updated and adapted without interrupting service availability. This supports continuous service delivery and therefore, sustainable delivery of capability.

The NEC initiative recognises that offering functionality is the main requirement in supporting military capability, and that functionality can be delivered without ownership of the delivery mechanism.

In NECTISE, design for change is being addressed by exploring architectures that can both adapt to changes and themselves can evolve within carefully defined limits. Traditional design methodologies (e.g. waterfall) address the issues of design for change by planning future changes in detail. However, modelling future changes is an extremely difficult task for complex system development. Agility in NEC represents an ability to adapt to changes occurring in through-life provision of capability not only at the design time but also at runtime. The frequent changes made in the short life-cycles could cause significant modifications of the interfaces of these military services, which lead to serious versioning and compatibility problems. Services need to be frequently re-integrated and re-configured to provide a reliable and sustainable capability. When factoring the agility of the systems that need to be evolved in order to satisfy new requirements of services then if some of these systems are legacy systems potential impedance to this redevelopment agility should be considered properly. By comparison to product delivery, service delivery is a continuous process, assuring reliability by maintaining the service provision and evolving the service implementation to respond to changes in environment, situation, supply, information and ongoing development.

SOA enables systems to operate across organizational boundaries and, therefore, contributes towards the ability for services to be discovered from different organizations and bound just before runtime [3]. This is supported by mechanisms of discovery to match services during composition. Architectural characteristics required by NEC, such as flexible interoperability and future proof evolvability, can be for the most part provided by SOA, where organisations, systems and computing each have defined service interfaces.

It should be noted, however, that services may be information services and are dedicated to providing data as opposed to purely computational services or those that control a physical resource - an example of an information service is exposing weather data or sensor data through a Web Service. The rapid growth of information services and resources in military systems makes it difficult and challenging to manage dynamic information and resources efficiently. In military systems, information is usually obtained from multiple data sources (e.g. database systems, file systems) to be integrated into a usable format for decision-making by battlefield commanders. For example, to analyse the implications of an aerial attack, background information could include data on meteorology, topography, settlements, population, infrastructure, sociology, and even hydrology. The premise of SOA, therefore, works towards standardising the mechanism by which data and information are described; for instance through XML Schema.

2.4 SOA and Workflows to Realize NEC

One advantage of SOA is that the functionality provided by system components can be exposed as services and reused in different contexts. A way to achieve this integration of services to realize business processes is through the use of a workflow. In the case of Web Service based SOA implementations, a standard service integration workflow technology is the Business Process Execution Language (BPEL)[1].

Workflow is defined by Hollingsworth [28] as *"the computerised facilitation or automation of a business process, in whole or part."* Business processes are descriptions of an organisation's activities designed to fulfil a customer's need, whereas similarly information processes are tasks performed by systems (people or computers) that process or provide information [27].

In the NEC context, as described by the UK Defence Industrial Strategy, a real-world example for capability enabled by networks is to achieve sensor data fusion through *"rationalising the masses of incoming sensor data"*. Our research work was based around a Software Demonstrator to illustrate a situational awareness capability realized by dynamically discovering and aggregating sensor data. An example SOA workflow would be the integration of data from sensors, which can be discovered at run-time using ultra-late binding. This example forms the basis of the practical Software Demonstrator described in Section 4 and was introduced briefly in our previous work

[1] http://docs.oasis-open.org/wsbpel/2.0/wsbpel-v2.0.eps

[43]. **Figure 2** shows a simple illustrative example of a target tracking capability. In this example, the capability is to detect and track a moving target within an area of interest. This is realized by integrating sensors that have their functionality shared on the network as services. The benefit of this approach is that these sensor services can be discovered and configured both at mission planning time and at run-time. Using a workflow approach, the sensors can be discovered and then configured based on functionality and quality of service attributes. The resultant data returned from the sensors is then aggregated through a data fusion algorithm. In the illustrated example the target (which for the case of our chapter's example is an automobile) can be seen travelling through grassland valley terrain between two hills.

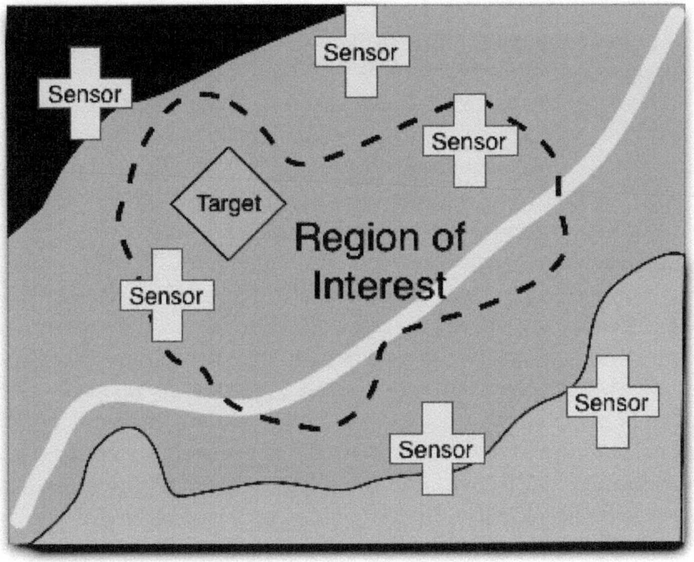

Fig. 2. High-level illustration of sensor integration scenario for target tracking

2.5 An SOA Integration Model for Realizing NEC

The abstract SOA integration model for NEC has been developed through the NECTISE architectures research work. It has been presented both within the NEC and academic communities [33,34,35]. It helps understanding of the reuse of system/platform functionality and data flow through services to realize a capability. **Figure 3** visually represents our model for the abstraction of SOA in the provision of network-enabled military capability within NEC.

In the capability layer, new capability requirements are determined through long to medium term capability planning. In the integration layer, configurations and specifications are defined based on these requirements. Configuration defines the actual combination of services used to implement the capability. This allows the abstract concept of the capability to be defined in terms of a set of abstract specifications.

In this model, functions (*'F'*) provided by concrete platform hosted systems are exposed and abstracted as services (*'FA ... FD'*) in order to be reused. In the integration layer these services are integrated close to (or at) run-time to realize a capability; in the diagram this is shown through the use of a workflow to invoke the individual services. As discussed earlier in this chapter, workflows are one way to integrate information services and can be implemented at the integration layer. The effects from the integration within the workflow are then synchronised to implement military capability.

To give a concrete example of this integration in action (based on the scenario presented in *Section 2.4*), a capability definition is created whereby a moving target needs to be tracked across a terrain. In order to realize this capability an integration workflow is used based on an SOA architectural style. Surveillance functionality is provided by an [open] network of sensors of different types.

The integration workflow submits real-time requests to a sensor service registry that dynamically discovers sensors, retrieving attributes such as position and range. A selection algorithm determines which sensors can 'see' the region of interest. The relevant sensors are contacted, which return information about the detected points of interest. This sensor information is processed to eliminate duplicates and points outside of region of interest and the detected feature positions are displayed on a map. This sensor information along with human interpretation combines to realize a target tracking capability.

In order to realize military capability, the following life-cycle description can be applied:

1. At the Capability Management level, there is a life-cycle (for example, owned by the MOD), which defines the need for capabilities, defines them, evolves them, uses them strategically and eventually ends their life if needed.
2. Within each capability, there will be a life-cycle which takes the definition of the capability from (1) and creates that capability. In the case of SOA, this would be by composing a capability from services. This process will involve defining the individual services needed to fulfil the capability,

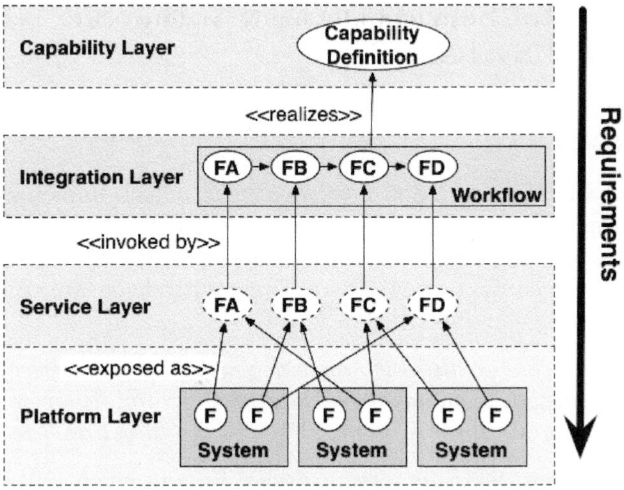

Fig. 3. An SOA integration model for realizing NEC

constructing the capability, maintaining/evolving the capability, and ending the life of the capability. This is where new services would be defined, if necessary, or selected for the workflow from existing services.
3. There is also a life-cycle for the individual services themselves. This will go along the line of create, maintain/evolve, end of life, as will happen with the capability itself similarly.

Summarising, the System of Systems to realize NEC will in itself contain a number of asynchronous life-cycles for individual systems. SOA has been proposed as an architectural approach to implementing this System of Systems. The rationale behind this is that the impact of changes to system internals should not affect other systems due their encapsulation through well defined interface specifications.

To recapulate, in this chapter the realization of applying SOA to an NEC environment in light of real-world constraints of military systems is investigated. Whilst SOA based integration is fairly well established in the research literature at a high-level, this chapter aims to add to this knowledge by focussing on how to implement an SOA system of systems in an NEC context. Specifically the challenge investigated is based on a fundamental characteristic of NEC in that its underlying systems will not be built from scratch as SOA-enabled, but rather will evolve from a whole host of existing legacy assets. Examples of concrete assets are a sensor or a database. This investigation is conducted by looking at SOA-based integration from both the capability integrator and service provider's perspectives and is further broken down in terms of the SOA-based service and system platform delivery later on.

3 Incremental Service Delivery within the NEC System of Systems

3.1 Introduction to Constraints of Developing Service Systems for NEC

The delivery of NEC will not involve a clean sheet implementation, but will come largely from a migration of existing assets. As the UK MOD Defence Equipment and Support (de&s) Handbook on Systems Engineering [19] states, in the context of MOD's working with disparate commercial approaches to systems engineering:

> *"MODs open architecture approach needs to acknowledge existing legacy system boundaries, while over time and where practicable seeking to migrate these systems closer to the target architecture."*

Incremental delivery is an important design feature when building complex systems. As has been acknowledged for single software systems, in order to deliver capability, full capability can be delivered through a single step, or alternatively through a number of evolutionary steps [9]. Brooks [23] describes this pattern of building a system (as opposed to specifying a satisfactory system in advance) as growth. The UK Defence Industrial Strategy [38] recommends incremental acquisition life-cycles to reduce project management risk and to enable the flexible insertion of new technology to respond to evolving requirements.

Whilst dynamic architectural styles (eg., SOA) have been proposed for the realization on NEC by supporting ongoing development without dependence on previous implementations [43], the *'growing'* nature of military Systems of Systems to realize NEC will involve legacy assets (systems) that will need to be factored in to the development process of services and their integration. The large volume of existing equipment and often the data bound to these provides a major constraint in the migration constraint of military systems to the NEC environment . A legacy system can be described as an old system that still provides essential business services [47]. Most of this equipment was not designed as SOA-enabled; additionally most legacy equipment is no longer actively being further developed. The clean-sheet ideal of developing SOA services that can be integrated and reused is not as simple as presented at the outset when legacy systems are factored in.

3.2 Existing Approaches to Reuse Legacy Systems as (Web) Services

In this section existing literature on recovering and reusing existing legacy systems as services is discussed. These techniques can be broadly categories into *'white box'*, *'black box'* and *'grey box'* techniques, generally described as follows:

- **Black box** - Legacy system has no source code available and cannot be decomposed.

- **White box** - Legacy system has source code available. This allows for refactoring and re-architecting of the system.
- **Grey box** - As with black box, but the system can be decomposed, for example the system uses a component architecture.

A black-box approach to reusing legacy software systems across hosted processor execution environments is proposed by Corman [16]. Here the description of a software wrapper given here is of a *"software adapter or shell that isolates a software component from other components and its processing environment (its context)"*. The options he gives for achieving this wrapping include:

- **Re-host** - Translate/recompile source code for new target host processor - this assumes either access to source code or the ability to transcode existing executable code.
- **Hybrid** - Spread system functions between old and new processors. Migrate over time.
- **Emulate** - Emulate the original target processor architecture. This type of approach has been described elsewhere by Booch [10] as putting it *'on life support'*.

The use of software emulation of an execution environment allows for a growth path for legacy and useful system executables that previously depended on now obsolete hardware. The approach presented in the paper describes a graphical tool to express data interfaces and constraints of the legacy system to determine whether requirements of the system on a new execution environment can be met. This approach based on the premise that legacy development tools can be used to maintain a computing system executable. The third-party maintainers of software systems, however may be forced (due to contracting) into using prescribed tools for generating artifacts that only useful within a specific suite of development tools, according to Kennedy [30]. These tools will also be subject to a servicing life-cycle and will become legacy systems themselves.

In Corman's paper the emulated approach was tested on an F-15 tactical fighter to allow the reuse of an Ada 83 Operational Flight Program component with a COTS PowerPC microprocessor. Through a demonstration activity, the emulated component was able to operate within given execution timeliness requirements despite the emulation overhead. Whilst this approach has demonstrated that the legacy system can be reused to extend its service life in a new processing environment, new requirements (functional and non-functional) over time acting on the legacy system as a result of operating an a new context are not yet considered.

The creation of wrappers in a workflow integration context has been used to mediate between legacy C code and Object-Oriented Java code for use within Triana workflows for scientific communities, as documented by Huang [29]. In this situation, tools have been developed to aid in the semi-automatic conversion between C and Java interfaces and then for the wrapping of Java components within the Triana workflow environment. This technique, however, requires access to the original C code and assumes that the C code will recompile on the target execution environment.

The approach proposed by Zhang [53] can be classified as a grey box approach. Here the assumption is that a particular legacy system can be broken up into smaller components. Following this process these components are wrapped as reusable information services of various granularities based on a domain model analysis to bridge the difference between the domain model and the legacy system component model. A decision tree based on specific criteria is used to determine if a legacy system is suitable for decomposition within this approach. These criteria involve:

- Is reusable and reliable functionality embedded within the system with valuable business logic available?
- Are reuse components from the legacy system reasonably maintainable compared to maintenance effort on the whole legacy system?
- Can the functionality be exposed as independent services and are they useful from the requirements point of view?

If these are satisfied then this approach takes legacy systems with source (or at least individual binary components/classes) available and partitions them up to expose certain parts as services. Here the legacy code components are modelled as UML models to aid in understanding of the legacy system and the linkage between components. Once particular components have been decided as mapping closely to the business defined service components (in this case Web Services) then the glue code is developed - in the case of Zhang's paper, the approach is using Java to mediate between existing C++ code and an Axis SOAP processor.

In summary, whilst this approach attempts to reduce the system complexity compared with wrapping black-box legacy systems, a shortcoming of the approach is that the maintenance associated with the 'glue code' of the mediator components is not given due consideration. Furthermore, this approach has limited applicability to black box *'monolithic'* systems that cannot be decomposed unless they provide available source code. For NEC there are a number of challenges that complicate this situation. Firstly as described, NEC initiative recognises that for service providers and contractors offering functionality, this service functionality can be delivered without ownership of the delivery mechanism. This means that it is not possible to make general assumptions for the state of legacy systems used to provide this service - for instance, whether they are black box monolithic systems or whether they are fully documented systems with source code available in order to be modified and customised by any party within the supply chain.

The approach described by Kotonya [31] aims to modernise legacy systems and provide a means for requirements mapping. To achieve this, domain entities/actors associated with the system are identified and secondly their needs are mapped to the reusable legacy systems. By performing this the impact of change from the requirements domain can be assessed on the legacy systems. Part of this analysis involves assessing the impact on system life-cycle planning as a result of new enhancements, for example the extensions to legacy systems required to allow operation on new platforms. To achieve this requirement, sources are modelled as viewpoints from various stakeholders within a UML model - this allows further mapping onto modelled legacy systems through a service intermediary.

This approach offers an advantage over Zhang [53] as systems can be modelled in a way that allows for the expression of component cost, certification and dependability requirements. Similarly as with Zhang [53] this paper acknowledges and suffers from the problem that system degradation will occur manifested as patches and *"increasingly complex glue code"*.

3.3 Abstract Decision Process Model for Wrapping Legacy Components

An abstract decision process model for wrapping legacy components is proposed here and illustrated in **Figure 4**. This provides a guide to how existing system components can be selected for integration into the System of Systems that will be used to realize NEC. For instance, this model is to be used to assist in the integration process of system components when migrating to or between execution architectures or in new operating contexts. Wrappers are a standard technique for reusing a legacy software component in a new originally unintended context and can be used to reuse legacy systems in new contexts. The concept of wrappers will be explored further in Section 5.1.

The decisions that the model will guide involves whether to wrap existing components based on the characteristics of modifiability for these components. Such guidance will inform the developer/integrator when new system life-cycles are created as a side-effect of implementing a wrapper - these are shown in **Figure 4**.

As discussed later in Section 5.1 the characteristics of the legacy system that will guide this process model's paths involve whether existing system components are modifiable. In the case of software components this can be through the inability to gain access to source code or that the component may not be modifiable (or at least difficult or costly) due to its phase in the servicing life-cycle. In the case of a wrapper being used, there may be no need to modify the underlying component.

The main decision points of the process shown in **Figure 4** can be broken down as follows:

- The ideal case is where a system component does not need to be wrapped and can be integrated as is. In the scope of this chapter we do not focus on the evaluation and certification process of integrating systems in NEC for space, but do point the reader to our previous work conducted within the NEC context [52].
- If the source code or development tools for the system component are available (and ideally documented) then the option is available to modify

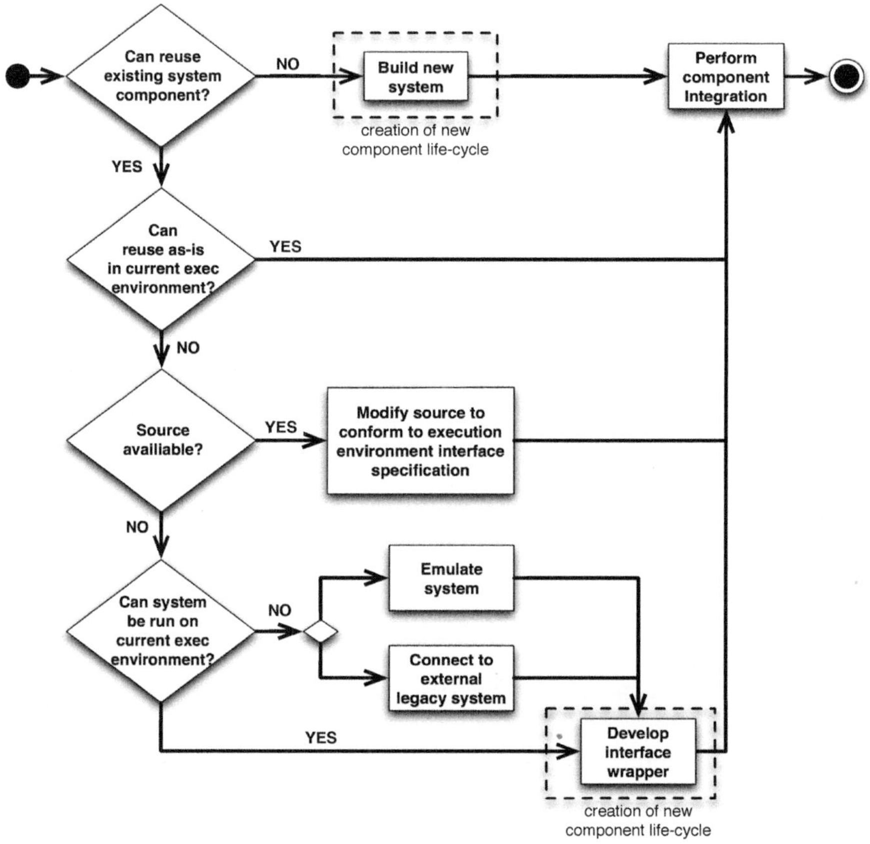

Fig. 4. UML activity diagram for wrapping options when integrating a component into a system or workflow

the component to operate with the interface standard required for integration in the System of Systems. This maintenance may be costly due to a lack of developer expertise or documentation within the organization to perform this modification of legacy code.
- In the case where the system component is non-modifiable, then the decision to make is whether the (software) system can be executed on the target execution environment or not.
  - Where the component can be executed on the target execution environment, but does not conform to the correct interface, then an interface-mediating wrapper can be developed.
  - Where the component cannot be executed on the target execution environment - for example due to relying upon a different microprocessor architecture for execution or incompatible service container - then there are two possible options:
    1) emulate the microprocessor architecture;
    2) reuse the legacy execution environment and provide an interface (possibly physically) between the two execution environments.

    An interface-mediating wrapper can then be developed to mediate between the differing system interfaces.

A key hypothesis to state here is that the creation of wrapping interfaces will increase the overall complexity of the wrapped systems as a whole and, therefore, the System of Systems. The wrapped interfaces will need to evolve in accordance to the interface requirements of integrating with the System of Systems environment over time. The process model, therefore, provides decision support to trade-off reimplementation cost against complexity and maintenance responsibility.

4 NECTISE SOA Demonstration and Critical Evaluation

In this section we discuss our experience with the NEC Software Demonstrator and discuss the legacy problems it can help to expose both above and below the Service Layer. The NEC Software Demonstrator was used to illustrate aspects of the research into systems architecture and through-life systems management (TLSM) within the NECTISE project and was implemented using Web based tools and technologies; for instance WSDL, SOAP, BPEL, XML Schema, HTML+AJAX.

The scenario aim of the Software Demonstrator was to model a Region Surveillance capability using dynamic service integration of sensor networks in the NEC battlefield - this allows a comprehensive ‘picture’ to be formed of the geographical region based on data communicated to a controller from deployed mobile sensors. Due to the high cost and confidential nature of testing our research work on real military systems a small but informed simulation environment has been developed for use within this research work. The surveillance capability was based on Points of Interest (PoIs), physical and military features within a geographical area that are detected by a group of simulated sensors. The integration of sensors was achieved by using a workflow to contact a sensor service registry and to dynamically discover sensors for a given region. The workflow uses simulated sensors that access a simulation of PoIs and then exposes this data as services. The sensor data is then processed to eliminate duplicates and points outside the region of interest and the detected feature positions are displayed on a map. The workflow can be illustrated at a high level in **Figure 5**. This diagram can be directly mapped onto the SOA integration model shown in **Figure 3** and illustrates the implementation technology used in the lower boxes. The workflow integrates Web Services which were the chosen demonstration implementation technology for SOA. The implementation of region surveillance used Google Maps to display the results from the feature detection workflow. A screenshot from this interface can be shown in **Figure 6**. This screenshot shows the output of the sensor integration workflow plotted onto a Google Maps interface within a web browser. The blue rectangle is the region of interest. The lists on the right of the image show the detected features and the sensors that have been accessed in the workflow.

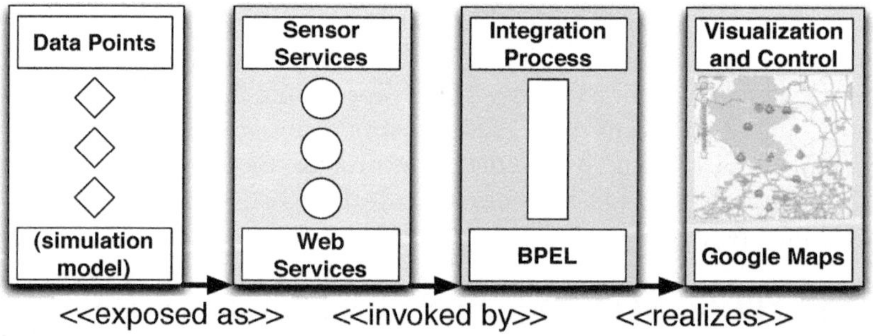

Fig. 5. A high-level overview of the sensor integration workflow

The Software Demonstrator was developed in NetBeans 6.1 IDE on a GlassFish application server, which enabled our research group members to write, deploy, test and debug SOA applications using the Extensible Markup Language (XML), Business Process Execution Language (BPEL) and Java Web Services.

4.1 Makeup of the Demonstrator Implementation

The Software Demonstrator consists of three main modules:

1. Web Services to simulate capabilities of different systems;
2. A workflow module for dynamic web services selection and composition;

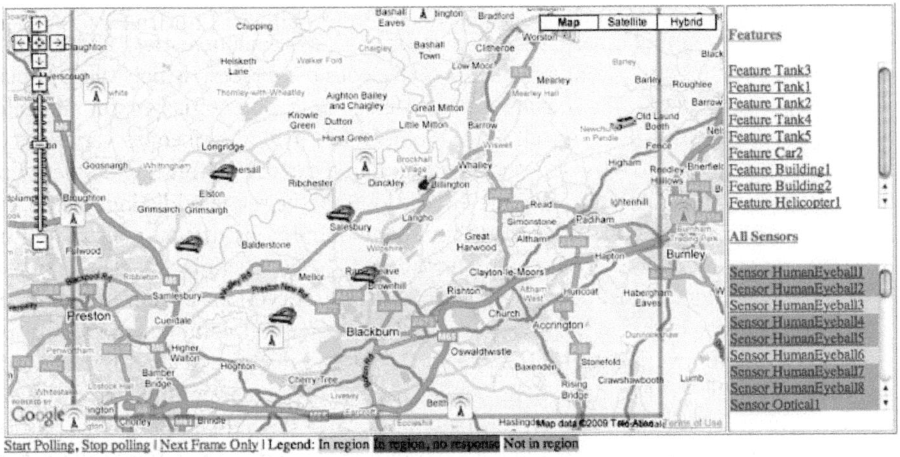

Fig. 6. Sensor surveillance Software Demonstrator screenshot

3. A client interface with sensor information display and user input.

The workflow dynamically discovers and integrates date from the sensor services and is implemented in a Java written Web Service itself. An Apache Derby SQL database server contains a model of the sensor and feature positions and their other attributes. There are five types of sensors maintained in the database including human observer, optical, infrared, and long-range and short-range radar. The detectable features include bridges, buildings, vehicles, helicopters, and humans. Each feature has a coordinate position xn, yn and movement is simulated through a path of coordinates. One or more of the sensors can detect each type of feature. Sensors are assumed to be static. **Figure 7** illustrates a user querying information about a sensor in the field.

Three types of web services have been created: Sensor Registry, Sensor, and Data Filter. The Sensor Registry was implemented as an Enterprise Java Bean (EJB). The registrys function is to return a list of sensors that can 'see' the region of interest. This square region is defined by the request with two pairs of coordinates x1, y1 and x2, y2 and Sensor services are selected from their attributes about position and range. A Sensor service is called for each entry in the list returned by the registry. The Sensor returns a list of all features that sensor can detect. The Data Filter service reduces the detected features from all the sensors by eliminating duplicate feature points and those outside the region of interest.

The Map Client allows the user to define the coordinate pairs for the ROI along with required response time. The user interface, shown in **Figure 6**, displays a map of the ROI utilizing Google Maps and POI overlay. In order to integrate the Map Client application with the rest of the components of

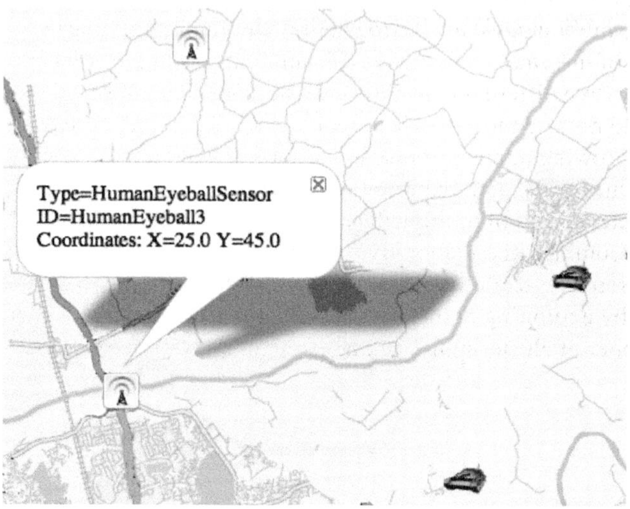

Fig. 7. Example of a user querying a sensor node within the Map Client

the demonstrator system as a whole, we needed the application to act as a client component of the Sensor Workflow component.

Due to the 'Same Origin Policy' [42] built into most web browsers, this technical limitation required a mediating component to connect the web browser client-side application with the Sensor Workflow. This policy prevents the properties of a document loaded from one origin being modified from another origin. Whilst we could have implemented a Web Services client within the web browser application, the allocation of this responsibility to the Servlet meant that the implementation architecture was simplified and allowed us to build upon existing and well-proven technologies from the Java Framework for Web Services.

In order to achieve this connectivity, an intermediate component was created to mediate requests from the user interface to the Sensor Workflow. To implement this mediator, a Java Servlet was created to mediate requests from the HTML+JavaScript part of the Map Client to the Web Service interface of the Sensor Workflow. Whilst we could have implemented a web services client within the JavaScript part of the user interface, the allocation of this responsibility to the Servlet meant that the implementation architecture was simplified and allowed us to build upon existing and well proven technologies from the Java Framework for Web Services in addition to easing the debugging process.

The implementation of the user interface of the Map Client consisted of an HTML+Javascript application that makes asynchronous requests to the mediator servlet commonly known as the AJAX technique (Asynchronous JavaScript And XML). In order to connect the client application to the mediator Servlet a simple XML schema was created to pass region and QoS requests to the Sensor Workflow through the mediator and consequently, return sensor and feature information from the Sensor Workflow. The features were then overlaid into a Google Maps display of the ROI. A UML deployment diagram is illustrated in **Figure 8**.

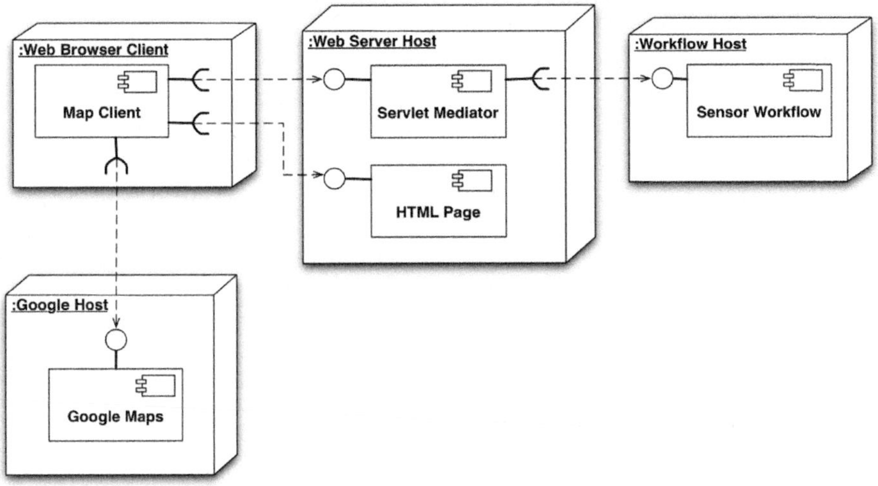

Fig. 8. UML Deployment Diagram for Map Client

Whilst the primary aim of the Software Demonstrator was to demonstrate architectural and dependability aspects of NEC (based on our previous experience [34]), the life-cycle aspects are focussed on in this chapter and demonstrate the challenges of migrating legacy assets to an SOA NEC environment as introduced previously in the chapter.

4.2 Life-Cycle Aspects of the Demonstrator

The following life-cycle related use-cases were illustrated through the Software Demonstrator:

1. Show Platform Upgrade
2. Show Service Life-cycles (Evolution)

Each of these use-cases is described in the context of the life-cycle work and illustrate the SOA implementation challenges when reusing legacy systems in SOA. These use-cases help expose evolution challenges above and below the Service Layer.

Show Platform Upgrade. This use-case illustrates a service upgrade. The upgrade to a service may be by changing the underlying system implementation without changing the service definition. The upgrade may improve quality of service attribute values, or it may improve efficiency of service delivery, thereby providing an internal benefit without affecting service delivery. This is found at the Platform Layer in **Figure 3**. The process can be described as having foundation in the concept of Refactoring [1].

Show Service Life-cycles (Evolution). This use-case illustrates the life-cycle of a service and its underlying system(s). The service life-cycle would include:

- Conceptual definition using service discovery to identify capability gaps;
- Development using evaluation to identify and trade-off possible solutions;
- In-service, by adding and removing the service from deployment;
- Disposal, where a service implementation is retired, causing:

  1. A change in service execution from one system to another by being replaced by another system;
  2. A change to the integration due to upgrading to new version, requiring all users of the existing service to migrate to the new one;
  3. End of life for service, requiring users of existing service to find alternative (newer) service or cease use of that function entirely.

The lesson learned here was that whilst it is possible to view the wrapping of a legacy system to provide an SOA interface as a solution for controlling the description of and access to its functionality, there are problems regarding the evolution and versioning of this interface description. Here is a research challenge whereby there is a dependancy between the service clients and service providers based on the agreed functionality of a particular version of a service. For example, if a service provider updates the interface of a service in a non backwards compatible manner and does not maintain the old version of the service interface, then the old client will not be able to access the service without reengineering. This shows that the problem of 'legacy' can occur at the Integration Layer.

By supporting the existing interface, backwards compatibility can be ensured. An example of this is to add an additional versioned interface to a service or by extending the existing interface with new parameters or functions in a backwards-compatible manner. A second option is for the service provider to provide multiple interfaces [4,13]. Whilst it is possible to still maintain interfaces for older versions of clients, this creates a maintenance responsibility.

As stated, in the case where backwards compatibility cannot be maintained and a new interface is defined, then interface mismatch will occur and will require upgrading of clients. In this case, older clients need to be migrated to operate with the new service interface. Here a wrapping solution may resolve the interface mismatch between the old client and new service, with the associated issues of increasing the complexity of the System of Systems as a whole and the creation of maintenance responsibility for the wrapping solution at the SOA Interface Layer level. There are many situations in the real-world where a client not being able to upgrade to a new service interface, for instance:

1. client systems no longer being maintained (legacy systems);
2. timeliness of upgrade cycles.

Within the development of the Software Demonstrator the concept of service evolution was demonstrated during its development. Two versions of a sensor access component were developed with both versions corresponding to the same client interface. This meant that the client was able to access the second version of the component with the same interface despite it being evolved to provide a different underlying behavior. To realize this, two versions of sensor workflow were implemented to represent serial and parallel processing of sensor aggregation as shown in **Figure 9**.

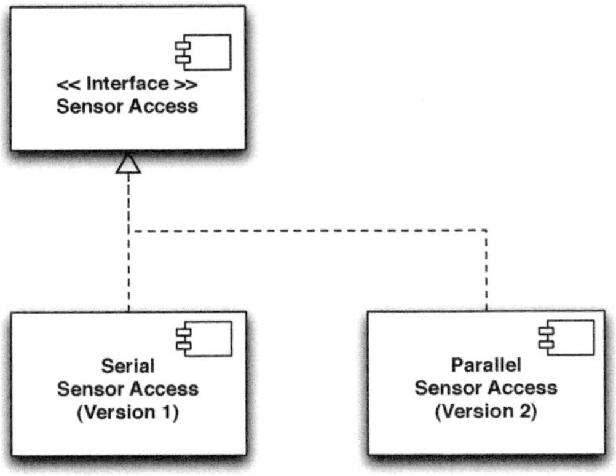

Fig. 9. UML diagram of serial and parallel sensor access components

Version 1 - Serial Sensor Access - Version 1 of the Sensor Access component implements a serial process of sensor access, thus the sensor accessing is sequentially processed. In implementation, there is only one thread in the program. Each sensor name returned from the aggregated sensors list is handled sequentially.

Version 2 - Parallel Sensor Access - Version 2 of the Sensor Access component used parallel sensor access. For each sensor name returned from Sensor List, a thread was launched to access the sensor Web Service. This provided the following advantages:

– Improved workflow performance by launching network requests in parallel, rather than waiting for one to finish before launching the next. In related research work in the GRID context, Quan [17] successfully demonstrated that parallel matching of GRID processing resources and

their subsequent configuration allowed that approach to meet SLA deadlines and achieve cost optimization.
- Improved resilience to failure of any Web Service requests. Web Service technology usually has a timeout for lost responses, typically 30 seconds. By issuing requests in parallel, the overall timeout would be a maximum of one failed response.
- The additional enhancement was to model variable availability in sensor accesses. Each sensor has a random ability to respond to a request. This is a basic model of two aspects:
  - network traffic delays in either request or response
  - request or response message loss - in particular using Web Services there is no guaranteed messaging, unless WS-ReliableMessaging is used.

The implementation of both the Version 1 (serial) and Version 2 (parallel) versions of the sensor access services that correspond to the same interface demonstrated the following advantages:

1. The two versions of the software services could be developed independently and orthogonally.
2. The versions of the software services could be swapped at run-time, meaning that they could be used for demonstration purposes to demonstrate the effects of serial and parallel access to sensors from a workflow. However, this meant that the interface could not be changed, leaving the only option of extending the interface and require clients to possess knowledge to utilize this (advanced) interface.

4.3 Exposing a Legacy Sensor Application to an SOA Network

The example presented here demonstrates wrapping legacy systems to operate within an SOA environment. In this example an unmodifiable legacy sensor application is exposed as a Web Service in order to allow it be integrated into an SOA enabled business process. This system model can be summarized with UML as in **Figure 10**.

In this model a Service provider ('**Sensor Service Provider**') and client '**Sensor Client**' are defined. The Web Service ('**Sensor Wrapper**') server is hosted within the '**Sensor Service Provider**' along with the legacy application ('**readsensor**') that communicates with the physical sensor hardware. '**Sensor Wrapper**' acts as both an external facing Web Service and a wrapper to the '**readsensor**' application.

In this example '**readsensor**' is a legacy executable that reads in an image from an optical sensor, performs some image processing and then directs a textual encoding of the image in base64 to the standard output stream. We acknowledge that this is a contrived example as the Demonstrator was simulation based and did not use any physical equipment; however, it demonstrates our argument. A mockup screenshot of an example client of the service can be seen in **Figure 11**.

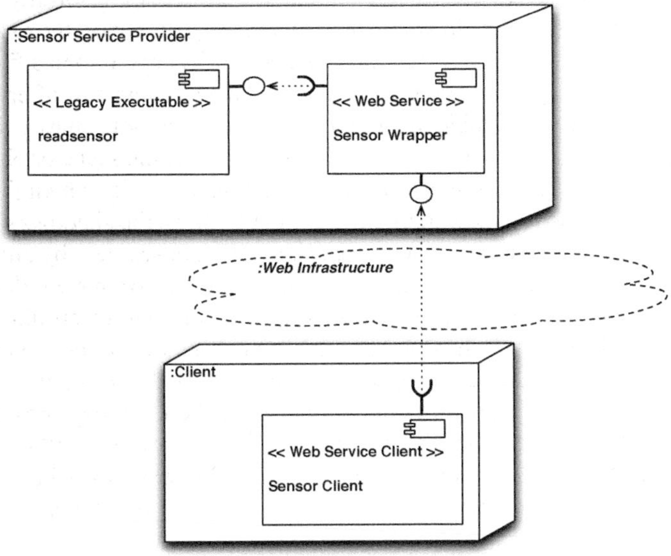

Fig. 10. UML diagram of wrapped sensor System of Systems

Fig. 11. Mockup screenshot showing a typical client decoded optical sensor image of the tracked target

The wrapping of the '**readsensor**' application to a GlassFish hosted Web Service was a relatively simple task to implement and test. The reader is directed towards the code listing in **Program 1** to see the implementation of this wrapper. However, this example exposes the fragile nature of this kind of wrapping solution as there are limitations on the value the wrapping component can add to the legacy application. To give an example; the business process utilizing this exposed sensor application has a requirement to not only retrieve the sensor image data, but also to provide a timestamp and geo-location coordinates for the sensor. Given that the original '**readsensor**' application cannot be modified; this may be achieved by recording the time that the sensor was accessed and attaching a GPS to the service provider with the assumption that the sensor reading is taken immediately following the call to the '**readsensor**' application and the server is located at the same location of the sensor. This implementation can be summarized as the *'decorator design pattern'* [24].

Whilst the above example demonstrates that it is possible to bridge the mismatch between capability requirements on an unmodifiable legacy system and the functionality that the system exposes, there will be a limitation to how far this approach can be taken. To give an example, if a business requirement is that the sensor imagery needs to be exposed along with directional data (for example, as a compass would provide), then this functionality cannot be provided from '**readsensor**' as-is. Likewise, if the sensor is a remote sensor and not physically attached to the Web Service mediatior, then it may not be possible to determine the geo-location of the sensor imagery.

Program 1 Simple Sensor Wrapper program in Java

```
/**
 * Simple Sensor Wrapper
 */
@WebMethod(operationName = "getSensorImage")
public String getSensorImage()
{
  /**
   * Launch 'readsensor' program and read the output
   * represented by plain text with base64 encoding.
   */
  Runtime runtime = Runtime.getRuntime();
  Process proc;
  StringBuffer programOutput
            = new StringBuffer();
   try
   {
      proc = runtime.exec("readsensor");

      InputStream inputstream = proc.getInputStream();
      InputStreamReader inputstreamreader
            = new InputStreamReader(inputstream);
      BufferedReader bufferedreader
                = new BufferedReader(inputstreamreader);

      // read the program output
      String programOutputLine;
      while (
        (programOutputLine = bufferedreader.readLine())
            != null )
      {
          programOutput.append(programOutputLine);
      }

   } catch (Exception ex) { return null; }

   /**
    * Return the wrapped program output to the
    * Web Service client.
    */
   return programOutput.toString();
}
```

5 Conclusions of Legacy System Migration towards SOA

Wrappers are a standard technique for reusing a legacy software component in a new originally unintended context and have been discussed in this chapter as a solution for this purpose. Whilst in this chapter software wrappers have been discusses as the primary technique for reusing legacy components there are, however, other techniques that are covered here. Aside from reimplementing the component from scratch, an alternative technique is to 'recover' legacy software through reverse engineering of its business logic [53]; this also helps to document its organisation and functionality [47]. The drawback of this approach is that it is invasive to the legacy system and that the accurate recovery of the business logic is difficult. Furthermore, reverse engineering of software systems may be impractical if source code is not available [26].

We identified and discussed the NEC need to investigate the wrapping of legacy equipment into an SOA as a response to the requirements for NEC discussed in **Section 2.3**. In this chapter, the realization of this process has been explored in more detail to expose hidden factors (for instance, System of Systems complexity, coupling, and maintenance costs) when migrating to an SOA enabled NEC environment from existing legacy assets.

The point to focus on here is that suppliers and system providers transferring to SOA based delivery of functionality will undertake the process of wrapping existing legacy assets into an SOA in order to conform to an SOA implementation's communication infrastructure, for instance a Web Services middleware. We can characterize this wrapping of legacy systems to an SOA middleware interface as the Adapter design pattern [53,25], which is sometimes referred to as a software wrapper. The software wrapper is a form of encapsulation whereby a software component is exposed with an alternative abstraction manifested as a software interface [6]. The purpose of the wrapper, therefore, is to allow system components that would be otherwise incompatible to be able to communicate with each other. Whilst the wrapper can be a complex piece of software, the cost of wrapping is often less than reimplementing the legacy system/component from scratch [47].

5.1 Challenges in Legacy System Migration towards SOA for NEC

The frequent changes made in the short life-cycles for new requirements on services to realize ongoing capability requirements for NEC could cause significant modifications of the interfaces of these military services, which lead to serious versioning and compatibility problems. Services need to be frequently re-integrated and re-configured to provide a reliable and sustainable capability. When legacy systems are factored into the System of Systems the problems that occur here, therefore, are twofold:

- 1) When new requirements are required of services, the systems that provide them need to be re-engineered for them to evolve to meet their new requirements. If the underlying systems are *'healthy'* systems that are currently in a evolutionary or servicing phase of their life-cycle (as can be illustrated in a Staged Life-cycle [40]) then this is just a reengineering process, where current approaches to agile development can be applied where appropriate. If there is a legacy system that is being wrapped to the service interface, then a danger with the practice of wrapping legacy assets is that the software system underneath the SOA interface will be difficult or costly to modify once in the late servicing and phase-out phases of a staged life-cycle model. As stated one attribute of this problem is the *'health'* of the development of that underlying system. For example if the system is in a phaseout phase of its life-cycle then it will be costly to modify, particularly if it relies upon diminished manufacturing sources (DMS) [32] both for hardware and software development.

- 2) A wrapping process could be used to satisfy an initial requirement to use a particular legacy system through an SOA integration middleware 'as-is', for example taking a black box approach. However, there will be challenges in evolving that system in order to adapt to ongoing capability requirements as a result of exposing the asset in a new environment through the service infrastructure. Examples of such challenges are having to deal with the increased information processing throughput over time; and using a system (or system component) in a domain or context that it was not intended or designed for.

Reflecting on the discussion of existing approaches to wrap legacy systems to service interfaces in **Section 3.2** - (with the exception of legacy systems which are either highly composable or have full source code available) - these systems suffer from the specific problems that they involve glue code to bridge the gap between the service requirements and the functionality and throughput ability of the legacy system. Challenges that still remain from existing work is the management of complexity over time with new requirements demanded from services - this has the potential to become unmanageable in a long running System of Systems over time. In the NEC context, the major problems here are the large number of stakeholders with differing capability requirements that will demand a high tempo for agile changes to be made services.

Using agile development methods for NEC, it is very important to check whether the current change will affect the interface of the service or QoS defined in the Service Layer Agreement. Good advice is that a comprehensive analysis and evaluation of the impact on dependencies must be done before any change is put into effect. However, where there are demands from multiple stakeholders in NEC that demand rapid response to ongoing capability requirements, this is not an easy process to go through when factoring in the 'legacy' challenge.

5.2 Maintenance Life-Cycle of Wrappers

Just as software components/systems need to be maintained, wrappers will possess their own servicing life-cycle since they will wrap between one or more evolving interfaces. Wrappers will need to maintain interoperability despite ongoing interface evolution. **Figure 12** provides a simple model to illustrate the relationship of the wrapper component against the two component interfaces that it is mediating between.

Fig. 12. UML component diagram representing an abstract model of component interface wrapping

In the UNIX programming domain, programmers are advised to avoid using wrapping *'glue'* layers to hide cracks and unevenness in software layers [41]. The problem here is that this process is responsible creating an additional layer of glue with the caveat of too many layers being the loss of transparency within a system. Whilst the technology in the SOA domain is different, the concept still applies, as the network integration of services will include wrapping components that will have to be managed and maintained. Therefore, the life-cycle complexity of the System of Systems is increased through the use of wrappers. We can logically conject that since a wrapper of a system mediates between two moving standards (which may be proprietary) the wrapper will also need to be maintained and evolved in accordance with its wrapped interfaces.

Figure 13 demonstrates the placement of the legacy system wrapper in the context of NEC service system integration. As with the capability life-cycle description in Section 2.5, requirements can be viewed as affecting the underlying system through the interface. These changes, as discussed earlier, can be functionality changes or performance requirement changes for instance an increase in processing throughput.

6 Conclusion and Opportunities for Future Work

This chapter reports our understanding of how to implement and realize NEC Systems of Systems when developing from existing assets rather than a purely clean-sheet position. The NECTISE program included the investigation of sustainable system maintainability with respects to the NEC problem and this work has explored a number of challenges to be faced in realizing NEC when migrating legacy assets to an SOA based System of Systems over time.

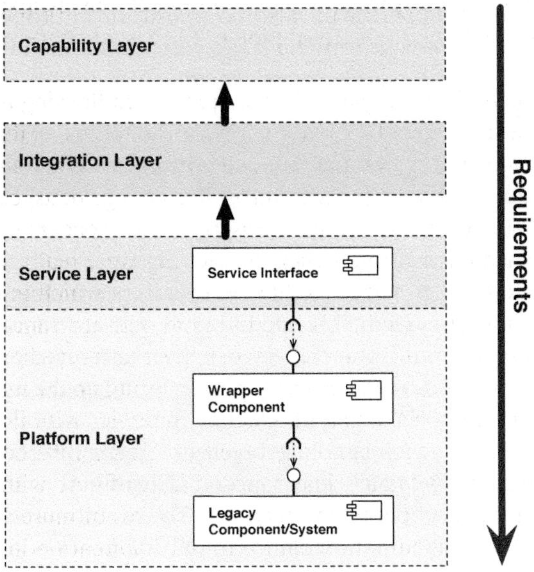

Fig. 13. Wrapping a legacy system in the context of capability integration

The challenges and constraints that have guided this research in the NEC domain include:

- System of Systems will be grown over time and not comprehensively designed upfront.
- Factors such as cost will drive the migration of existing/legacy assets to an SOA infrastructure rather than developing solutions from scratch.
- Systems that realize NEC possess varying timescales for their life-cycles and need to provide integrated operation at their current stage. In military systems, long life systems are common; for example a ship may have a service life of forty years. During this time the operational environment and operational concepts will both evolve.

This work shows that whilst it is possible to produce an abstract system model for SOA integration by abstracting the functionality of platform hosted systems into services which can be recombined (as in **Figure 3**), there are still *'devils in the detail'*. Application of SOA to an NEC environment has been investigated in light of real-world constraints of military systems. Specifically the reuse of legacy systems in an SOA-enable NEC environment has been considered. SOA-based integration has been viewed from both the capability integrator and service provider's perspectives and is further broken down in terms of the SOA-based service and system platform. We have explored problems from both legacy systems (changing underneath and extra requirements due to a context change) and evolution and mismatch at the Interface Layer. Problems of 'legacy' can occur at both the Platform Layer and at the SOA Integration Layer. These will occur, not just at the stage where non-SOA legacy assets are integrated, but when SOA-enabled services are evolved over time.

A summary of the understanding for the use of legacy systems is as follows:

1. The wrapping of legacy systems has been illustrated through the diagram shown in **Figure 13** and the worked sensor example in Section 4.3. Ongoing capability requirements affect wrapped legacy systems through the Service Interface layer. Limitations to how far this approach can be taken have been discussed, in particular for performance limitations and the inability to modify the system. The mediator/wrapper will need to bridge the gap of requirements put upon that system in its new context and the functionality implemented by the system in its original context. As discussed in Section 5.2 a hypothesis is presented that maintenance life-cycles need to be associated with wrappers created to overcoming interface mismatch. A decision support model has been proposed (**Figure 4**) to inform the developer of the creation of these life-cycles and the maintenance responsibility. A more structured documentation and analysis process is planned for future work. Part of this documentation includes the tracking of the creation of and life-cycles associated with wrapping components.
2. Whilst it is possible to view the wrapping of a legacy system to an SOA interface as a solution for controlling the description of and access to its functionality, there are still problems regarding the evolution and versioning of this interface description. Whilst it will be possible for service providers to support older versions or at least provide backwards-compatible versions of interfaces, semantic changes or the maintenance of a single interface version will prompt a reengineering task for existing clients. The evolution of the dependancy between service clients and service providers based on the agreed functionality of a particular service interface version is a challenge for future research.
3. Use of wrapper/adapter technologies is a potential solution to the problem of architectural and interface mismatch. This occurs below the Interface Layer for replacing replacing traditional systems whilst still maintaining the SOA interface definition and also above the Service Layer when providing functionally equivalent services from different service providers. The Software Demonstrator has illustrated that changing a serial sensor aggregation component with a parallel sensor aggregation component to improve performance is possible whilst encapsulating this change behind the same service interface.

Future work has been considered which would include formally analyzing the complexity, maintainability and interoperability for wrapping legacy systems into SOA infrastructures used in a System of Systems to realize NEC. Modeling of this work would be investigated in the context of documenting and reverse engineering the wrapping and legacy architectures through the use of MODAF (MOD Architecture Framework).

A structured approach to evaluating the replacement of components is required for future development of the Software Demonstrator. An approach for the design of a component behaviour test suite has been proposed by Flores [22] for equivalent components. This approach can be used as a starting point and will be key to evaluating the through-life upgrading, refactoring and replacement of components whilst still maintaining a System of Systems to realize capability.

Finally, whilst only wrappers have been suggested here, there is scope for other techniques for mitigating architectural mismatch to be investigated and evaluated.

Acknowledgements

The work reported in this paper has been supported by the NECTISE programme, jointly funded by BAE Systems and the UK Engineering and Physical Sciences Research Council Grant EP/D505461/1. The authors would like to thank John Davies from BAE Systems (INSYTE) for his input into this research work.

References

1. Refactoring: improving the design of existing code. Addison-Wesley Longman Publishing Co., Inc., Boston (1999)
2. Allen, P.: Service Orientation: Winning Strategies and Best Practices. Cambridge University Press, Cambridge (May 2006)
3. Alonso, G., Casati, F., Kuno, H., Machiraju, V.: Web Services - Concepts. In: Architectures and Applications, Springer, Heidelberg (November 2003)
4. Andrikopoulos, V., Benbernou, S., Papazoglou, M.P.: Evolving services from a contractual perspective. In: van Eck, P., Gordijn, J., Wieringa, R. (eds.) CAiSE 2009. LNCS, vol. 5565, pp. 290–304. Springer, Heidelberg (2009)
5. Baresi, L., Di Nitto, E., Ghezzi, C.: Toward open-world software: Issue and challenges. IEEE Computer 39(10), 36–43 (2006)
6. Bass, L., Clements, P., Kazman, R.: Software Architecture in Practice, 2nd edn. Addison-Wesley Professional, Reading (April 2003)
7. Bass, L., Clements, P., Kazman, R.: Software Architecture in Practice, 2nd edn. Addison-Wesley Professional, Reading (April 2003)
8. Bernstein, P.A.: Middleware: a model for distributed system services. Commun. ACM 39(2), 86–98 (1996)
9. Boehm, B., Hansen, W.: The spiral model as a tool for evolutionary acquisition. In: Bowers, P. (ed.) Journal of Defense Software Engineering, Crosstalk, vol. 14.5, pp. 4–11 (May 2001)
10. Booch, G.: Nine things you can do with old software. IEEE Software 25(5), 93–94 (2008)
11. Booth, D., Haas, H., McCabe, F., Newcomer, E., Champion, M., Ferris, C., Orchard, D.: Web services architecture. In: W3C Note NOTE-ws-arch-20040211, World Wide Web Consortium (February 2004)

12. Brooks Jr., F.P.: The mythical man-month. In: Proceedings of the International Conference on Reliable Software, page 193. ACM Press, New York (1975)
13. Cisco Systems. Service virtualization: Managing change in a service-oriented architecture. Technical report (2007), http://www.cisco.com/en/US/prod/collateral/contnetw/ps5719/ps7314/prod%5fwhite%5fpaper0900aecd806693c2.html - (last accessed October 13th, 2009)
14. Cohen, D., Larson, G., McDougal, D., Ware, B.: Extending life cycle of legacy systems. In: International Conference on Computer Networks and Mobile Computing, vol. 0, p. 291 (2003)
15. Comella-Dorda, S., Wallnau, K.C., Seacord, R.C., Robert, J.E.: A survey of legacy system modernization approaches. Technical report, Report CMU/SEI-2000-TN-003, Software Engineering Institute, Carnegie Mellon University, Pittsburgh PA (2000)
16. Corman, D.: The iuls approach to software wrapper technology for upgrading legacy systems. In: Bowers, P. (ed.) Journal of Defense Software Engineering, Crosstalk, vol. 14.12, pp. 9–13 (December 2001)
17. Altmann, J., Quan, D.M., Yang, L.T.: Improving the capability of the sla workflow broker with parallel processing technology. International Journal of Computer Systems Science and Engineering 24 (September 2009)
18. de&s.: The systems engineering handbook: Principles, practices and techniques. In: Draft D. UK Ministry of Defence (2007)
19. Director General Safety & Engineering. Implementing systems engineering in defence. Technical report, UK Ministry of Defence. de&s (2008)
20. Erl, T.: SOA Principles of Service Design (The Prentice Hall Service-Oriented Computing Series from Thomas Erl). Prentice Hall PTR, Upper Saddle River (2007)
21. Erl, T., Karmarkar, A., Walmsley, P., Haas, H., Umit Yalcinalp, L., Liu, K., Orchard, D., Tost, A., Pasley, J.: Web Service Contract Design and Versioning for SOA. Prentice Hall PTR, Upper Saddle River (2009)
22. Flores, A., Usaola, M.P.: Testing-based assessment process for upgrading component systems. In: IEEE International Conference on Software Maintenance, ICSM 2008, October 28, vol. 4, pp. 327–336 (2008)
23. Brooks Jr., F.P.: No silver bullet essence and accidents of software engineering. Computer 20(4), 10–19 (1987)
24. Gamma, E., Helm, R., Johnson, R., Vlissides, J.: Design Patterns. Addison-Wesley, Boston (1995)
25. Gamma, E., Helm, R., Johnson, R., Vlissides, J.: Design patterns: elements of reusable object-oriented software. Addison-Wesley Longman Publishing Co., Inc, Boston (1995)
26. Garlan, D., Allen, R., Ockerbloom, J.: Architectural mismatch: Why reuse is so hard. IEEE Software 12, 17–26 (1995)
27. Georgakopoulos, D., Hornick, M., Sheth, A.: An overview of workflow management: from process modeling to workflow automation infrastructure. Distrib. Parallel Databases 3(2), 119–153 (1995)
28. Hollingsworth, D.: Workflow management coalition - the workflow reference model. Technical report, Workflow Management Coalition (January 1995)
29. Huang, Y., Taylor, I., Walker, D.W., Davies, R.: Wrapping legacy codes for grid-based applications. In: Parallel and Distributed Processing Symposium, International, vol. 0, p. 139b (2003)

30. Kennedy, D., Nesterov, S.: Issues with third party maintenance of software intensive legacy systems: A case study – avionic mission systems. In: 18th Annual International Symposium of INCOSE Utrecht, The Netherlands (2008)
31. Kotonya, G., Hutchinson, J.: A component-based process for modelling and evolving legacy systems. Softw. Process 13(2), 113–125 (2008)
32. Littlejohn, K., DelPrincipe, M.V., Preston, J.D., Calloni Dr., B.A.: Reengineering: An affordable approach for embedded software upgrade. In: Bowers, P. (ed.) Journal of Defense Software Engineering, Crosstalk, vol. 14.12, pp. 4–8 (December 2001)
33. Liu, L., Russell, D., Looker, N., Webster, D., Xu, J., Davies, J., Irvin, K.: Evolutionary service-oriented architecture for network enabled capability. In: International Workshop on Verification and Evaluation of Computer and Communication Systems (VECoS), Leeds, UK, Published by the eWiC series of the British Computer Society (BCS) (2008)
34. Liu, L., Russell, D., Webster, D., Luo, Z., Venters, C., Xu, J., Davies, J.K.: Delivering sustainable capability on evolutionary service-oriented architecture. In: IEEE International Symposium on Object-Oriented Real-Time Distributed Computing, vol. 0, pp. 12–19 (2009)
35. Liu, L., Russell, D., Xu, J., Davies, J., Irvin, K.: Agile properties of service oriented architectures for network enabled capability. In: Realising Network Enabled Capability (RNEC 2008), Leeds, UK (2008)
36. Microsoft. Soa in the real world. Online Book. (August 2007), `http://www.microsoft.com/DOWNLOADS/details.aspx?FamilyID=cb2a8e49-bb3b-49b6-b296-a2dfbbe042d8&displaylang=en` (retrieved September 21, 2009)
37. NECTISE. Network enabled capability through innovative systems engineering (nectise) (2005), `http://nectise.com/`
38. UK Ministry of Defence. Defence industrial strategy: Defence white paper (cm6697). Technical report, UK Ministry of Defence (2005)
39. UK Ministry of Defence. Joint Services Publication 777, 1st edn. (2005), `http://www.mod.uk/NR/rdonlyres/E1403E7F-96FA-4550-AE14-4C7FF610FE3E/0/nec_jsp777.pdf`
40. Rajlich, V.T., Bennett, K.H.: A staged model for the software life cycle. Computer 33(7), 66–71 (2000)
41. Raymond, E.S.: The Art of UNIX Programming. Pearson Education, London (2003)
42. Ruderman, J.: Same origin policy for javascript. Technical report (2009), `https://developer.mozilla.org/En/Same_origin_policy_for_JavaScript`
43. Russell, D., Looker, N., Liu, L., Xu, J.: Service-oriented integration of systems for military capability. In: IEEE International Symposium on Object-Oriented Real-Time Distributed Computing, vol. 0, pp. 33–41 (2008)
44. Russell, D., Xu, J.: Service oriented architectures in the delivery of capability. In: Proc. of Systems Engineering for Future Capability (2007)
45. Russell, D., Xu, J.: Service oriented architectures in the provision of military capability. In: UK e-Science All Hands Meeting 2007, Nottingham, UK (2007)
46. Shaw, M., Garlan, D.: Software architecture: perspectives on an emerging discipline. Prentice-Hall, Inc., Upper Saddle River (1996)
47. Sommerville, I.: Software Engineering (International Computer Science), 8th edn. Addison-Wesley Longman Publishing Co., Inc., Boston (2006)

48. Szyperski, C.: Component Software: Beyond Object-Oriented Programming. Addison-Wesley Longman Publishing Co., Inc., Boston (2002)
49. uddi.org. UDDI Executive White Paper (2001)
50. W3C. Soap version 1.2 part 0: Primer 2nd edn. W3C Recommendation (April 2007)
51. Web Services Description Working Group. Web services description language (wsdl) version 2.0 part 1: Core language. Technical report, W3C (2007)
52. Webster, D., Looker, N., Russell, D., Liu, L., Xu, J.: An ontology for evaluation of network enabled capability architectures. In: Realising Network Enabled Capability (RNEC 2008), Leeds, UK (2008)
53. Zhang, Z., Yang, H.: Incubating services in legacy systems for architectural migration. In: Asia-Pacific Software Engineering Conference, vol. 0, pp. 196–203 (2004)

Author Index

Batch number: 10614940

Printed by Printforce, the Netherlands